The Blessings of Disaster

The Lessons That Catastrophes Teach Us and Why Our Future Depends on It

MICHEL BRUNEAU

Prometheus Books

Essex, Connecticut

Prometheus Books

An imprint of Globe Pequot, the trade division of
The Rowman & Littlefield Publishing Group, Inc.
4501 Forbes Blvd., Ste. 200
Lanham, MD 20706
www.rowman.com

Distributed by NATIONAL BOOK NETWORK

British Library Cataloguing in Publication Information Available

Library of Congress Cataloging-in-Publication Data Available

ISBN 978-1-63388-823-4 (cloth : alk. paper) | ISBN 978-1-63388-824-1 (ebook)

♾™ The paper used in this publication meets the minimum requirements of
American National Standard for Information Sciences—Permanence of Paper
for Printed Library Materials, ANSI/NISO Z39.48-1992.

To Parize,
who, during the snowstorm on our wedding day,
never imagined that our journey would involve
sleeping on an active volcano,
living in earthquake countries,
and strolling on beaches stripped by hurricanes.

Some subjects are so serious that one can only joke about them.

—Niels Henrik David Bohr
Nobel Prize in Physics 1922

Contents

MEET THE FUTURE

Selected Acronyms

CBD: Central Business District (of Christchurch, New Zealand)
DWR: California Department of Water Resources
EPA: US Environmental Protection Agency
FEMA: Federal Emergency Management Agency
NOAA: National Oceanic and Atmospheric Administration
SWP: California State Water Project
USGS: US Geological Survey

Preface

If you are reading these words right now, you are most probably not dead. At least, not yet.

You most likely consider this to be a good thing (and so are authors and book publishers, because dead readers constitute a notoriously bad market). Being alive might not seem to be anything exceptional, as most rational, reasonable, mature adults believe that they are generally cautious, place value on safety, and would not purposely expose themselves to danger. Rational people would not drive through a red light, would not operate a hair dryer while in a bathtub, would not drink bleach, would not pet a twenty-foot-long alligator, would not walk their dog while holding a lightning rod in the middle of a thunderstorm, would not stab a horse in the butt while standing behind it, or perform other stunts that would compete for the Darwin Awards (bestowed to those who kill themselves in creatively stupid ways and, in doing so, by the virtues of natural selection, significantly improve the human race by progressively eliminating the moron-gene from its DNA pool).[1] Most rational humans do take sound actions and make sound decisions when facing the risk of immediate death or—at best—death in the short term.

What is more interesting, and intriguing, is that most rational adults typically do little, or sometimes nothing at all, to protect themselves from things that could kill them years from now, and particularly from events that could suddenly and simultaneously kill a large number of people. The world is full of such looming disasters. Yet the further away in the future that deadly event might be, or the lower its probability of occurring, the less is done. As if death a few years or decades away is not a "less imminent danger," but rather a "nonexistent danger"—at best, a blurry possibility that might be acknowledged with the best of intentions, but not necessarily acted upon.

Why is that?

To what extent is this the case and how can this foretell the future of our civilization?

It is the purpose of this book to answer these questions. This is done by exploring how our societies deal with low-probability high-consequence events, by bringing together in counterintuitive (and hopefully entertaining) ways information from history, science, engineering, economics, risk, politics, and human behavior, addressing how knowledge, interpretations, reality, misconceptions, resourcefulness, and beliefs may condition certain actions, inactions, decisions, indecisions, preparedness, and responses—effective or not—to cope with these events.

This exploration proceeds by providing in-depth perspectives on the causes of disasters, on how humankind copes and learns from successive disasters, and the implications of this process on how civilization will face other future hazards and existential threats. This linkage between the past and the future can be done because, be it for natural hazards or anthropogenic ones (that is, made by humans), for current threats or existential ones, the patterns of human behavior before and after a disaster are the same. Therefore, those looking at the future with eco-anxiety, existential dread, and other fears calling them to action can gain valuable insights by reviewing how humanity has addressed or ignored other hazards.

With that mindset, a working title for this book was "Earthquakes Are Good for You—If You Survive." "Earthquakes," in that context, is a metaphor for a large number of life-changing events. The book's title could just as well have been "Earthquakes, Hurricanes, Tornadoes, Floods, Volcanic Eruptions, Terrorist Attacks, Technological Catastrophes, and All Kinds of other Disasters Are Good for You—If You Survive," but this would have required the cover page to be typeset in a font too small for marketing purposes. Alternatively—and more bluntly—it could also have been titled "Evolution by Disasters and Tombstones." But "The Blessings of Disaster" paints a crisper picture, because, in the end, catastrophes serve as wake-up calls for survivors, who then improve things for the better. Sadly, a disaster is a necessary rite of passage to a next step along the ride up to a more robust and resilient society.

The journey through this world unfolds here in three steps.

The first part of the book focuses on hazards that have created past disasters—and that will continue to do so for the foreseeable future. Not only to understand the physics at play in each case, but also to witness how mundane it can be to live on a volcano, to work in a building behind a levee, to retire in hurricane country, to trust technology that is certain to fail, and more—until a disaster happens, of course. This is important because coexistence with disaster-prone conditions is universal. To put it plainly, there are no disasters without humans. Disasters "R" Us. When it comes to the simple matter of preparing, not preparing, and coping with various hazards, human behaviors range from baffling to amazing.

The second part of the book seeks to explain why John and Jane bought a bungalow straddling the San Andreas Fault, Jim and Janet retired to a beach villa that will fly away in the next hurricane, Julio and Juliet reside on the slope of Mount Vesuvius, and Jack and Jill went down the hill and built their dream home there, in a flood zone. It may not make sense, but at the same time, it makes perfect sense. Confusing predictions and statistics, imperfect building codes, disinterested politicians that are simply mirrors, more pressing priorities, and the brain's biases in interpreting reality and beliefs, all play a part—with potentially deadly consequences.

The last part of the book is an extrapolation of the previous observations that makes it possible to address the issues of monetary collapse, climate change, overpopulation, nuclear holocaust, and other similarly joyous topics. In other words, it seeks to answer the question: "Are we doomed?"

Examples provided throughout the book span over a range of disasters. If it seems that points are made by referring to earthquakes more often than for other disasters, it is because in a world with fifty shades of disasters, earthquakes provide the perfect black-and-white contrast needed in case studies: they are sudden and devastating, but their threat is rapidly and conveniently forgotten during their long periods of quiescence.

As a word of warning, this book is not intended to be a rigorous treatise. By any scientific standards, it is not a dissertation resulting from years of research in public policy, sociology, psychology, economics, philosophy, theology, or witchcraft. According to his official bio, the author is recognized for his decades of work in disaster mitigation from the perspective of . . . earthquake engineering. Oops! Admitting to be an engineer can be an effective repellent in a party, so this is maybe where some will close the book for good. Yet, an earthquake engineer may be good company here, because, beyond the fascination with the extreme power of earthquakes, everybody wants to be reassured that our infrastructure will not collapse and that we will not suffer or die following a big one—as if for once, we could be protected from the inescapable flaws of human nature.

For those who did not close the book, it is hoped that the pages that follow will provide an enjoyable ride—a roller coaster one that will start from the world of earthquakes and loop (roll, dive, twist, and corkscrew unpredictably) through many other similarly captivating types of disasters and domains of human nature. However, as you proceed forward, be warned that you are taking a risk. This is best explained using an end-of-chapter capper—a format that will be used throughout the book to insert (under the heading "On the Disaster Trail") some of the author's personal anecdotes that could be skipped by those who prefer their books written in a third-person point of view.

Finally, two realities must be acknowledged before going any further.

First, this book does not promise to change you or make you better. In fact, if anything, it is the exact thesis of this book that human nature cannot be changed easily—or, arguably, in some aspects, cannot be changed at all, barring a disaster. Those rummaging through this book in search of a life-altering experience will not find such a secret recipe (there are religions, cults, and drugs better suited for that quest). At best, this book may lead some to reflect a little bit on certain perspectives and interpretations of reality, and maybe spark a few ideas. It may lead to some introspection and some New Year resolutions, but the chances that it will change anything over the long term are slim. It takes a disaster to create the conditions leading to long-lasting changes for the better (which, again, is the thesis presented here). However, if the book can offer a glimmer of hope and show that eventually things come out right, and sometimes better, for the lucky ones who survive earthquakes and other disasters, then that is already a decent accomplishment.

Second, it is important to acknowledge that there are thousands of professionals that have been working—and continue to work—behind the scenes to improve things, such that future disasters are lesser disasters for each subsequent generation. Sometimes they succeed, sometimes not, but they are tirelessly championing the goal of achieving a safer world, and it is largely because of these invisible heroes that many people will actually survive disasters, and do so facing a lesser hardship than would have been the case otherwise. Engineers, scientists, policy makers, and many more have performed this thankless job—and are exceptionally thanked here for that. People rarely get medals or rewards for averting a disaster because disasters that have not happened are unknown. Yet safety is not magically happening by itself. Some parts of this book will briefly—superficially at best—highlight how these heroes have contributed to make the world less disaster prone, safer, and more resilient, but providing a thorough review of all these successes, and of the sustained and coordinated efforts that have been necessary to get there, is way, way, way beyond the scope of this work. As a consequence, some professionals with decades of achievement in the many disciplines that have contributed to abate the risk of disasters, or that have provided deep academic treatments to relevant subjects, may find that this small book does not fully recognize the extensive contributions of their disciplines to the subject at hand, and is thereby only a crude and overly simplified view of the bigger picture. They may feel that scholarly works on many relevant topics and subtopics have not been given due credit. These omissions, simplifications, and short-cuts will hopefully be forgiven, with the understanding that this book is not targeting a limited academic or specialized audience but rather is intended to connect with the broadest possible audience in a direct and thought-provoking way.

Therefore, when all is said and done—and read—it is the author's sincere hope that the book will increase awareness, trigger fruitful discussions, and broaden the conversation beyond the realm of academia, to embrace all constituencies and lead to a better future. Beyond that, all other concerns that may arise might be best appeased by sharing a glass of wine.

ON THE DISASTER TRAIL

Who to Trust?

Some years back, I spent a few weeks in Taipei as part of an international, collaborative, structural engineering research project investigating the seismic performance of steel plate shear walls—which is the long and formal way to say I was having a good time doing full-scale experiments destroying huge chunks of cleverly designed steel structures with a good friend and colleague at the National Taiwan University. One evening, back at my hotel room after a long day of testing in the lab, while busy answering emails, the hotel started to sway. "Earthquake!" shouted the little seismograph hardwired into my brain (I am getting good at recognizing those). The whole room started to move back and forth in space and the first period of vibration of the entire building was driving the response. Severely. My first reaction usually when exposed to earthquake shaking is to try to guess the magnitude of the earthquake and the distance to the epicenter (it is a stupid game and I have explained how it works elsewhere, in a different context).[2] This time however, the motions kept amplifying, the sway was becoming more severe, and I realized that I was on the tenth floor of a reinforced concrete building in a country where some of those have spectacularly collapsed in past earthquakes. My first thought was: "Damn it! If it is the 'Big One' for Taipei, I hope the structural engineer who designed that building knew what he/she was doing." Evidently, I survived—and largely, in this case, because it was far from being the "Big One." However, the point of that story is that this thought that arose at the start of an earthquake should exactly match your thought right now, at the start of this book: "Damn it! I hope the structural engineer who designed this book knew what he was doing."

MEET (SOME OF) THE HAZARDS

Which Little Pig Are You?

A MATTER OF RELATIVITY

In response to a moderate earthquake that had struck a foreign country and damaged a significant number of engineered bridges and buildings, the government had dispatched a team of experts to perform earthquake reconnaissance activities. The team's mandate was to travel across the affected region, to document the extent of the damage, to determine to the extent possible what caused the extensive damage suffered by the infrastructure, and, most importantly, to report on whether such a disaster could happen at home.

Sad but true, there is always much to learn from disasters that kill and injure thousands of people and produce billions of dollars in damage and losses. Teams of engineers have conducted earthquake reconnaissance visits after damaging earthquakes all over the globe in the past decades, so this was by itself not an unusual thing to do. However, this time, it was different. Not only had this earthquake caused $40 billion in damage, but it did so by striking a country that considered itself a leader in the development and implementation of modern seismic design codes and standards; a nation that was part of that elite club of players whose members represent more than 60 percent of the world's net wealth; and, most significantly, a friendly nation with whom the government had long-standing close ties. On that account, there was potentially much more to learn than usual from the earthquake damage, and the findings from the earthquake reconnaissance mission were sure to be valuable.

So, the government got its report.

What happens once a technical report is printed and submitted very much depends on the political forces at play. Countless reports "rest in peace" on library shelves; less fortunate others are "filed" in bankers boxes buried in archives. Yet, when the stars align and the timely words fall in receptive ears, lucky reports serve their purpose and can lead to changes in building codes, enhance design specifications, or fulfill some other noble purposes mostly invisible to the public. In some rare instances, parts of these reports that have had an impact are quoted by politicians.[1] For whatever the reason may have been, in one such moment, an elected government official[2] deemed it appropriate to reassure the public and commented on the many bridges that collapsed during the earthquake in that other country. He emphasized that the way bridges were designed there was different than at home and that, contrary to what had been observed in that other country, the nation's bridges were safe. "Safe!"

That elected official was a representative of the Japanese government commenting on the damage from the January 17, 1994, magnitude 6.7 Northridge earthquake that struck at 4:30 a.m. near Los Angeles and during which many bridge spans and overpasses either fell off their support or collapsed due to column failures. Most significantly, one of those collapsed spans occurred along the busiest freeway in the U.S. (Interstate 10), which, as a result, was closed to traffic for almost three months.[3] These collapses could have been of dramatic consequence had the earthquake not happened so early that day, but instead during rush hour with cars zooming by at more than seventy miles per hour.

The Japanese official's confident statement undoubtedly reassured the population at the time. Most unfortunately, he was wrong.

Exactly one year to the day after the Northridge earthquake, the Great Hanshin earthquake struck near Kobe at 5:46 a.m. on January 17, 1995. Beyond demonstrating that, statistically speaking, large earthquakes have a propensity to strike on a January 17 before business hours (one can really make statistics say anything), this magnitude 6.9 earthquake destroyed many bridges across the region,[4] including large segments of the elevated railway for Japan's iconic bullet train—which could have been of dramatic consequence had the earthquake not happened so early that day, but instead during rush hour with trains zooming by at more than two hundred miles per hour. Also noteworthy, the elevated Hanshin Expressway that ran though Kobe suffered massive damage over nearly its entire length, and a segment of it collapsed in such a spectacular manner that photos of it made the front page of magazines and newspapers worldwide—and, not to forget, provided sensational opening footage for the television evening news. Beyond bridges, the earthquake also damaged more than one hundred thousand buildings,[5] triggered about three hundred fires that burned for days, and debilitated Kobe's entire infrastructure. And Kobe's port, one of the world's busiest at the time of the earthquake, was ravaged by the earthquake and never recovered its stature.

So much for "safe."

However, that was not the worst of it. This earthquake was also embarrassing for the Japanese government for a number of other reasons.

First in the list of embarrassments[6] is the fact that the prime minister learned that the Kobe earthquake had occurred, not from government agencies, but rather from broadcast news. Unlike some other countries, where the president wants to be (or already is) the most powerful man/woman on earth, in Japan, the prime minister only aspires to be a good leader for his nation—sometimes acting as a manager of timid transformational forces, most often plenty happy to tend to the day-to-day business of presiding over the government. Nonetheless, it remains that the prime minister is the top decision maker in the country.[7] It is the prime minister's office where the proverbial "buck" is supposed to stop, so from the head-of-state perspective, being the last one to know amounted to losing face—a particularly painful thing in many parts of Asia. Comparing the prime minister to a father (or mother, but Japan has not yet elected a woman to that position), it would be like having dad watch television in the den upstairs and learn on the news that a few hours earlier, somebody broke into the kids' bedroom in the basement, thrashed the place and set the curtains on fire, while the kids huddled in their closet. In 1995, the World Wide Web did not exist, but emails, telephones, CB radios, telegraphs, and smoke signals all existed; the fact that the prime minister's office did not receive an official notification of the earthquake and of its severity before news crews could be dispatched illustrates how local authorities were taken by surprise and the extent of their disarray following the earthquake. Part of this confusion can be explained by the fact that, in spite of evidence to the contrary, the general belief was that future large earthquakes were certain to strike soon in the Kanto area of Japan, close to the more than thirty million people that lived in and around Tokyo, but certainly not so in the Kansai area, close to the more than twenty million people that lived in the contiguous towns of Kobe, Osaka, and Kyoto (see end of chapter).

Second subject of embarrassment: the post-earthquake response was aggravatingly slow. For example, it took nine hours before the military was ordered to assist, and, thereafter, army vehicles spent hours stuck in massive traffic jams created as the population either tried to leave or to return to Kobe, winding around town to circumvent the collapsed bridges. In many instances, when firefighters finally reached burning homes (and sometimes entire burning neighborhoods), it was often to no avail as waterlines had been ruptured. To top it off, as a matter of national pride, the government reflex was to decline international assistance until shamed to do otherwise. The US offer of a nearby aircraft carrier that could have provided a floating two-thousand-bed medical facility was declined, and Swiss dogs specially trained in post-earthquake search-and-rescue operations were held for days in quarantine at the Kansai airport.[8] In fact, the government's response was so poor that the Yakuza (which is the Japanese mafia) reportedly took it upon itself to provide food and water to residents in some neighborhoods (which is, after all, not totally surprising given that a large part of organized crime's activities are about providing services that the government does not). Part of the problem was attributed to the fact that there was not an equivalent to the US Federal Emergency Management Agency (FEMA) in Japan at the time—although FEMA has had its problems with disasters too, as will be addressed later.

Third embarrassing matter: recovery was uneven. On the positive side, repair of bridges progressed at an impressively brisk pace; in fact, train service throughout Kobe was completely restored in four months, which is impressive considering that Kobe's three separate train lines (JR, Hanshin, and Hankyu), in addition to the Shinkansen (bullet train), each suffered damage over more than 20 miles of elevated tracks. However, the number of gas lines failures was so large that it took approximately three months to fully restore service to all customers. Three months can be an untenable delay for those who depend on gas for heat and cooking; contrary to transportation, where there are alternate roads and/or means of transportation that can be taken, there is only one line supplying gas per residence, and no alternative. Furthermore, because of earthquake damage or fires, a lot of people became homeless overnight. In rural settings, people can literally camp in their backyards, but in one of the most densely populated countries in the world, life is far from pastoral, and a lot of people were displaced. As a gauge of the problem, seventy thousand people spent two months or more living in temporary shelters, and some temporary housing units provided by the government remained needed for over five years. Of course, some optimistic spirits have professed that the post-earthquake recovery went smoothly considering the circumstances, but had this truly been the case, maybe Kobe's deputy mayor in charge of reconstruction would not have doused himself in kerosene and set himself on fire fourteen months after the earthquake,[9] becoming in the process one of the many victims of post-disaster stress—although not all those so afflicted have set themselves ablaze.

Now, at this point, it is important to emphasize that the Japanese were not ignorant of their exposure to damaging earthquakes. Quite on the contrary, and this is best explained by a department store analogy.

In Japan's crowded urban environment, department stores are typically multistory buildings often located at major train stations (not coincidentally, since some of the stores were originally owned by railway companies). Although the practice has tended to disappear in recent years, some of the bigger stores still employ young "elevator girls," dressed in the company uniform, whose purpose is to cheerfully greet customers entering the elevator, graciously thank them as they leave, and in-between call out the services provided on the floors at every stop. Going up, they would announce:

- Basement: Food department—the quintessential Japanese grocery store, providing everything one may desire, from blemish-free perfectly shaped fruits (a national obsession) to live octopus.
- Ground level: Cosmetics and beauty products—yes, that, by itself can fill an entire floor.
- First floor: Women's fashion.
- Second floor: More women's fashion—yes, that, by itself can fill more than an entire floor.
- Third floor: Men's fashion.
- Fourth floor: Sporting goods.
- Fifth floor: Home furnishing—from water purifiers to waterbeds.
- Sixth floor: Kids clothes, stationery, and toy department—the only place in the world where one can buy an Ultraman figurine (not your typical superhero—readers will have to Google it to appreciate).
- Top level: Restaurants of all kinds, which display, next to their entrance, plastic replicas of their main menu items—most convenient for the locals or international visitors who cannot quite decipher the Kanji, Katakana, and Hiragana symbols that combine to create the Japanese written language.

So, Japan, in a nutshell (and arguably in its collective subconscious), is a department store of disasters, where one would find:

- Ground floor: Floods—for example, thousands died in the summer of 1953 when dikes along rivers failed due to heavy rain.
- First floor: Landslides—another consequence of downpours, and a widespread problem given that 73 percent of Japan is covered by mountains.
- Second floor: Earthquakes.
- Third floor: Earthquakes—yes, that by itself can fill more than an entire floor.
- Fourth floor: Fires and conflagrations—for example, 143,000 people died in Tokyo in 1923, when the fire caused by the earthquake

caused more destruction and death than the earthquake itself. Note that, given the predominance of timber in Japanese residential construction, other cities there burned down on their own without the triggering effect of an earthquake, such as Hakodate, Hokkaido, in 1934.

- Fifth floor: Tsunamis, which are massive waves triggered by off-shore earthquakes. It is reported that 21,959 people died during the 1828 Sanriku earthquake and tsunami that struck Tohoku, the largest to hit Japan until the 2011 Tohoku earthquake and tsunami during which fewer people died (15,896), but which is considered to be the costliest natural disaster in recorded history (some estimates reach $360 billion),[10] and that created a meltdown at the Fukushima Daaichi Nuclear Power Plant complex.
- Sixth floor: Typhoons and storm surges, which go hand-in-hand—more than 19,000 people died in the 1828 typhoon that made landfall on Kyushu with 180-miles-per-hour winds.
- Seventh floor: Volcanoes—the Japanese islands are dotted with 110 active volcanoes, and while Mount Fuji near Tokyo has been dormant since 1707, a 2018 government study[11] indicates that a repeat of that event would paralyze the economic capital of Japan indefinitely.
- Eighth floor: Nuclear Bombs—Hiroshima and Nagasaki are the only two cities ever leveled by nuclear weapons. On August 6, 1945, 70% of Hiroshima's buildings were destroyed, and over 100,000 people died instantly or in the subsequent months from the effects of radiation.
- Top floor: Godzilla—this is actually where the collective subconscious part comes into the picture. As world record holder for the longest continuously running movie franchise, with thirty-one Japanese full-length features since 1954, a bad actor in a latex Godzilla suit, scaled as needed to always be taller than Japan's tallest building of the day, has trampled and destroyed countless scale models of Japanese cities, as the embodiment of the subconscious conviction that the country is perpetually on the brink of being annihilated by forces beyond its control. In short, Godzilla means: "If you live in Japan, beware that some disaster is always lurking around the corner."

No one spending time browsing through the Japanese department store of disasters can miss the fact that disasters—including earthquakes—have consistently occurred throughout Japan's history. Hence, prior to the 1995 Kobe earthquake, everybody in Japan knew that they lived in earthquake country. How they acted in the years prior to that earthquake very much depended on their perception of the risk of it happening soon, in their very own backyard, and of the possible consequences of that event. For each organization and each individual, the level of preparedness very much depended on the complex juggling of relative priorities that takes place in the human psyche. And that is not unique to Japan, but rather universal.

Which brings up the Three Little Pigs.

There have been many versions of this classic nursery rhyme, from 1853's England to today's YouTube (with Disney's Silly Symphony adaptation about halfway), so everybody is presumably familiar with the story. In a nutshell, a first Little Pig builds a straw hut and the second one a house of sticks. This is expeditious and leaves plenty of time for frivolous play and to ridicule the third Little Pig who labors to build a brick home. Then, out of the blue, as if nobody saw it coming, a "Big Bad Wolf" shows up and blows away the first two huts. Depending on the age of the audience, the first two pigs either find refuge in the fortress of the compassionate third Little Pig or end up in a delicious pulled-pork sandwich. The moral of the story is that hard work pays off and—evidently, here—that the third Little Pig is a better engineer: It pays to build a more resilient structure.[12]

Indeed, the Three Little Pigs story is most relevant here, when dealing with extreme events and circumstances, because not all houses are created equal and some will suffer more damage than others during any disaster. However, one of the most important points of the story, but one that is not stated even though it is at the root of countless decisions, is that if no wolf ever came, the first two Little Pigs would have won, so to speak, with more free time to enjoy life and dollars to spare—which is essentially counter to the lesson underscored by the nursery rhyme. Likewise, when it comes to earthquakes, investments in earthquake protection measures may never actually provide any return on investment in the lifetime of the investor if no damaging earthquake occurs. The same is true for all other extreme hazards or life-impacting conditions. Therein lies the dilemma. One can invest resources and energy in hope of maybe reaping a benefit in some distant future. However, even when fully aware of the risk, betting on the probability that no disaster will occur, to spend time and money instead on things providing immediate rewards, is always an option. And a most attractive one at that.

As such, at any point in time, depending on circumstances and timing, everybody can be any one of the three Little Pigs, which makes preventing disasters an uphill battle.

This is what is explored at length in this book.

ON THE DISASTER TRAIL

Kobe or Tokyo?

After a long day spent visiting a number of research labs in the Tokyo area, some Japanese engineers and I were in a small family-owned restaurant— the type where a television is on a shelf along the wall behind the chef. We were enjoying a glass of sake while waiting for the sushi, when all were jolted.

"Earthquake. Nice!"

"Hai!" they replied.

The shaking was mild and the entire room moved for only a few seconds, with a bang, as if a small truck had rammed the building. It was not scary but it is always impressive to feel an earthquake—and fun too when they are small and there is no damage.

"You haven't seen anything yet," they added, smiling, looking at their watch. "Minutes."

Sure enough, within three minutes, on the television, a news anchor interrupted the ongoing programming to report that a magnitude 4 earthquake had just happened, showing on a map the epicenter where it occurred and a map of the area over which it was felt. Wow. Being able to report in such "near-real-time," in mid-1994, was state-of-the-art, and as impressive as the earthquake itself.

I mentioned to them that I was going to be in Kyoto during the first six months of 1995, with my family, and that I hoped they would also get the chance to feel such small earthquakes and live a similar experience. They laughed and replied, "There are no earthquakes in Kyoto. That is in the southern half of Japan. If you want to experience earthquakes in Japan, you have to be near the Tokyo area, in the northern half of Japan."

Of course, six month later, on January 17, 1995, the magnitude 6.9 Great Hanshin earthquake shook the entire Kansai area that includes Kobe-Osaka-Kyoto (in the southern half of Japan), killing more than five thousand people and creating over $100 billion in damage. It was the largest Japanese earthquake since the 1923 Kanto earthquake that devastated Tokyo and the surrounding prefectures.

The "Big Bad Wolf" had visited Southern Japan.

Earthquakes Happen Because . . .

THE DNA OF AN EARTHQUAKE

Those who fundamentally believe that the earth is flat, or that the 1969 moon landing is an elaborate hoax filmed in a Hollywood studio, or that earthquakes are deliberate actions of God (or gods) to destroys sinful cities that have attained the "Sodom and Gomorrah" elite status, are likely to find this section offensive and can easily skip it without detrimental effects on their reading enjoyment (however, note that requests for pro-rated refunds of the book's cover price are non-receivable).

Early on in the history of humanity, the best brains of the world were quick to figure out the mechanisms that, every now and then, unpredictably, made the solid bedrock of all civilizations shake, rattle, and roll frantically for a few terrifying seconds. When these great thinkers experienced shaking strong enough to move all frames of reference into an infernal dance and frightening enough to violate their beliefs in the immovability of the world, when the surge in adrenalin amplified the acuteness of their senses and distorted their perception of time and space, they understood it all. They were able to explain what had happened and thus re-stabilize a world that had been tilted out of balance for a moment, although unaware that various nations had established similarly accurate, competing models of how the universe triggered such tremors.

In India's model, the Earth was supported by four gigantic elephants standing on the back of an even bigger turtle, itself standing on a cobra quite sizeable in itself. Understandably, any twitch by any one of those pillars of the world created terrifying ground motions.

The Japanese model, of superior simplicity, understood that all of the nation's islands rested on the back of Namazu,[1] the giant catfish pinned at the bottom of the sea by a demigod holding a stone on its head. Whenever the demigod got tired—as all demigods do every now and then—the pressure released and the catfish managed to wiggle, generating all those

11

vibrations that wreaked havoc at the surface.[2] Today, in a society where cute characters that look like childish-artwork (known as Kawaii)[3] are ingrained in culture and marketing, cute little catfishes are still used to symbolize earthquakes, for example in the logo of the 2020 World Conference in Earthquake Engineering[4] and of the Japan Meteorological Agency Earthquake Early Warning System.[5] In the latter logo, a worried (but cute) yellow catfish is broadcasting warning waves from its antenna-like (but cute) tail.[6]

It goes without saying that Japan and India were the more advanced civilizations. Had all of the world's fine thinkers met at some Antiquity's International Conferences, the Japanese and Indian delegates would have had quite a laugh at the Siberian model of the earth resting on a giant god-driven dog-sled in which earthquakes occurred when dogs scratched their fleas, as well as other equally misguided models that relied on giant frogs or dragons.

Eventually, it became obvious to the sages and savants that the gods did not need fishes or dogs, and could perfectly wreak havoc on their own. Some were baby gods, in Mother Earthquake's womb, creating earthquakes every time they kicked her tummy, as any pregnant woman knows; others were full grown gods who carried the earth in their arms or on their shoulders and had a bad day or a fit of anger every now and then.

Other earthquake-creating gods had more complex personality issues, like the Scandinavian god Loki. There, as in all Norse mythology, the narrative gets complicated: in a nutshell, for having murdered his brother, Loki was sentenced to spend the rest of eternity tied to a bunch of big rocks deep within the earth, which so annoyingly happened to be located right under a poison-dripping snake. With each abrupt twist by Loki to avoid a falling drop of venom, came—what else—an earthquake.[7]

At the other end of the spectrum, for simplicity, were the super-specialists, like Rūaumoko, the Mauri god whose main job description is to create earthquakes and volcano eruptions, which he triggers by merely walking around New Zealand. Yes, that country is quite seismically active: one earthquake per Rūaumoko step to be exact.[8]

Eventually, some civilizations came to their senses and realized that gods were not needed to create earthquakes. Even the Greeks—who long held that earthquakes were caused by Poseidon striking the ground with his trident—eventually accepted Aristotle's more rational and elegant explanation that earthquakes were nothing more than the results of crazy winds trapped in underground caverns, struggling to escape.[9]

Yet, for most nations, moving away from the view that turtles, dogs, frogs, fishes, dragons, and annoyed gods were responsible for creating earthquakes was a hard transition—particularly when the prevailing theology wholeheartedly embraced the due rewards and punishments of divine justice. Which brings 1755 Lisbon into the fore.

At that time, the capital of Portugal was flush in gold imported from its Brazilian colony, and flush with religious orders. As ecclesiastic establishments of the era generally excelled at enriching themselves, their growing wealth attracted more devotees, and by 1750, Portugal counted more than 200,000 priests, nuns, and friars, and no less than 409 monasteries and 129 convents of 12 monastic orders (ranging from the Franciscans, with approximately 200 monasteries, to the Carthusians with only two).[10] All of that while Portugal's entire population was approximately two million people,[11] and its size half of that of today's Florida. To say that Lisbon was the capital of what considered itself to be a pious nation is, by all counts, an understatement. With Franciscan, Bernardine, Augustinian Dominican, Jesuit, Benedictine, Carmelite, Paulist Hieronimite, Loios, Trinitarian, and Carthusian orders running the show, Catholicism had a monopolistic control of the Portuguese faith.

Presumably holding the record for largest number of prayer candles burning per square miles (outside of the Vatican), to the eyes of all these devout Catholics, if a city had to be in God's good graces, it had to be Lisbon. Or so they might have thought, until that fateful All Saints' Day in 1755, when many of Lisbon's quarter of a million people were in church celebrating the spiritual bridge between heaven and earth provided by all of the Church's known and unknown saints.

Suddenly, in the Basilica de Santa Maria, chandeliers began to swing madly, the walls of the cathedral rocked, and the congregation rushed out to the streets. From the square in front of the church, where they were joined by other residents fleeing from all nearby buildings, some reported seeing the spires of the Basilica wave "like a cornfield in a breeze." Then it stopped. Terrified, most stayed there and prayed, asking for God's protection. The response came promptly. Shaking resumed, with more power than before. The shock was so severe that the heavy masonry façades of the Basilica and of adjacent grand buildings collapsed toward the square, burying everybody in stone rubble—or, symbolically, in piled up tombstones.

Forty minutes later, a tsunami hit Portugal's shore; it flooded the harbor and the downtown area. To top it off, throughout all that violent shaking, some of the burning votive candles uncontrollably spilled their vows all around, igniting all nearby combustible materials, including the tapestries ubiquitous in those days to both insulate and decorate churches. In other buildings, collapsing wood floors and roofs sometimes landed in burning fireplaces, which added more fuel to the problem. From there, it grew into a conflagration that burned for three days. It is estimated that 85 percent of Lisbon's buildings were destroyed and that 15 to 30 percent of Lisbon's population died from the earthquake-tsunami-fire triple whammy.[12]

All these deaths and damage did not make the Lisbon earthquake that different from other similar catastrophes prior. What was remarkable, rather, was its timing: smack dab into the Age of Enlightenment.

It did not matter that Catholicism focused on afterlife, with heaven and hell being the essential outcome as reward or punishment for present life, providing a meaning for the current existence. Those embracing a life of religious devotion were hard pressed to explain why a God of mercy and justice would decide, out of the blue, to unleash fire and fury on a city that was considered to be a bastion of the faith, to destroy its sacred temples, and to wipe out disciples by the thousands. Crushing them under collapsing basilicas, to make it worse. So, to some thinkers, either the faithful of Lisbon had been wrong all the time to believe in a merciful God, or earthquakes had to be caused by something else altogether that had nothing to do with God.

Before the earthquake itself, in the early stages of the scientific revolution, philosophical optimists, such as Leibniz, seeking to reconcile scientific understanding and religious beliefs, had advanced that God had created the best possible of all worlds.[13] They believed that the universe created by God was a Celestial Clockwork[14] that had to be, maybe not perfect, but the best of all possible outcomes—the premise being that since only God can be perfect, the universe could only be "nearly perfect." Leibniz, a world-renowned mathematician,[15] claimed to have logically proven this. As such, the philosophical optimists had painted themselves into a corner: Either God had created the best possible world, or God did not exist. As such, the Lisbon earthquake severely shook and fractured the foundations of that philosophy. The widespread destruction and death throughout the city could not possibly be understood as being the outcome of a near perfect divinely ordered clockwork, created by a God of goodness and mercy. Only an uncaring and dark force could have so harshly swept away so many Christian lives and churches—crushing the philosophical optimists in the process.

The irony of the Lisbon earthquake was not lost on the critics of religion. As one of the most highly regarded satirical polemicists at the time, Voltaire had a blast. The Lisbon earthquake led him to write the novel: *Candide: or, The Optimist*—his magnum opus. Not surprisingly, the book was officially banned by the Church, on account of religious blasphemy and political sedition, which must have made for great publicity, as it became the best-seller of 1759, and is alleged to have been the fastest selling book ever at that time.

While the contentious matters of philosophy and religion attracted all the attention then—as is often the case now—behind the scenes, a more scientifically inclined thinker looked at the 1755 Lisbon earthquake from a different perspective. Since earthquakes were not anymore, literally, acts of God, it made sense to try to understand what they are. By definition, this can effectively be considered the birth of seismology, even though the word itself did not exist at the time. Based on his studies of the Lisbon earthquake, the English geologist John Michell published in 1760 the first scientific article attempting to explain what created earthquakes.[16] Like most scientific publications, this one "flew under the radar," was not banned by the Church, and was most probably not a best seller. Evidently, it would have been a miracle of science for the first study on such a complex and difficult topic, on a phenomenon that cannot be directly observed or measured, to reach conclusions that were flawless. So, not surprisingly, he was totally wrong when he stated that earthquakes are caused by boiling water from nearby volcanoes, but it was a nice try, nonetheless. However, he hit a bullseye when he stated that earthquakes create shock waves and that, like any other traveling wave, the location where the earthquake was triggered could be determined by triangulation, knowing the time when the waves were felt at various locations—a brilliant insight given eighteenth-century technology. Note that it took almost a century after Michell—until 1846—before the words "seismology" and "epicenter" were coined by Robert Mallet.[17] Mallet was an engineer who detonated underground bombs, like mini earthquakes, seeking to measure the speed of shock waves in various kinds of soils.

From there on, the most broadly known achievements of seismology were to develop: (1) the Mercalli Intensity scale to measure the severity of the earthquake by correlation to the damage it produced, and (2) the Richter Magnitude scale to measure the inherent power of an earthquake as a physical phenomenon by itself, independently of the damage it produces (more on that later). Development of the knowledge to understand what geophysical process actually produces earthquakes took much longer, and that story follows a tortuous path, because before getting there, one first had to recognize that the surface of the earth was not static—a fact that was hard to swallow. In other words, a map of planet earth purchased three billion years ago at the Galactic Hitchhiker's store is nowadays completely obsolete, because every continent and many other parts of the earth's surface have moved over time.[18]

Once upon a time, all of the earth's continents slept together, as in a hippie commune. Africa was spooning with South America, and North America embraced both at the same time.[19] This consensual intimate relationship between all partners lasted for a while, but eventually, like many similar polyamorous relationships where equilibrium is difficult to maintain, it did not last. Since their "break-up," the continents have been moving away from each other, like partners from a violent divorce. In other words, not only is the earth shaking every now and then, but it does so because its continents are traveling. Some take their time, moving about half an inch per year, while others race through at four inches per year.[20]

The idea that continents were once snugging together can be traced back to 1596, when the Dutch cartographer Abraham Ortelius—who is credited for having created the first modern world atlas—noticed how well South America and Africa would fit together if they were two pieces of a worldwide jigsaw puzzle. He even went as far as suggesting that they might have been torn-away by earthquakes and flood.[21] History does not say if any of his contemporaries took the idea seriously, or dismissed it as the senile ramblings of a sixty-nine-year-old dinosaur in an era when life expectancy was about forty years. Irrespectively, many respected individuals reflected on this over the course of the following centuries, and most notably through the nineteenth century,[22] with the geographer Antonio Snider-Pellegrini going as far as showing in an 1858 map how Africa and South America could have fit together. Yet, while some other geographers probably thought in similar terms, it is the thirty-two-year-old German *meteorologist* Alfred Lothar Wegener who apparently first coined the term "continental drift" in two papers published in 1912.

Even though a 1912 meteorologist who publishes in scientific journals is more credible—albeit sometimes less well paid[23]—than one who presents the weather forecasts during the evening news, in the eyes of geologists that credibility was apparently not worth much. To put it mildly, Wegener's theory was not met with much enthusiasm. He certainly had supported his theory with evidence of matching geologic structures and fossils where South America and Africa would have been once glued together, but he had no explanation as to what force could have propelled the continents apart from each other.

Yet, over the fifty years that followed in the twentieth century, it was found that the movement of continents over time was the only possible explanation to a number of other scientific observations. For example, when rocks are formed, some of the minerals that are embedded in the rocks lock in the orientation of the earth's magnetic field. When rocks of different ages were compared, the orientation of the magnetic field was different, so either the magnetic north pole drifted over time, or the continents did. If the magnetic north pole had drifted while all continents remained where we know them to be today, then at any given point in time, the magnetic pole location would have been recorded to be the same by all rocks on all continents. It was not the case, so only the drifting and rotation of continents in different directions could explain this discrepancy in the data. Likewise, early studies that surveyed the deep ocean floors provided tons of evidence that the seafloor of the Atlantic Ocean was spreading in both directions from the mid-oceanic ridges,[24] expanding the width of the ocean at those locations. Imaging techniques decades later showed that, in the Pacific Ocean, a layer of the oceanic crust was pushed under other layers, reducing the width of the ocean there. This makes sense since the circumference and diameter of the earth, as a sphere, was not observed to grow.

Rapidly, it was possible to map the boundaries of all the moving "plates" at the surface of the earth, where either "new earth" was pushed out at the bottom of the oceans, or where "old-earth" was pushed in.[25] This rapidly became the plate tectonics revolution.[26] Furthermore, the birth of space-age satellite-based geodesy in the late twentieth century made it possible to measure the drift of each of these individual plates using the Global Positioning System (GPS).[27] As a result, plate tectonics is nowadays a widely recognized and accepted phenomenon.

The plate tectonic revolution was also timely, as it tied together nicely with "on the ground" observations following earthquakes. This all helped explain what creates earthquakes—finally, getting to a credible explanation.

Thirty miles northwest of San Francisco, the San Andreas Fault crosses the Point Reyes National Seashore. The long and narrow Tomales Bay at the northeast edge of the National Seashore makes it clear on a map where Point Reyes is sawed away by the San Andreas Fault. This straight line cut, clearly visible from aerial photographs, is inconspicuous at ground level, particularly when the fault is inland. Hence, when an Olema Valley farmer built a picket fence, years prior to 1906, unbeknown to him was the fact that it was perpendicularly crossing the fault—and that he had built a monument that would decades later belong to the federal government. All because when the 1906 earthquake hit, the land on the west side of the San Andreas Fault moved north. The part of the fence sitting on that side of the fault came along for the ride, sheared off from the rest of the fence, and traveled a grand total of sixteen feet (a road crossing the fault further north close to Tomales Bay shifted twenty feet).[28] A replica of the west and east remnants of that fence are now part of the Earthquake Trail in the National Seashore, as a reminder of the 1906 earthquake.[29]

Geologists knew that earthquakes occurred along faults, and the San Andreas Fault had already been identified eleven years earlier.[30] Events such as the 1906 earthquake made it possible for them to witness the slippage that developed along the fault during that earthquake and formulate theories—such as the elastic rebound theory—to explain those locally observed permanent shifts. In essence, the elastic rebound theory can be explained as follows. If the surfaces that meet at the fault line were perfectly lubricated, like two pieces of polished steel sprayed with 10W40, the relative movement between both sides of the fault would occur continuously, at a slow rate imperceptible to the human eye. However, rough rocks are not polished, but rather locked together along surfaces going miles deep in the ground and hundreds of miles horizontally. They do not slip easily. The continental plates may be stuck together at their edges (that is, at the fault line), but that does not prevent them from continuing to displace relative to each other. Like two dog owners moving away from each other while their dogs dig their paws determined to stay put together. If the dog leashes are rubber bands, as the owners move further away and progressively more energy accumulates in the stretching leashes, Fido and Fluffy are setting themselves up for quite a rebound.

The energy locked in a major fault is enormous, and when that energy is released, earthquakes are produced by the rocks along the fault grinding on each other during that rebound. The sheer magnitude of this energy makes nuclear bombs look puny. For example, the 2011 Tohoku earthquake released energy equivalent to 9.32 million megatons of TNT, which is roughly six hundred million times the energy of the "Little Boy" nuclear bomb that exploded over Hiroshima in 1945.[31] To be clear, the plan hatched by the zany sociopathic villain in one of the worst-ever James Bond movies, which consisted of blowing up a single bomb inside the San Andreas Fault to wipe out Silicon Valley, was ludicrous, a severe case of bad math, and bad movie making.

It takes time to build up all that energy along a fault before it releases. If a fault always slipped twenty feet during each earthquake, and the relative movements between two plates were recorded to be one inch per year, this would imply that there would be an earthquake every 240 years. Unfortunately, there are at least two reasons why this will not happen as simply as that. First, seismic faults are not Swiss Cuckoo clock parades of wooden characters recurring at perfectly regular interval. Across a segment of fault miles deep and hundreds of miles long, rocks not only lock together in a random fashion, but there is no way to see how they are locked together. Second, humans have only been recording useful information about specific ground motions for a few hundred years—or less in some cases.

Earthquakes that have happened before human settlements are not forever lost though. Seismologists have dug trenches across faults in search of geological evidence of prehistorical earthquakes in the rock layers, which has made it possible to determine that the San Andreas Fault near San Francisco has ruptured at an average interval of two hundred to three hundred years.[32] Over a million years, the fact that the San Andreas Fault moves at a relative rate of roughly one inch per year[33] adds up to a total displacement of one million inches, but in fits and starts. At that rate, given that the fault slices California north to south, all the way down to Southern California, Los Angeles will become a suburb of San Francisco in only twenty-four million years. By the way, it is always great fun to see the bewilderment of those who watch an accelerated video animation of that phenomenon and think that it will happen in twenty-four years because they fail to see the "x10⁶" in small print next to the numbers on the counter (which is a mathematical shortcut to say 24,000,000).

Nonetheless, while everybody can visit the sixteen-foot offset in the fence at Point Reyes, there is no way to know if the 1906 earthquake released all of its locked strains, or only some of it, leaving a lot of it temporarily locked and accumulating more stresses for another imminent release of energy. This means that the next "Big One" there could be anytime (and, to make it more exciting for San Franciscans, keep in mind that the city is surrounded by other equally dangerous faults expected to rupture sooner).[34]

All of that is to say that, bottom line, earthquakes have happened for as long as the earth has existed, and most probably will continue to do so for as long as it will exist. This process cannot be stopped.

Might as well learn how to live with them.

CAN'T ESCAPE THEM IS BAD; CAN'T PREDICT THEM IS WORSE

DISCLAIMER: Some scientists underscore that there exists an important difference between a prediction and a forecast. Some argue that a prediction must define at what specific time and location an event will occur, whereas a forecast is expressed as probabilities over a short time window.[35] Others argue that a forecast relies on calculations using past and present data to extrapolate into the future, whereas a prediction can be based on facts, judgment, experience, opinion, or a crystal ball.[36] As interesting a nuance as this may be, to the general public (and throughout this book), the two words are used interchangeably and context will be sufficient to determine whether the words refer to hours of calculations by an egghead using complex models running on a cluster of supercomputers, or an opinion from Grandpa after sniffing the air for a few seconds.

Anyone who believes that scientists will soon be able to provide a warning of exactly when and where an earthquake of a specific size will strike has been watching "way too many" bad Hollywood disaster movies.

Barring an explosion of amazing scientific or engineering breakthroughs (each of the $E = mc^2$ jaw-dropping kind), this is not likely to happen in this lifetime. Pessimists will argue that funding research focused on short-term earthquake prediction is like throwing dollars into a money pit, with little expectation of return on investment. Hard-core pessimists will add that it is more of a black hole than a money pit, sucking resources into oblivion. Optimists (and most definitely those who thrive on this kind of research funding) will argue that breakthroughs are imminent and that the benefits that society will reap by being able to make such predictions will far outweigh the costs of the research. Whether or not such predictions really can be of any value will be examined later, but it remains true that the inability to make short-term credible predictions has been frustrating to seismologists, particularly because it often seems to be the only thing that people want to know about earthquakes.[37]

In the long history of failed earthquake predictions, the "Parkfield Experiment" is a special example of optimism blown to pieces. Parkfield is a world famous, minuscule village located near the middle of the 750 mile-long San Andreas fault with a population of eighteen persons (in 2007).[38] World famous in the seismological community that is. It happens to be located along a part of the San Andreas Fault that creeps at a constant rate of roughly an inch per year.[39] In fact, the ends of a bridge built in 1936 across the fault[40] have continuously moved laterally with respect to each other, bending the bridge by more than five feet since.[41]

Moderate earthquakes of roughly magnitude 6 are known to have occurred there previously in 1857, 1881, 1901, 1922, 1934, and 1966. At some point in the early 1980s, someone brought up the fact that not only did this correspond to an average of twenty-two years, but it also seemed to happen like clockwork about every twenty-two years—as long as the geological clock can be forgiven for not having Swiss-watch accuracy, because the 1934 event clearly fell out of line by a substantial margin. Nonetheless, this unusual regularity in observed seismicity, coupled with the fact that seismographic records (available only for the last two in that series) showed identical epicenter locations along the fault, was alluring. If Mother Nature was to throw a wild earthquake party in Parkfield twenty-two years after 1966, seismologists surely did not want to miss it—as far as an earthquake can be an exhilarating proxy for sex and booze when it comes to a seismological wild party. As such, in 1985, the US Geological Survey (USGS) issued a bold but official prediction that the next magnitude 6 earthquake in Parkfield would occur in 1988—the very first such prediction by the USGS. The USGS is a serious agency of the federal government whose mandate is to "provide science about the natural hazards that threaten lives and livelihoods."[42] It not only published this prediction in *Science* magazine,[43] but also officially informed the governor of California—evidently catching the media's immediate attention. Wisely recognizing that the creeping motion of the San Andreas Fault did not operate as perfect clockwork, it qualified its prediction by stating that there was a 95 percent chance that a magnitude 6 earthquake would occur there in the ten-year window between 1983 and 1993. It was a fairly safe bet to make, since it called for a relatively small earthquake in an area where only a few ranchers lived, thus avoiding the risk of a panic and economic depression due to fleeing residents—in case anyone there trusted the government's prediction. Given that the 1985 prediction was already encroaching, two years into that ten-year window, $20 million[44] were spent in a rush to "pepper" the Parkfield area with instruments, such as seismometers, strainmeters, creepmeters, GPS receivers, and other state-of-the-art gadgets to record everything physical and chemical that could occur in rocks, soil, and ground water, during and after the event.[45]

Then, 1988 came and went without any magnitude 6 earthquake occurring in Parkfield. So did 1989, 1990, 1991, 1992, and, finally, closing the window with a big disappointment, 1993. In short, Parkfield could have been to seismologists what Woodstock was to hippies, if the main attraction had actually showed up to perform—but it did not.

A magnitude 6 earthquake eventually occurred in Parkfield in 2004, which prompted some to victoriously proclaim that the predicted earthquake had occurred after all, "as predicted," only a bit later than originally foreseen.[46] Those likely to find solace in that argument are probably the same ones who would not mind if the groom or bride showed up sixteen years late to their wedding, expecting guests to have waited in the aisles.

As a consolation prize, all those still dreaming of an earthquake-Woodstock can travel to the Parkfield café, rated by TripAdvisor as the best (and only) restaurant in Parkfield,[47] to take a selfie with the café's big outdoor steel water tank where is inscribed in vivid letters: "Earthquake Capitol [*sic*] of the World—Be Here When It Happens." Admittedly, the best and the brightest might have been seduced by the simplicity of the measured constant creep-rate along the fault and the perception of a steady return period in Parkfield. Critics have described the assumptions underlying the prediction as simplistic and possibly politically motivated.[48]

In fact, Parkfield is nothing but one more unsuccessful experiment. Actually, hundreds of millions of dollars have been spent by various government agencies and research funding agencies throughout the twentieth century in support of research on earthquake prediction—with, arguably, nothing to show for it.[49] This was often done in spite of repeated testimony by engineers that it would be more pragmatic instead to fund research on how to design buildings to perform better during earthquakes.

Funding for research on earthquake prediction peaked in the 1960s and 70s, and proponents of this research used a battery of scientific approaches and measurement techniques to prop their optimism in this wild goose chase. Many felt they had to catch up to the Russians and most particularly the Chinese who claimed great successes in earthquake prediction.

Indeed, in the mid-1970s, the Chinese government claimed that they had nailed the science of earthquake prediction. In one such success story, they had issued an official warning at 10 a.m. on February 4, 1975, predicting that an earthquake was to occur soon in the Liaoning Province.[50] The fact that this is a large province covering 56,332 square miles—being more than 350 miles at its widest point[51]—somehow got lost in the official narrative. When a magnitude 7.3 earthquake actually occurred nine hours later, near Haicheng, located right in the middle of that province, wild claims were made to the effect that the scientific foresight precisely had pinpointed the location of the earthquake ahead of time. It was also claimed that the government accordingly had ordered the city to be evacuated, but no such evidence has ever been produced. In any case, the earthquake killed 1,328 people in Haicheng—hardly a success story by any standard. The truth of the matter, though, is that hundreds of foreshocks occurred over a three-day period before the large devastating earthquake hit, including hundreds of shocks in the early hours of February 4—foreshocks being small earthquakes that damage nothing but usually scare people enough to make them sleep outside out of precaution, particularly when foreshocks come in swarms. Therefore, while it is true that the Chinese seismologists made a successful prediction, anybody with an IQ above 50 might have also successfully called this one.

The problem is that most large earthquakes are more sneaky than that and do not provide the luxury of foreshocks. That was the case for the tragedy of the Tangshan earthquake that hit on July 28, 1976. In spite of the Chinese claims (emboldened by the "success" of the Haicheng prediction eighteen months earlier) that seismologists there had nailed the science of earthquake prediction by monitoring water levels in wells, concentration of radon and other chemicals in groundwater, inclination of the land using tilt meters, changes in magnetic fields, temperature anomalies, and changes in the behavior of dozens of species of animals, the plain fact remains that nobody—absolutely nobody—foresaw the Tangshan earthquake. By the official count of the Chinese government, it killed two hundred forty thousand people, making it one of the deadliest earthquakes in recorded history. Although some unofficial estimates report six hundred fifty-five thousand victims, the true number will likely never be known because one plus one equals whatever the government needs in some regimes.

In an entire textbook devoted to the topic of earthquake prediction written by the German-Chilean-Mexican renowned seismologist Cinna Lomnitz,[52] each scientific method that has once been considered valid or promising for earthquake prediction is reviewed, explained, and—more importantly—debunked. As stated loud and clear in the book's concluding chapter, earthquake predictions are about as successful as predictions of economic recessions or the outcome of elections. Simply put, earthquakes cannot be predicted.

Today, posted on the USGS website is the definitive statement: "Neither the USGS nor any other scientists have ever predicted a major earthquake. We do not know how, and we do not expect to know how any time in the foreseeable future. USGS scientists can only calculate the probability that a significant earthquake will occur in a specific area within a certain number of years." On that last point, the USGS typically does so to provide the values (and maps) of seismic parameters used for the design of buildings and other infrastructure, using statistical modeling to forecast over a fifty-year period.

In summary, none of the prediction methods proposed in the past work. Still, facts have never prevented colorful characters from making claims to the contrary.

QUACKS AND CLOWNS, OR GENIUSES?

As the geologist Jim Berkland used to say, "It's really very simple to predict earthquakes, and anyone could do it. The hard part is being right."[53] Following the principle that most broken clocks can give the right time twice per day, if one predicts earthquakes often enough, a successful prediction is likely to occur at some point.[54] Since people are inclined to remember only successes, the red carpet is usually ready for charlatans, pseudo-scientists, prophets, and attention seekers.

Sometimes it is all about packaging. For example, there is apparently a guru out there advertising his services to expecting parents, claiming that he has the mind power to control the universe such that they will get a kid of the gender of their choice. In fact, he is so confident in his special "skill" that should his intervention fail in their specific case, he will fully refund them his $5,000 fee. With such an ironclad money back guarantee, what is there to lose? Inevitably, the guru is bound to be right half the time, and thus to keep his fee half the time, which makes for a pretty lucrative business. Earthquake prediction by quacks, clowns, and prophets can be somewhat similar in concept, except that it is not always clear if there is any monetary gain to be made in this case. In this era of social media, some have been ingenious in turning prediction websites into dollars, but most seem motivated by love of the limelight or by an eagerness to save lives using their sixth sense, extrasensory perception, psychic gifts, or super-powers. Many have also been skillful in getting free publicity by attracting media attention.

In a world where "infotainment"[55] is king and "information" is not, clicks and likes and eyeballs matters more than substance. What does it say when all international media outlets considered it newsworthy to report the story of a ten-year old grilled cheese sandwich that sold for $28,000 on eBay because the burn patterns on its toasted bread purportedly showed the image of the Virgin Mary?[56] Although some argued that the face looked more like Greta Garbo or Marlene Dietrich than the Virgin Mary,[57] it made perfect sense to hard-core believers that only a miracle from Jesus's mother could keep a ten-year old sandwich free of mold. Some have ventured scientific explanations as to why a ten-year old grilled cheese can be mold-free,[58] some have taken advantage of the frenzy to explain the brain's ten-dency to formulate familiar images in clouds, ink blots, or other random patterns (a phenomenon called pareidolia),[59] and others have simply called it a hoax.[60] Not to forget those who took advantage of the news coverage to advertise toasters that can burn a perfect image of Jesus on the morning daily bread.[61]

If that entire circus is considered legitimate news, it then stands to reason that any atypical or free-spirited earthquake prediction will get its fifteen minutes of fame. Examples include the prediction by a self-proclaimed Dutch "quake-mystic" that an 8.8+ magnitude earthquake was to strike California on May 28, 2015 (it did not happen),[62] and (broad-ening the horizons a bit) that a "megaquake" of magnitude 8 or greater was to occur somewhere on the planet between December 21 and 25, 2018 (it did not happen), and that a magnitude 7 earthquake was to occur on the planet on April 12, 2019 (it did not happen either).[63] Considering that at least one magnitude 8, eighteen magnitude 7, and 150 magnitude 6 earthquakes occur each year on the planet,[64] lowering the size of the pre-dicted earthquake with each prediction is an excellent strategy to eventu-ally hit the jackpot. Another brilliant strategy would consist of enlisting 365 Nostradamus-minded friends to each pick a day of the year and post on Facebook their prediction of a magnitude 8 earthquake somewhere on the planet on that day; one of them is sure to be crowned a genius by all media when the yearly magnitude 8 happens. Even more so if it happens near a populated area rather than in the middle of nowhere as is more frequently the case.

Also, since making a prediction on a hunch costs nothing, might as well aim high: a Pakistani visionary with a "sixth sense" alerted the governments of seven Asian countries on November 2017 that a massive tsunami was expected within two months in the Indian Ocean (it did not happen).[65] This benevolent warning forced all the emergency response agencies in those countries to be on alert just in case—not because anyone believed it, but rather because nobody wanted to be blamed if by sheer coincidence such a tsunami did happen that month.

What matters for a prediction method to be credible is repeated successes under close, independent, expert scrutiny, and predictions by "free-spirited theorists" have not scored well on that account. Predicting earthquakes by looking at the shape of clouds, lights, electric signals,[66] vapors, snowpacks, aching bunions, and other unrelated phenomena has happened consistently over the years and has—not surprisingly—been consistently ignored by the scientific community.[67]

It is amazing (and possibly disturbing) that brash forecasts based on nothing more than hunches or pseudoscience receive massive media attention and have even prompted many to stock-up on supplies and brace themselves for the upcoming jolts, while forecasts anchored in real science often fail to generate excitement—or often any response at all. Or maybe it is not so surprising, given that some people pay fortune tellers to "read" their future in tarot cards (cartomancy), lines in the palms of their hands (chiromancy), crystal balls (crystallomancy), shapes of clouds (nephelomancy), numbers (numerology), tea leaves (tasseomancy), and even the wrinkles, cracks, bruises and dimples on butt cheeks (rumpology).[68] For instance, in 1968, clairvoyant Elizabeth Steen predicted that a giant earthquake would destroy San Francisco that year, and moved with her family to Spokane, Washington[69]—followed by thirty-five other families equally moved by her mystical prophecy. Evidently, nothing happened in 1968, but others built on that prediction after the fact. Combining it with an earlier prediction by another "prophet" and a fictional account in a book of California falling into the sea in 1969, some people became convinced that the fateful event would occur in April 1969. Arguably, these were the "groovy" sixties, when hard drugs facilitated prophecies and conspiracy theories were plentiful—after all, if the government lied about JFK's assassination, the moon landing, and flying saucers in Area 51, why not earthquakes? This might explain why hundreds left the state, sometimes incited by preachers leading the flock to safety, in spite of all the public announcements and newspaper articles attempting to counter the rumor by providing factual information. Again, nothing cataclysmic happened— San Francisco made it through the year.

There is, in principle, no limit to the possible escalation of media madness and nonsense driven by a "precise" earthquake prediction. This was well illustrated by the dozens of national and international media news crews that drove satellite and radio vans to New Madrid, Missouri, on December 3, 1990, waiting for the earthquake predicted by Dr. Iben Browning to happen. Dr. Browning's PhD was in the field of zoology, with minors in genetics and bacteriology.[70] He became a business consultant in various scientific fields. His prediction of a magnitude 6.5 to 7.5 earthquake to strike in New Madrid on December 3 was based on the expectation that the alignment of the Moon and Sun with the Earth on that date, combined with a few other astronomical factors, would create record tides last seen 179 years ago. Given that a series of large earthquakes of up to magnitude 8 had occurred in the New Madrid area from December 16, 1811, to February 7, 1812 (that is, 179 years before 1990), it apparently all added up. To put things in perspective, Browning predicted that bridges across the Mississippi River would collapse and that a third of the buildings in Chicago would suffer damage from the upcoming earthquake. No matter how many experts discredited the prediction, nor how many scientists explained that none of it made sense, media could always find someone to make comments that left enough uncertainty in the air.

Prior to the D-day, members of emergency response agencies in Missouri and nearby states attended seminars to learn about earthquakes and made sure all their equipment was ready to deploy. Many of these agencies conducted earthquake drills and prepared emergency shelters equipped with food, water, and medical supplies. More than a thousand emergency responders participated in a mock disaster drill in St. Louis. The Arkansas National Guard performed a drill for a scenario that assumed a magnitude 7.6 earthquake with 4,950 dead, 25,097 seriously injured, and 98,020 left homeless. At least sixteen hundred members of the Kentucky National Guard were on call. Leaflets on how to mitigate losses from earthquakes were widely distributed.

Schools around the New Madrid area in Arkansas, Illinois, Indiana, Missouri, and Tennessee closed on December 3, some for three whole days just to be safe. Some businesses closed, including an aluminum plant with fifteen hundred employees. Some people decided it was best to leave town for that day. Christmas parades and seasonal concerts were canceled. Religious activities abounded, and a preacher used a van equipped with megaphones to preach about—why not?—the end of the world. Of course, for many people, December 3 turned out to be a great outdoor event, with rock bands, "quake-burgers," "Shake Rattle and Roll T-shirts," and all the stuff typically found in a tailgating party or a holiday celebration.

As expected, December 3 passed, the earthquake did not happen, and the news crews went back home. All that remains from that temporary insanity is some old T-shirts, maybe some survival toolkits, and a fascinating USGS report that was produced in an attempt to understand how such an unfounded prediction attracted so much attention.[71] The report contains nearly two hundred pages of press clipping (as a small but representative fraction of the total number of news articles published) that "built-up" the frenzy by promoting the prediction and informing people on what actions to take to prepare, survive, and live after the earthquake. The report concluded that, as one of its many findings, "what began as an interesting story, but one not given particular credibility, evolved into a contagion that promoted Browning's prediction. Many educated and otherwise sensible people were caught up in believing that perhaps Browning was right. New York's finest ad agency could not have done it better."

Once all is said and done, there is only one method to predict earthquakes that is 100 percent infallible. It is known as the Richter Law of earthquake prediction—from the same Richter who invented the earthquake Richter scale and the seismograph, no less. This best-kept secret states that every earthquake takes place within three months of an equinox,[72] and to this day, no earthquake has ever occurred outside of this range (it usually takes a few seconds to do the math on that one).

More seriously, the only possible way to predict the future occurrence of earthquakes is to review information from all sound scientific bases—fault-slip rates, average return periods of past earthquakes, and other seismological data—and provide a probabilistic assessment. Notably, for the San Francisco–Oakland–San Jose Bay Area, a region crisscrossed by multiple faults that have been extensively studied for decades by the best minds in the field, the official prediction is that there is a 72 percent chance for one or more earthquakes of magnitude greater than 6.7 to occur between 2014 and 2043.[73] That is informative in some quantitative way, and hopefully helpful to incite people to prepare, but it is probably no more useful than if the Three Little Pigs were told that there is a 72 percent chance that the Big Bad Wolf will show up some time in the next thirty years.

ON THE DISASTER TRAIL

Make It Stop
Three days after the Christchurch earthquake in New Zealand, I was awakened at 3 a.m. by a rather strong aftershock. There had been a few more during the night strong enough for me to open an eye, think "aftershock," and fall back to sleep, but that one was much stronger. It also lasted longer. It said, "are you sure it was a good idea to fly down here, dummy?"

I got out of bed and walked to the window. Across the street, the masonry building that had collapsed the day of the big shock was there, in the dark, as a grim reminder of what happens to those poorly equipped to face the forces of nature. I remembered that a good friend once told me, "Some people chase butterflies, others chase earthquakes." Not quite. You cannot drive around to "catch" an earthquake when it happens, like the storm chasers who follow dark clouds to catch tornados, but you can walk through the rubble it creates. Actually, more importantly, you can walk in the buildings still standing up to diagnose why they suffered specific types of damage. In research labs, we spend hundreds of thousands of dollars to tests specimens that are parts and pieces of buildings—or even to test scale models of entire buildings on shaking tables—in well controlled experiments, to investigate how things will behave during future earthquakes. Here, nature gave us all the results of many full-scale experiments, leaving damaged "specimens" everywhere. All that is needed is to figure out what the actual properties of the tested specimens were—as a sort of "reverse engineering" problem.

Staring at the collapsed building in the dark, I remembered another evening, almost twenty years earlier, in Los Angeles following the Northridge earthquake, in another hotel whose architecture suggested that it was built with a decent amount of reinforced concrete walls and was therefore likely to be safe for a few nights during a post-earthquake visit. In the penthouse restaurant of that other hotel, an aftershock started to shake the room. A woman at a table nearby started to shout: "Not again! Make it stop, make it stop!" She was quite frazzled. She might have been one of the twenty thousand people whose house got damaged, forcing her to stay in a hotel. There had been more than $20 billion worth of damage to single-family homes north of Los Angeles, in Northridge—only half of it insured.[74] Here in Christchurch, thousands of homes were similarly damaged (the newspapers mentioned one hundred thousand homes needing repairs, ten thousand to be demolished and rebuilt),[75] but most of it was insured to some degree.

Looking at the collapsed masonry building across the street. Its damage. So predictable. Everybody knows how unreinforced masonry buildings collapse during earthquakes. Not much new to learn there. So predictable. As the saying goes, it is not a matter of "if," but only a matter of "when." The big "when" question for which there might never be an answer.

That was when, lost in thought, I remembered the expert from NASA whom I had met fifteen years earlier. Without doubt, a credible scientist. I had completely forgotten about him. He had told me then—with a certain excitement—that NASA had totally figured out how to aggregate data from various satellite-based sensors to be able to predict earthquakes accurately. This breakthrough discovery was soon to be announced. Of course, such an announcement from NASA never came in the end. NASA had not predicted any of the world's earthquakes in the fifteen-year span since our encounter either—Christchurch included. For sure, if anybody ever finds the Holy Grail that makes earthquake prediction possible—and especially NASA—it will not be kept secret.

Earthquakes to Tsunamis

JAPAN—A REDEEMING VIEW

The National Railroad Passenger Corporation, known as Amtrak, receives roughly $1.5 billion per year in subsidies from the federal government (corresponding to approximately 30 percent of its total income)[1] to provide passenger train services to the nation.[2] In return for this small investment, Amtrak's customers have the pleasure of playing the popular game "Guess what time your train will arrive." The Amtrak "report card" sets the punctuality goal as follows: If trains running on a given line arrive within fifteen minutes of their scheduled arrival time at least 80 percent of the time, it is a success.[3] To be clear, this means that if a train scheduled to arrive at 8:00 a.m. actually arrives at 8:14 a.m., it is considered to be "on time." However, for trains running on lines for which Amtrak does not receive subsidies from states (in addition to its massive federal government subsidies), the passing grade is set at 70 percent. On the basis of these generous passing-grade definitions, in 2019, seventeen of the twenty-eight state-supported Amtrak routes failed to achieve the self-imposed 80 percent standard, and fourteen of fifteen of the other routes failed to achieve the 70 percent standard.[4] According to Amtrak's Office of Inspector General, in 2018, passengers traveling to Atlanta on the Crescent line arrived "on schedule" (recall that this is defined as meaning no more than 15 minutes late) 3 percent of the time. These were the lucky folks; everyone else suffered an average delay of 124 minutes, and 46 percent of the trains were delayed by more than two hours.[5]

Further north, Canada's ViaRail does not fare any better. On some regional lines, federal government subsidies amount to approximately $600 per passenger, for arrivals late by two hours on average.[6] On January 20, 2019, passengers traveling from Montréal to Halifax (775 miles away) arrived at their destination thirty-three hours later than scheduled,[7] for a total travel time of roughly fifty-five hours (equivalent to fourteen miles per hour, the average speed of someone riding a bicycle—but still an impressive four times faster than if riding a donkey at steady pace).[8]

In both countries, delays are largely because passenger trains travel on corridors owned by freight railroads. Amtrak owns only 3 percent of the tracks on which it travels;[9] ViaRail none. A train with three hundred containers generates $600,000 of income to those who own the tracks, while ViaRail pays them $25,000 to run a passenger train on the same line.[10] It does not take a degree in accounting to figure out why cows on a freight train to the slaughterhouse have priority over folks on their way to the daily grinder. Strangely, this state-of-things seems to be considered normal—or, at least, not enough of a concern nationwide to trigger riots in the streets by angry mobs of passengers demanding change.

In Japan, the picture is a little bit different. At a typical suburban train station, on a weekday at rush hour, where trains are scheduled to stop by every twenty minutes, there will be nobody on the railway platforms. No adults, no kids, no dogs. Nobody. That is, until two minutes before the train's scheduled arrival time, when suddenly a mob storms the station and fills the platform, packing it elbow to elbow. Two minutes later, as expected, the train arrives on time, the doors open, and the mob rushes in—squeezing everybody already on board a bit more in the process. The doors close and departure is *on time*.

Seeing a deserted station become so crowded at once may be disturbing to the casual tourist, but it makes complete sense. With the train arrival and departure times being known to the minute—and reliably so—what is the point of waiting on a railway platform when there are more enjoyable or important things to do in life?

Japanese train conductors have a clock on the dashboard and specific checkpoints that must be met—almost to the second—along the route, and anyone can lean on the window behind the conductor and watch the process unfold (like clockwork, literally). This is critical given that, for Japanese railways, a train arriving one minute behind schedule is considered officially delayed.[11] This typically happens only when an earthquake causes operation stoppage or when people commit suicide by jumping on the tracks in front of an arriving train—and when that happens, apology notes are immediately distributed to all passengers so that they can give these notes to their bosses to in turn excuse their tardiness showing up to work.[12] In fact, in two instances, Japan's Metropolitan Intercity Railway Company issued an apology because its Tsukuba Express connecting Tokyo and Tsukuba departed the station twenty seconds early (November 2017) and twenty-five seconds early (May 2018).[13] The official apology[14] stated, "The great inconvenience we placed upon our customers was truly inexcusable."

Such a fact may be mindboggling to the average North American (even more so when stuck in a US train that is hours late), but in a country where crowds are used to storming the train station seconds before the departure time, it all makes sense. Since 1984, the average delay over a year for the Japanese bullet train (Shinkansen) has exceeded one minute only twice: it was 1.1 minutes in 1985 and 1.3 minutes in 1990.[15]

As they do for their train system, the Japanese take the construction business very seriously. As such, it is worthwhile here to provide a balanced view of the Japanese's relationship with earthquake preparedness, response, and recovery—good and bad. This relationship is valuable to appreciate because, nowadays, the Japanese viscerally know that their homeland is earthquake country—maybe more so than any other nation. Nowhere in Japan can one hide from earthquakes. In fact, every year since 2005—maybe in some ways to be more proactive and make amends for the Kobe earthquake—the Japanese government has released a colorful earthquake map indicating the probability of powerful damaging earthquakes occurring within the next thirty years in the various parts of the country. From red in Tokyo and Yokohama where the odds are 85 percent and 82 percent respectively, down to orange, dark yellow, and light yellow for the other progressively less risky regions. No green. The government is making it clear that there is no such thing as a region with zero-probability of earthquakes in Japan.[16]

For sure, as indicated earlier, the Japanese were caught off-guard by the Kobe earthquake, in part because the bulk of their attention and preparedness efforts were focused on the Tokyo region to prevent a dreaded repeat of the Great Kanto earthquake and fire of 1923 that killed more than one hundred forty thousand people.[17] As the Kobe disaster unfolded, all other governments worldwide watched humbly, because none could have done any better if similarly caught with "their pants down" (which is arguably the correct scientific term for when people who should have known better look clueless after an earthquake, and must deal with the resulting total mess). Nowhere in modern times had such a catastrophic earthquake hit a major, highly developed urban city that had all the sophisticated, complex, interdependent, and intertwined infrastructures needed to support today's fast-paced economic life. Naturally, the outcome was not pretty. The graphic details of how this earthquake had wreaked havoc on Kobe were broadcast all over the world 24/7, underscoring the fragile dependence of urban life and global economy on the vulnerable built environment and its entangled transportation, gas, power, water, and communication networks, and all the ensuing problems that arise when these systems fail.[18] For sure, 6,279 deaths, 136,000 housing units lost,

300,000 people turned homeless, and $100 billion in damage[19] definitely helped make that point.

On the positive side, the Kobe earthquake gave the Japanese government a solid "kick in the butt" (another highly technical term used to describe the jump-starting of important, new initiatives). Earthquakes can be good to those that survive and many post–Kobe earthquake achievements are worth highlighting.

First, as mentioned earlier, many aspects of Kobe's reconstruction were fast. "Bullet-train" fast. The bullet train runs though most of the length of Kobe in tunnels under the mountains along the north edge of the city, emerging at the Sannomiya downtown area for the length of a train station, and then again in Nishinomiya at the east end of the city, where it continues on elevated bridges all the way to Osaka. There, the bullet train's elevated lines suffered damage over thirty-six spans, totally collapsing in some places.[20] Lightning fast, all these bridges were repaired and reconstructed within eighty-one days.[21] In some cases, this involved picking up parts and pieces of the collapsed reinforced concrete spans, slapping them together and wrapping them in steel plates, as if reassembling the pieces of a jigsaw puzzle, and injecting concrete inside the steel boxes to fill-in where concrete was missing.[22] It helped a bit that the Kobe earthquake happened at 5:46 a.m., because no trains were running that early. So no bullet train emerged out of the tunnel at 175 miles per hour to fly over where no bridge existed anymore. Otherwise it might have taken weeks (or months) to respectfully recover all the body parts scattered among the rubble—a typical Shinkansen train has sixteen cars and about thirteen hundred passengers.[23] The odds of such a mess happening at rush hours would have been high, with a train running in each direction in as little as five minutes,[24] and considering that emergency braking of a bullet train at full speed still takes one mile before full stop.[25] In truth, not everything in Kobe was repaired that promptly, as was mentioned in a prior chapter, but things got done—which is more than can be said for some countries where the urban scars of an earthquake remain visible for years, sometimes decades. Nonetheless, in spite of the Japanese's efficiency with some specific construction activities, it remains that Kobe is a narrow, densely built urban strip wedged between the sea and mountains and packed with 1.5 million people, which can impede effectiveness. As such, it took six months to remove all the rubble, just as long to restore all six train lines, two years to rebuild Highway 1, which crossed town along its length, and two years to rebuild the port.[26]

Second, the Japanese government took proactive steps to ensure that the collective memory would not forget. Going beyond the usual commemorative plaque, statue, sculpture, trinket, or monument that governments often plunk down somewhere in sincere (or fake) contrition, the Japanese built a full museum to "prevent memories of the Great Hanshin-Awaji Earthquake" (as the Japanese call it) "from fading and to pass on to future generations the thoughts of the survivors and the lessons learned from the disaster."[27] The museum is also part of the "Disaster Reduction and Human Renovation Institute." The term *human renovation* (not defined anywhere in its website, and possibly a typical case of Japanglish) likely refers to the institute's broader mission of training disaster managers and broadly sharing knowledge with specialists in many related disciplines, as well as providing assistance when major disasters occur. Strategically located next to Kobe's Art Museum and a waterfront park, it offers multiple floors of exhibits, 3D multimedia simulations, and full-scale reproductions of damage scenes.[28]

Third, the country invested. Some of it, evidently, could be counted as the typical politically astute way to respond to a disaster—the classic recipe that calls for scattering money all over the place to appease the grumbling population after the earthquake rumble. Maybe so. Nonetheless, the Japanese did not pull punches there either. As one interesting example, the government funded the construction of the world's largest shake table testing facility, and had it built at the northern edge of the Kobe Prefecture.

A shake table is a platform that can be moved in space by servo-controlled hydraulic actuators, pretty much like a flight simulator or some large amusement park rides—think of Star Tour in Disney World, for example.[29] Except that instead of being a box carrying forty people with mouse ears through a CGI field of asteroids in a fake starship, a shake table is a large flat platform on which anything that is constructed will be moved the same way the ground moves during an earthquake—vertically and horizontally at the same time. Like a giant jukebox, one can "play" the "time signature" of any past earthquake for which a strong motion recording is available from the world's database of ground motions, which makes it possible to investigate how anything placed on the shake table would have performed during that earthquake. However, whereas the Disney ride only has to move forty Star Wars fans collectively weighing maybe as much as twelve thousand pounds (in an extreme case), a building and its contents can typically weigh between fifty to two hundred pounds per square foot of floor, depending on the type of construction. For a small 50' x 50' floor area, that equals one hundred twenty-five thousand to five hundred thousand pounds per story times the number of stories. In other words, to move a five-story building, it takes roughly the power of more than two hundred Star Tour simulators—and probably the force of as many Obi-Wan Kenobi's.

Not surprisingly, before the Kobe earthquake, the largest shake tables in operation in research facilities across the world were roughly 20' × 20' when capable of full 3D motion (including pitch, roll, and yaw), which means that they could only support scale-models of buildings or small parts of buildings. Somewhat larger tables that could only move horizontally existed in Japan. After Kobe, the Japanese government literally put half-a-billion dollars "on the table" to build a full motion 65' × 45' monster platform capable of testing structures weighing as much as 2.6 million pounds, with servo-controlled actuators capable of imposing 5 million pounds of thrust in each horizontal direction, and 14 million pounds of thrust vertically. This testing facility also has its own power plant to generate the energy needed to dynamically move that table such as to recreate the movements of the Kobe earthquake at the base of full-size structures.[30] Named "E-Defense" (to defend against the attacks of earthquakes—presumably like one defends against attacks of Godzilla), the facility opened in 2005. Having "built the monster," the Japanese then proceeded to "feed the monster," providing research funding to build and test nearly one hundred multiple full-scale buildings of up to seven stories over the subsequent decade.[31] If what separates kids from adults is the size of their toys, in this case, the Japanese definitely built the biggest playground—a gigantic Star Tour, for the exclusive pleasure of earthquake engineering researchers, albeit without music, smoke, mirrors, lasers, and sound effects.[32]

Fourth, Japan increased earthquake awareness countrywide. If earthquake awareness existed in Kobe prior to the 1995 Kobe earthquake, it was at best somnolent. As mentioned earlier, the government's response was late, emergency relief was inadequate, and many would have starved if not for the more rapid and more effective food distribution by the Japanese mafia.[33] When an organization profits from extortion, blackmail, financial market manipulation, and protection rackets, then it makes sense that the parasite must ensure survival of the host—speaking of the Yakuza here, not of the government, although some would argue that the description best fits the latter in some countries. In such dire circumstances, it is remarkable that there was no reported looting following the earthquake.[34] Some have argued that the media were encouraged by authorities to close an eye—or, actually, a lens—to the few instances of fistfights, looting, and price gouging that were witnessed, but these misdemeanors at least did not occur at wholesale level.[35]

Some Japanese experts have alleged that the Kobe disaster is not the consequence of a deficiency in seismic awareness prior to the earthquake, but rather the results of a smug overconfidence in the superior quality of

Japan's infrastructure compared to everybody else worldwide.[36] Either way, the reality of the 1995 earthquake hit home and provided the impetus for the government and engineering community to revise the seismic design and detailing requirements of all standards, and to mandate more stringent inspection rules for construction.[37]

Evidently, awareness by itself does not solve all problems. For example, from 1988 to 1998, the $3.6 billion Akashi Kaikyo suspension bridge was being built between Kobe and Awaji Island. It was to become the longest suspension bridge in the world. At the time of the 1995 earthquake, only the towers and cables had been constructed. Unfortunately, it so happened that the Nojima fault that ruptured for over twenty-five miles during the Kobe earthquake ran right between the two towers. As a result of the offset created by that rupture, after the earthquake, the towers found themselves one meter further apart than before. Fortunately, pulling the ends of a hanging cable is not too dramatic. The plans for the supported spans of the bridge were redrawn to accommodate the new site geometry and construction resumed. When the bridge opened in 1998, it not only was the longest suspension bridge in the world (with a 6,532 feet center span), but it was one meter longer than originally planned.[38] Arguably, the engineers must have been aware of the presence of the fault and possibly expected that the bridge could accommodate the motion at the cost of some repairs after a rare earthquake at some point in the distant future. They probably just did not expect the distant future in question to occur so soon. Therefore, awareness in this case would likely not have changed the outcome, because it was physically impossible to bridge from Awaji Island to Kobe without crossing the dreaded fault line.

However, better disaster awareness can make a big difference for government emergency agencies planning their response to future catastrophic events. And, this being Japan, it did not take too long for all this practice to be put to good use, as the next massive earthquake happened only sixteen years later.

THE GREAT EAST JAPAN EARTHQUAKE OF 2011

Having learned hard lessons from the 1995 Kobe earthquake, the Japanese were definitely more ready when a magnitude 9.0 earthquake struck on March 11, 2011, off the coast of the Tohoku region. Officially named the "Great East Japan earthquake," it was said to produce damage over 3,800 square miles and in thirty-seven cities—an area six times larger than the one affected by the Kobe earthquake.[39] Contrary to the government's initial confusion and slow response after the Kobe earthquake, this time the government was immediately informed of the disaster as it struck, and Japan's Self-Defense Forces were mobilized within minutes. The first troops arrived within hours and more than one hundred thousand were deployed to the disaster area within three days. While there had been no coordination between the Japanese government and the few nongovernmental organizations (such as the Red Cross) that volunteered to help after the Kobe earthquake, this time coordination was immediately engaged with more than one hundred recognized nonprofit, private organizations and half a million registered and prepared volunteers. And while the Japanese government had rejected all of the 70 international offers of international assistance in Kobe, it accepted all of the 170 offers received following the 2011 Great East Japan earthquake.

Total cost of damage from the Great East Japan earthquake was estimated to be between $200 and $300 billion. More than two hundred miles of rail lines and seventy-seven highway bridges were closed, and as was the case following the Kobe earthquake, much of the infrastructure suffered damage due to the ground shaking, but many aspects of response improved. Gas and electric utilities were reconnected in one week instead of the six months it took in Kobe; railway service and roads all reopened in fifteen days, instead of the seven months in took in Kobe.[40] And so on.

This was largely due to the fact that, while the earthquake itself was one of the largest ever recorded, it occurred offshore. As a result, the ground shaking itself on shore was typically less than considered for the design of new buildings. Also, the smaller Sanriku-Minami earthquake that had occurred in 2003 in the same region a few years earlier, like a wake-up call, made it possible to better prepare. For example, a number of columns of the elevated Shinkansen train line in the Tohoku region were damaged during the 2003 event. A retrofit program was rapidly implemented to enhance the ability of that infrastructure to resist earthquakes. That work was completed prior to the 2011 earthquake and, consequently, these columns experienced no damage during the 2011 Great East Japan earthquake.[41]

Yet, the official reported death toll turned out to be 15,897, with 2,534 additional people reported missing[42]—more than three times the human losses that occurred in Kobe. Only 6 percent of all the casualties occurred in buildings that collapsed during the earthquake shaking itself[43]—still a large number, but not the dominant factor. The big killer in this case was the tsunami created by the earthquake.

When it comes to depicting what a tsunami is, Hollywood somehow is stuck in this groove of portraying it as a single one-hundred-foot-tall wave crashing on a city (or a thousand-foot-tall one, in some of the worst movies). Maybe this artistic vision helps with the box-office receipts, but this is not how tsunamis work.

First, it must be acknowledged that waves hundreds of feet tall are possible, but this type of mega-tsunami only happens when massive landslides fall into water bodies. The largest wave ever recorded in history was generated by one such landslide when a magnitude 7.9 earthquake occurred in Alaska close to Lituya Bay, on July 9, 1958. Mountains up to 3,400 feet tall, with steep walls similar to what is found in fiords, are located at the head of that bay. During the earthquake, a piece of mountain measuring forty million cubic yards got loose and plunged from a height of two thousand feet above sea level. At roughly 2,500 pounds per cubic yard, that made for a 100-billion-pound rock hitting water at 250 miles per hour,[44] like a mega-size pool bomb. The resulting splash wave ramped up along the surrounding mountains with such velocity that it stripped them of all vegetation. The resulting exposed bare rock provided evidence of how high the water reached at various locations. On the mountain immediately facing the splash, the water rose up 1,720 feet. By the time the wave reached the narrow bay's outlet into the Pacific Ocean, the run-up was "only" 75 feet (beyond that point, the wave rapidly dissipated, absorbed by the wider sea).[45] Computer simulations were able to reproduce the phenomenon. Not to be outdone by Hollywood, the researchers produced a "Director's cut" of the actual computer simulation[46]—except that this one was anchored in scientific models, not artistic license. Regular tsunamis born from fault movement in the ocean do not even come close to producing such insanely high waves.

The mechanism that gives birth to most tsunamis is displacement along a fault under water. In simplistic terms, it can be explained as follows. If both sides of the fault move sideways from each other, no wave is generated because water cannot be sheared. However, if one side of the fault moves up with respect to the other, all the water above the side that moves up is also moved up.[47] That upward energy creates the tsunami wave. The bigger the earthquake, the bigger the vertical movement, and the bigger the size of the wave.

While a wind-generated wave typically moves at speeds ranging from five to sixty miles per hour, with a wavelength of three hundred to six hundred feet, a tsunami wave travels at five hundred to six hundred miles per hour over deep water, with a wavelength of sixty to three hundred miles, slowing down to about thirty miles per hour in shallower waters when approaching the coast. As such, tsunami waves (yes, plural, because they arrive in bunches) will arrive from 10 to 120 minutes apart from each other. Because of those unique characteristics, when a tsunami wave travels across deep sea, it is barely noticeable. Once it gets close to shore, as it slows down, as all waves do, it starts to form a more noticeable wave. As the energy piles up water to create a wave, water is usually first drained away from the shore. This massive recess of the water from the beach is usually an unmistakable hint to start running the other way and uphill. Curious folks who decide to go seashell hunting and explore up-close the shore exposed by this "unexplained" sudden, extremely low tide unknowingly walk to their death as the tsunami wave will arrive shortly after. Amateur videos have captured the fate of such unfortunate curious strollers during the 2004 Indian Ocean tsunami.

When a tsunami wave arrives, it is typically in the form of a tidal bore rather than a giant wave;[48] the water level rises within minutes, and it is the sudden inundation of the coast and the volume of water that comes with it, due to the long period wave, that creates the problem. It is like a flash flood rising from the ocean, engulfing the entire visible coast, and spreading miles inland while pushing tons of debris along the way. Considering that a cubic yard of water weighs almost a ton, when that cubic yard is filled with floating debris, it is clear that one can only go with the flow when caught by it. And hope to survive. Most do not. It is estimated that two hundred twenty thousand people died across fourteen countries during the December 26, 2004, tsunami created by the magnitude 9.1 earthquake that occurred in the Indian Ocean.

In response to that 2004 catastrophe, the United Nations helped coordinate international efforts to install an Indian Ocean Tsunami Warning System that became operational in 2006.[49] It consists of a network of seismograph and tsunami reporting buoys that can detect a tsunami wave offshore, so that an alarm can be issued ahead of its arrival to shore, making it possible for people in nearby countries to run to higher elevations—which had not happened in 2004. Operating such a network seems sensible, but unfortunately, when a magnitude 7.5 earthquake struck offshore of Indonesia on September 28, 2018, no data was received from the twenty-two buoys near Indonesia because they had stopped being operational in 2012 due to vandalism and lack of maintenance (vandalism of buoys puts truancy in a league of its own).[50] The government still issued a tsunami warning and evacuation order as a precaution, but since cell phone towers along the coast had been destroyed by the earthquake and there were no emergency sirens installed in the cities along the coast, nobody received the evacuation order. The tsunami struck Indonesia and killed twelve hundred people.[51]

Returning to Japan, the tsunami created by the Great East Japan earthquake of 2011 made quite a mess across the Tohoku region. Immediately after the earthquake, Japan's Meteorological Agency issued a tsunami warning, estimating the height of the water run-up to reach eighteen feet in some regions along the coast. However, given that seawalls from fifteen to thirty feet tall had been constructed along the entire Tohoku shore decades before, half the population in some of these coastal cities apparently did some mathematical guesswork, concluded that the sea walls would provide the needed protection, and ignored the warning.[52]

This being the first major tsunami happening in the smartphone era, it became the most recorded one in history. Hours of video are available showing the progressive run-up of water in various bays and shores, until all the tsunami defense walls built along the coast were overtopped by rushing water and drifting boats started to ride through the resulting waterfalls. One stranded driver even filmed the whole thing from inside his car as the street flooded and his car was swept away, thrashed about along with other debris—calmly running his windshield wipers through the entire ordeal.[53]

According to the Japanese agency tasked with the $250 billion reconstruction effort, as the ocean flowed inland—reportedly as far as six miles from the coast—more than 120,000 buildings were completely destroyed and over a million were half- or partially destroyed.[54] In fact, when it was mentioned earlier how fast utilities, railroads, and roads returned to service, that did not include the infrastructure that was permanently lost due to the tsunami. Likewise, most of the casualties caused by this earthquake are directly attributable to the massive tsunami that followed.

To make matters worse, the tsunami also flooded the site of a nuclear power plant and created a nuclear crisis—as will be described in more detail later. A total of 470,000 people were evacuated from the coastal region and by 2020, 77,500 had yet to return,[55] as the entire region that was affected by the tsunami was slowly being reconstructed. To make reoccupation of the region possible, one of the government's first tasks was to remove more than thirty-one million tons of debris from the disaster area.[56] By early 2018, new forty-one-foot-tall concrete seawalls had already been constructed along 245 miles of coastline, at a cost of $12 billion, to replace the shorter ones that failed in 2011, providing all coastal residents with a glimpse of what inmates experience, looking at a high wall all day long.[57] Beyond giving the term "ocean view" a whole new meaning, the Japanese government also emphasized that the seawalls could not be expected to stop all future tsunamis, that evacuation plans must be in place and promptly followed in case of overtopping, that lower elevation coastal zones should preferably not be redeveloped, and that new structures should be built with the ability to resist a certain level of tsunami wave forces.[58]

And not surprisingly, a museum was built in Tohoku in commemoration of the devastating tsunami that followed the 9.0 magnitude "Great East

Japan Earthquake" of March 11, 2011[59]—right next to the rebuilt, taller seawalls erected to protect the region from future damaging tsunamis.[60]

More protections, newer constructions, enhanced awareness. All good things.

And yet, earthquake disasters are still looming in Japan.

GETTING READY FOR THE BIG GAME

As damaging as the 1995 Kobe and 2011 Tohoku earthquakes were, these were small potatoes compared to the biggie that lies ahead for Japan. Yes, when it comes to disasters, Japan only plays in the big leagues.

The SwissRe reinsurance company reported that "the metropolitan area of Tokyo-Yokohama in Japan is by far the most earthquake-exposed community" of the 616 metropolitan areas it studied, worldwide.[61] It also stated that "the cities most exposed to tsunami risk are in Japan." Incidentally, beyond the earthquake-tsunami twofold blow, a 2014 report from the UN Office for Disaster Risk Reduction stated, "The exposure of the population in Japanese port cities to potential wind damage is extremely high. Tokyo ranks highest in the world for exposure to potential wind damage, with Osaka-Kobe in sixth place."[62] With all hazards aggregated, Tokyo is ranked number one for the predicted number of people affected and loss of productivity due to future disasters—a fun fact that was probably not highlighted in the city's bid to host the 2020 Olympics.

RMS, which brands itself as "The World's Leading Catastrophe Risk Modeling Company," has estimated that a repeat of the magnitude 8.2 1923 Tokyo earthquake alone—without a tsunami to boot—would cause insurance losses in excess of $100 billion today. Keeping in mind that not all the Tokyo infrastructure is insured, that implies a lot more than $100 billion in total damage—possibly even more than ever seen before. This is partly attributable to the fact that a lot of Tokyo is packed with old construction that is not expected to fare well in future earthquakes,[63] including old residential homes of the type that were badly damaged during the Kobe earthquake.

Japan has not always been the prosperous high-tech giant known today for its high-quality high-reliability products—overlooking some shameful recent incidents where quality has not been up-to-snuff, deliberately or not. Baby boomers will recall a postwar era where "made in Japan" meant cheap and of dubious quality, before the decades when SONY, Panasonic, Toyota, Honda, Canon, Nikon, Yamaha, Suzuki, and countless others "upped their game" and took over the world. In those early days, the country was hungry, raw materials were scarce, and the impact of this poverty was not only felt in low-quality exports. New buildings and infrastructure might be designed to the latest standards and equipped with the greatest technology—cost being no issue—but the overwhelming majority of the buildings, bridges, and other components of the Japanese urban landscape were designed and constructed in the 1950s–1980s. These are not necessarily "up to snuff" either. It was well demonstrated by the 1995 Kobe earthquake that older buildings and bridges were more prone to collapse and catastrophic failure. On the positive side, after the Kobe earthquake, massive efforts were undertaken to strengthen critical facilities, and particularly a lot of schools, but a study conducted after the Great Eastern Japan earthquake showed that while nobody was killed in the schools that had been strengthened prior to the earthquake—because none of them collapsed—they still suffered a lot of damage, had to be evacuated, and were to be demolished because repairing them would be too costly.[64] To the parents of the little kids who survived, that is good news. To the taxpayers or insurance companies who will foot the bill, it remains a painful outcome. And that is what was done for essential buildings deemed worth investing money in to achieve life safety—that is, life safety *alone*, nothing more. Far fewer old residential buildings in Tokyo have been strengthened in anticipation of the next large earthquake.

An old joke that poked fun at European stereotypes went like this: "Heaven is where the cooks are French, the lovers are Italian, the mechanics are German, the police are British, and the whole place is run by the Swiss. Hell is where the cooks are British, the lovers are Swiss, the mechanics are French, the police are German, and the whole place is run by the Italians." Not to be outdone, one of the Chinese adaptations moves it many notches up on the politically incorrect scale, stating, "Paradise is having Chinese food, a Japanese wife, and an American house, whereas hell is having a Chinese wife, a Japanese house, and American food." Leaving aside the sexist aspects here—where some horrible stereotypes are at play—and focusing on the infrastructure aspects only, since this is the topic as hand, the story suggests that Japanese housing is less than desirable. It may be true that a Japanese house, at roughly one thousand square feet,[65] is not spacious when compared to American houses, which average slightly more than two thousand square feet[66] (and more than twenty-six hundred for new homes[67]), but it is actually still twice more spacious than those in China[68]—which goes to show the Chinese jokers that, if anything, stereotypes are for fools.

Granted, the Japanese have not helped their reputation for cramped space, with real estate statistics reporting that 25 percent of the Tokyo population has 120 square feet of living space per person (down to as little as 50 square feet in some cases).[69] In fact, 70 percent of Tokyo's population is living in less space that the bare minimum recommended by Japan's own government for citizens to have a "healthy and culturally fulfilling life," pegged at 250 square feet for a single person, including bathroom and kitchen.[70] When space is at such a premium and every inch of space counts, it is surprising that Japanese music lovers have continued buying their music on compact discs and have been slower to embrace music streaming than those in other countries, but it shows that the Japanese psyche and practices are—well—uniquely Japanese.[71]

Beyond the discomfort of cramped space and high urban density, those lucky enough to own a detached home face the problem of high earthquake vulnerability if their house is in an older residential neighborhood. During the rebuilding period following World War II, up to 1981 when construction standards were substantially updated, the traditional Japanese house was built of timber posts and beams without much internal partitions that could provide bracing against the lateral forces imparted by earthquakes. Furthermore, to overcome the uplifting forces due to typhoon winds, roofs were typically built of heavy clay tiles. While the weight of the tiles works well to withstand wind forces, adding mass is the worst possible thing to do when it comes to earthquake forces. When the ground moves, it imparts an acceleration to the building. As Newtonian physics demonstrated in the seventeenth century, a mass that accelerates creates an inertia force. The bigger the mass, the bigger the force. In essence, this is why an eighteen-wheeler needs the force of a 600 hp engine to haul a trailer full of merchandise while a 123 hp Ford Fiesta is sufficient to lug around a small family with grocery bags in the trunk. In this case, the heavier the roof, the larger the force that the house must resist when its heavy roof accelerates sideways because of the horizontal ground motions. As a result, over 60 percent of the wooden structures in Kobe were heavily damaged or collapsed during the 1995 earthquake.[72] To add to the injury—in another pernicious "double whammy" effect—almost nothing burns better than a wooden house. In the days following the earthquake, broken gas pipes, toppled kerosene heaters, downed power lines, and damaged electric appliances and wiring combined to ignite 148 fires that spread over 163 acres (equivalent to the surface of 100 US city blocks), consuming 6,900 buildings and killing five hundred people.[73]

Note that these were genuine post-earthquake fires. Incidentally—as an aside—in many countries, homeowners do not know that standard insurance policies, by default, do not cover earthquake damage unless a special endorsement is added to the policy, at an extra premium, but that losses from fire following an earthquake are generally covered.[74] When that is the case, it would not seem to be in the best interest of insurance companies to publicize this loophole, as it could motivate the unscrupulous owners of homes hopelessly damaged by earthquakes, and uninsured for that damage, to set them on fire solely to collect the insurance money.

Yep, indeed, disasters are looming.

ON THE DISASTER TRAIL

The "Can't Escape It" Moment

Approximately 80 percent of Japan is covered by mountains and is therefore uninhabitable. The remaining 20 percent consists of crowded plains where industries and infrastructure are so densely packed together that it is also uninhabitable.

Although things are not nearly as dire as suggested by this case of bad Japanese humor, it remains that the Japanese have developed a distinct societal code of conduct that has made it possible for them to achieve some level of "harmony" while living in such close quarters—as far as harmony implies conformity to the group. Among the various social rituals needed to achieve this peaceful integration, getting down to some serious drinking with colleagues after work is a prized one. In an environment where many will perspire soaking wet at the mere thought of having to say "no" to someone, getting drunk allows everything to be said—and to be forgotten the day after.

Not for the drinking, but rather for the attraction of a serious culture shock, I had accepted an invitation to spend six months in Kyoto with my family, starting in mid-January 1995. Six months before departure, unable to purchase plane tickets to land in Japan on January 16—because the flight was sold out—we had to settle for the next available flight, arriving on the 18th.

The evening before our departure, my mother called, wondering if Kobe was close to Kyoto (yes, it is), because it was reported on the news that a big earthquake had just happened in Kobe, killing fifty people. More than twenty million people live in the Kansai area of Kobe, Osaka, and Kyoto, so fifty casualties (0.00025 percent of the total population) did not give rise to concerns that evening. Besides, the flight had not been canceled, so how bad could it be?

Amazingly, the brand-new Kansai airport was open and we landed safely on the 18th, but by then, the news reported that the death toll had exceeded five thousand. All concerns were legitimate now.

Some will call it luck, some disappointment—it is all a matter of point of view—but we missed the Kobe earthquake by twenty-four hours. We got the key to the rented house, opened a bank account, did a first grocery, and off I went to Kobe. I was fortunate enough to join the Architectural

Institute of Japan mega-team tasked to perform earthquake damage reconnaissance through the entire Kobe urban area, as the only "Gaijin" on the team—the jury is still out as to whether Gaijin is a pejorative, neutral, or positive way for the Japanese to refer to a foreigner.

This was a three-day expedition by 110 engineers driven by the desire to survey thousands of buildings throughout Kobe, from sunrise to sunset, to document the extent of the damage produced by the earthquake. As all hotels were evidently closed, the organizers had arranged for the team to sleep at a Buddhist monastery, as long as our entire team would be kept out of view of the monks—presumably so as not to disturb them in their meditations. The monastery was located at the top of a hill in Kobe, where no earthquake damage had occurred.

After grueling days of earthquake reconnaissance spent walking in dust—and sometimes, rubble—back to the monastery in the evening, some members of the reconnaissance team pulled out playing cards and bottles to partake in social rituals not quite in harmony with those of their hosts. As the monks were heard chanting their prayers in separate distant rooms—moving along on a path undisturbed by the earthquake—some of the engineers merrily violated the "Fifth Precept" of Buddhism ("do not take intoxicants").

Then, the ground started to move, and all that playing and drinking (and distant chanting) stopped. Time stopped. A time pregnant with the uncertainty of the future, that provided a glimpse into a universal human nature. An instant during which the desire to forget the weight, burden, and gravity of the situation, was robbed by a sudden, unwelcomed return to reality. That instant when everybody wondered if the tremor would grow or stop—or rather, whether the building would collapse, as all were engineers.

This is the moment when it becomes crystal-clear that there will be a point—anytime, anywhere—when it will be too late. A time when one must face the consequences of previous decisions made by others in the past: the "can't escape it" moment.

The Water Magnet

THE ATTRACTION

Like bugs that are hypnotically attracted to light,[1] humans are inherently attracted to water—it is in our DNA.[2] It is probably not that surprising given that life on earth first appeared in water; given that we spend the first nine months of our life floating in mom's amniotic fluid (which is mainly water enriched with electrolytes, proteins, and a bunch of other life-nourishing stuff); given that 50 to 80 percent of our body weight is water (depending on age); and given that we will not survive long without water. Not to forget all the joys that can be had with water, such as boating, fishing, surfing, snorkeling, swimming, and skinny dipping, to name a few. As any Florida real estate agent will attest, it seems like everybody wants a pool and/or a water view—be it of the ocean, the Intracoastal Waterway, a bay, a lake, a pond, or a bird pool. If it is possible, when precariously leaning over the railing at the very end of a balcony, standing on toes and twisting neck, to see a tiny bit of the sea through the tree leaves on a day when the wind conditions are just right and the horizon is clear, then it is a sure bet that this condo will be advertised as having an "ocean view."

Humans naturally gravitate toward water, period. Nearly 80 percent of the world's population lives within sixty miles of an ocean, lake, or river—the closer the better, as property prices unequivocally demonstrate.[3] That attraction is certainly fine, and it can be healthy. However, the danger of that vivid attraction to water arises when it hijacks all other considerations—when reality is distorted to fit a perception that the waterfront is a safe, peaceful, and idyllic location. Unfortunately, that distorted reality exists when it comes to hurricanes.

THE DNA OF A HURRICANE

Those who fundamentally believe that humans rode dinosaurs like horses a few thousand years ago, or that wet blanket wraps can cure tuberculosis as effectively as they can cure concrete (technically speaking, concrete does need to cure at a desired moisture level to gain strength), or that it was a bad idea for the gills-less residents of Atlantis to upset their gods, are likely to find this section offensive and can easily skip it without any impact on their longevity on this earth.

It was long believed that hurricanes were the product of a god with some anger management issues. Hurricanes and typhoons are essentially both cyclonic disturbances of the same nature. Cyclones over the Atlantic Ocean are called hurricanes,[4] because they come from the Mayan god Huracan,[5] whose single foot is actually a snake head blasting wind, water, and fire (lightning) from its open jaws.[6] Cyclones over the Pacific Ocean are called typhoons, because—well, it is less clear. In Hindi, Arabic, and Persian, the word "tufan" means "big cyclonic storm"; in Chinese, "tai fung" means "a great wind"; and there is a multi-headed fire-breathing dragon called "Typhoon" in Greek mythology.[7] While Typhoon's cute offspring included Cerberus, the three-headed mean dog that guarded the Underworld, and the Sphinx,[8] who killed all those who could not resolve its riddles, it is unknown if the reputation of that legendary Typhoon somehow traveled along the silk road to India and China,[9] but it is plausible.[10]

A tropical cyclone, be it a hurricane or a typhoon—or simply a cyclone, as folks in the southern hemisphere call them—occurs when a small weather disturbance "sucks-up" enough energy from warm tropical oceans to grow into a rotating, organized system of clouds and thunderstorms that produces maximum sustained winds of more than seventy-four miles per hour. At less than that, it is deemed a tropical storm. At less than thirty-nine miles per hour, it is a tropical depression.[11] Neither of those is a great day at the beach—except for extreme parachute surfers who think that the joys of huge waves and high winds are worth defying the hazards of thunderstorms.

The threshold for hurricane-force wind was set at seventy-five miles per hour by Herb Saffir, the structural engineer who developed in 1969 what was to become the Saffir-Simpson hurricane scale, because he felt, from his experience with building codes, that there was no way to quantitatively predict structural damage from winds greater than seventy-five miles per hour.[12] Saffir provided various descriptions of damage as wind speed increased, and the meteorologist Robert Simpson, director of the National Oceanic and Atmospheric Administration (NOAA) hurricane center, supplemented Saffir's scale by adding wind speeds threshold for each category of damage on the scale, as well as corresponding expected storm surge heights.

A storm surge can be defined as follows. As the high winds from a hurricane push water toward the shore, it can create a substantial temporary rise in the water level above normal tide levels. This can be one of the most dangerous parts of a hurricane. The storm surge produced by Hurricane Katrina in 2005 pushed the sea ten to twenty feet above tide level along the southeastern Louisiana coast, and twenty-five to twenty-eight feet above tide level along parts of the Mississippi coast.[13] That effectively brings the ocean into town. In other words, a small storm surge will bring water into the streets, a moderate one will bring waves crashing on building walls, a large one will swallow the neighborhood. This is the difference between "surprise, the beach is gone," "surprise, the beach is in the living room," and "surprise, the neighborhood is gone."

Hurricane wind speed obviously is the source of the storm surge problem, but the height of the surge is also affected by the speed and angle of approach of the storm, the slope of the ocean floor near the shore, and the topography of the shoreline and shape of coastal features—all features unrelated to wind speed.[14] As a result of a better scientific understanding of the above phenomena and evidence from past hurricanes, in 2010, the predicted storm surge levels were stripped from the Saffir-Simpson scale and it returned to be only a wind-speed scale.

The Saffir-Simpson scale arbitrarily sets wind speed thresholds of 74, 96, 111, 130, and 157 mph for the hurricane categories 1 to 5, beyond which various kind of wind damage are expected—but not guaranteed.[15] In that scale, the explicit descriptions of damage corresponding to hurricane categories were formulated in the 1960s and 1970s, and a lot of those were more "educated guesses" than hardcore engineering and science.[16] Furthermore, as building codes evolved and construction quality in coastal regions improved over time, the damage descriptions became less relevant, but the wind speed threshold limits remained. Incidentally, the specific numbers used to define the threshold levels of the scale—for example, using 74 mph instead of 75 mph to define Category 1—might suggest great accuracy, but it is not the case. Round numbers could have been used at the transition points, but there is no possible conversion of wind speeds in knots, mph, and km/h that could satisfactorily give round numbers in all three measurement units.[17]

Incidentally, there is no wind speed ceiling in that scale. The Category 5 label is assigned to all hurricanes having sustained winds of at least 157 mph, but given that Levels 1 to 5 are "spaced" about 20 mph from each other, and that nearly twenty hurricanes/typhoons to date have recorded sustained winds in excess of 175 mph, and a couple even reaching 195 mph, some have suggested the need to add Categories 6 and 7 to the scale.

So, to sum it up, hurricanes are a well-known natural phenomenon. A hurricane is simply a rotating low-pressure system that builds-up strength over warm ocean waters. The physics of how it forms and intensifies is relatively well understood.[18] In the Atlantic Ocean, this long-distance traveler often shoots off from the coast of Africa, meandering along westward with the trade winds[19] until it either: (1) dies for not finding the warm ocean waters it needs to feed upon; (2) veers off north as it bumps into high pressure systems; or (3) hits shore and moves inland, which is technically defined with scientific certainty as the point where "all hell breaks loose." At that point, three problems arise that can affect all those humans previously attracted by the beauty of the ocean at that very location of possible hurricane landfall: (1) winds speeds up to nearly 200 mph, (2) ocean levels surging up to tens of feet above normal, and (3) literally feet of accumulated rain.

Wind speed and storm surge are intertwined like the two twisted coils of a DNA strand. They both happen at the same time, and both can kill—so both are reviewed individually below. Floods due to rainfall will be the topic of the next chapter.

HUFF AND PUFF AND BLOW YOUR HOUSE

The cannon was facing an eight-inch-thick masonry block wall. The distance was small. Those in the room did not think much of it, as they had fired that cannon time and time again. Even though they knew firsthand the destruction it could wreak—execution style—it remains that firing a cannon is a job. Like any other job. It probably can be fun, but a deadly tool is still only a tool after a while. Yet, the blur of routine cannot be allowed to distract. Safety first. A cannon is powerful, and errors can be costly.

The cannon was loaded carefully, as always. The target was in the crosshair of the viewfinder. The trigger was depressed. The fifteen-pound projectile flew out at more than 100 mph, penetrated the wall, and went all the way through, as if the wall provided no defense.[20] Bullseye! Another wind impact test successfully completed. Propelled by the large missile pneumatic cannon at the research facility, the twelve-foot-long southern pine two-by-four that flew through the room was not stopped by the wall.[21] Hopefully, when that happens during a future hurricane, because debris are flying all over the place in 100–170 mph winds, there will not be people on the other side of that wall—people convinced that hurricanes are not a big deal and determined to "ride the storm," adamant that the news media always make things sound a lot worse than they really are.

Granted, the news media are guilty of doing just that—which does not help. If an entire town survives a storm unscathed, except for one poorly constructed roof that is blown off, all the networks will park their cameras and news anchors in front of the roofless home and fill the airwaves with it. They will show all the rain that fell in, interview neighbors for lack of experts, show the devastated owners—a bonus if they can catch them sobbing—and inflate the story beyond measure, describing the place like a war-torn disaster zone. Some will call this great journalism, but that is a matter of opinion.

The problem is that, over the long run, after calling wolf before each puny windstorm that rips a few shingles here and there, credibility erodes and there is the risk that the serious threats will be ignored. Those who rode the previous puny windstorms will be empowered to stay put for the real deal. Unknowingly, they will have to face the projectiles of the missile cannon—except that Mother Nature operates her own missile cannon recklessly, with lots of projectiles flying in all directions, because Mother Nature is a loose cannon. The energy of a projectile increases with the square of its velocity.[22] In simple terms, the impact of a projectile at 80 mph can produce four time more damage than one at 40 mph; one hitting at 160 mph produces sixteen times more damage. To make it worse, the number of things that can get ripped up—and thus become projectiles— also increase with the square of the velocity. This is because the pressure created on an obstacle by blowing wind quadruples when the wind speed doubles.[23] If construction is of good quality, it may be that no parts of a building will suffer damage in an 80 mph wind, but increase that wind to 160 mph and many things start to fly out. Roof shingles, parts of home siding and gutters, pool enclosure frames, outdoor furniture, traffic signs, tree branches, and all kinds of other debris. The kind of stuff that, at some point, is bound to kill one of those daredevil weather-anchors that feel compelled to broadcast their report while trying to stand up to hurricane wind and rain, for dramatic effect—and the kind of casualty that will end-up on YouTube for posterity.

In Florida, the latest building code requires impact-resistant windows, or shutters in all new homes built within one mile of the shore in regions where buildings must be designed to resist more than 100 mph winds.[24] Impact-resistant windows are so rated if they have been subjected to the missile cannon test and have successfully "survived"—which means that during the test, the glass panes can fracture, but the window should still remain in place and prevent water from penetrating. Windows can be certified for various speeds of the projectile, typically up to 55 mph.[25] Typically, only a quarter of the impact force of a 100 mph projectile, and one sixteenth that of a 200 mph projectile, but during a storm, not all projectiles ram windows head-on like the fifteen-pound board in the missile cannon test.

In some coastal areas across the United States, when building a new home, buyers are typically offered the option to install hurricane-resistant windows, albeit for a premium. Some may find it mindboggling that such a requirement is not mandatory in all coastal regions, but then again, in the land of the free and home of the brave, some towns do not even have a building code to start.[26] Such holdover towns typically are incentivized to change their mind after being devastated by a hurricane, or when insurance companies refuse to issue coverage to buildings not built in compliance with a specified building code.

In fact, as far as incentive is concerned, the insurance industry itself learned the hard way the fact that wind alone can produce massive losses during a hurricane—that is, even without any help from storm surge and floods. Every business exists to make a profit, and the insurance industry is no exception. Except that, unlike most businesses, the insurance industry collects money from its customers without providing any immediate good or service in return. Although an unflattering analogy, the insurance broker is a sort of bookmaker taking gambling money on long odds, as an intermediary placing bets and paying out the winnings on behalf of the gambler—except that in this case, the bookmaker takes the money upfront to avoid having to hire shady characters to collect from insolvent customers. Obviously, the long-odds bet in this case has nothing to do with predicting which horses will finish first, second, and third in a race, but the principle is the same. The insurance company collects the money from all, calculates the odds that the "winning combination" will occur, and works out the math such that the payout to the "winner" is less than the sum of all money collected. The only difference is that the insurance industry is dealing with unlucky outcomes—in other words, the "winning combination" in this case is the occurrence of an unfortunate event, be it a fire, a theft, a health problem, or any other type of infrequent circumstance that derails one's normal pursuit of happiness.

In principle, everybody understands the insurance principle—except, in some rare instances, the insurance industry itself. The whole principle of pooling everybody's money is to help out the unlucky few who run into trouble due to unforeseen circumstances, so the concept breaks down if everybody needs to collect at the same time because everybody ran into trouble due to the same unfortunate circumstances.

In 1992, Hurricane Andrew made landfall at the Biscayne National Park, about twenty miles straight south of downtown Miami. As the first Category 5 hurricane to hit the United States since 1960, it packed sustained winds of more than 140 mph,[27] and the corresponding punch delivered $26.5 billion in damage[28]—most of it from wind alone.[29] The wind forces wiped out sixty-three thousand homes; severely damaged more than a hundred thousand; left one hundred seventy-five thousand people home-less; and damaged eighty-two thousand businesses, schools, and hospitals, as well as thousands of traffic signals, and thirty-three hundred miles of power lines. As the state was busy trucking away the twenty million cubic yards of debris, it could only be forever grateful that it had been so lucky. Immensely lucky. Had the storm hit shore forty miles further north, it would have crossed the most heavily populated part of Florida, and would have left an even more significant trail of destruction.

Less lucky was the insurance industry. It was as if everybody at the horse races had bet on Secretariat in first, Sea Biscuit in second, and American Pharaoh in third, and that trifecta of thoroughbred lined up perfectly as it crossed the finish line. If such a racetrack catastrophe hap-pened, the bookies would run away, unable to pay all their clients. That is pretty much what many insurance companies wanted to do following Andrew. The hurricane triggered an insurance crisis in the state of Florida, mostly because the property and casualty insurance companies in the state had not realistically accounted for the possibility of a damaging hur-ricane hitting the state when calculating the insurance premiums. They collectively had "misevaluated" the risk—effectively living in denial of risk—to remain competitive on the market place. As a result, after Hurri-cane Andrew, they were left with $16 billion in insured losses, something they were unprepared to absorb. They were immensely lucky that the storm did not hit smack dab in the middle of Dade, Broward, and Palm Beach counties, where there was $370 billion in insured property—on top of much uninsured property—but it did not matter. Many insurance companies went belly-up.[30] Others wanted to withdraw from the state. Giving them further incentive to flee, builders and developers claimed that construction practices were unlikely to change because building hurricane-resistant houses would be too costly, positioning the housing market beyond the reach of too many people.[31]

To prevent a massive exodus of underwriters, the state legislated rules to allow limited and progressive disinvestment over successive years, such as not to leave state residents without coverage. In parallel, the state cre-ated an association that required participation of all private insurers to serve as an insurer of last resort for those abandoned by the insurance market. In spite of this, over the past decades, after each hurricane, some of the largest property insurance firms pulled out of Florida's insurance market,[32] not renewing the policies of millions of homeowners.[33] For those that remain, standard policies have a deductible for wind damage equal to 10 percent of the insured property values, meaning tens of thousands of dollars—although smaller deductibles of 2 percent or 5 percent are available as options, for higher premiums.[34]

Practically, hurricane-resistant homes can be built—and are being built all the time. An investigation conducted following Hurricane Andrew revealed that a lot of the damage occurred, first, because the South Florida Building Code might have called for buildings to be designed to resist 120 mph winds, but its provisions effectively did not match that claim, and; second, because buildings were either poorly designed or poorly constructed, and because building inspection was inadequate.[35] In other words, having great plans and design is only half the solution; implementation is key. For example, in many of the roofs that were ripped up by Hurricane Andrew, it was found that the nails driven by shoddy workers had missed the roof trusses altogether—the type of errors that could have been caught by building inspectors.[36] The best building plans are useless in the hands of a negligent contractor that cuts corners with the same disregard for safety as Napoli drivers who treat traffic lights like cute Christmas decorations—severe warning: it is very dangerous to go through a green light in Naples.

Constructions of the same vintage as those damaged by Hurricane Andrew still exist across all of Florida, including from Miami to Palm Beach. If that part of the coast narrowly escaped the blunt force of Hurricane Andrew, it is because proximity to the hurricane center makes a huge difference when it comes to damage. The spiral of clouds that constitutes a hurricane, as seen in satellite photos, can be gigantic and drop rain over multiple states at the same time, but wind speed is the highest near the center of the storm, along the eye-wall. The eye of the hurricane is the very middle of the cyclone. Somebody standing in the eye of the hurricane effectively is in a column, twenty to forty miles in diameter,[37] of relatively calm air under a clear blue sky. In fact, a great way to avoid hurricane winds would be to stay in its eye, walking/running/driving/flying with it all the way as it travels over the continent—a stunt impossible to accomplish, as one would have to go through half of the hurricane to reach its center in the first place. Right around the eye is the eye wall, which is where the strongest winds are developed. Then, moving farther away from the wall, the wind speeds progressively decrease. For Hurricane Andrew, peak sustained winds exceeded 140 mph along the wall at landfall, but were down to 110 mph in Miami and 75 mph in Fort Lauderdale,[38] respectively twenty and forty miles away. That also corresponds to nearly two- and four-times smaller wind forces acting on buildings. All of the old construction that still exists from Miami to Palm Beach got lucky in 1992, but it will eventually hit a wall someday—that of the eye of a future hurricane—and get somewhat less lucky then. So will those who believe themselves invincible, claiming they "survived the hurricane" because the clouds above their abode rained for a few hours in 1992, unaware that distance matters. Just like everybody can survive a magnitude 8 earthquake if standing far, far, far from the epicenter, everybody can survive a Category 5 hurricane if standing far, far, far from its eye—which is not a guarantee of survival in future events striking closer.

SURFING IN THE LIVING ROOM

On October 9, 2018, Hurricane Michael approached the coast of Florida, dead set on hitting the panhandle as a Category 4 hurricane.[39] Mandatory evacuation orders were issued that very morning for oceanfront counties lying within the projected path of the hurricane. Florida's governor had already declared a state of emergency days before, making it clear that this "monstrous storm"[40] was heading toward the coast, warning that it was going to be "the most destructive storm to hit the Florida panhandle in decades" and "life-threatening and extremely dangerous." He added: "You cannot hide from this storm. You can rebuild your home, you cannot rebuild your life."[41] Wind speeds of 155 miles per hour and twelve-foot storm surges were projected.[42] The only way to make it scarier would have been to forecast that the twenty-five-story-tall monsters of "Pacific Rim" would emerge from the sea at the same time to trample everything else that had not been already destroyed by wind and water alone.

Yet, in spite of these unambiguous statements, hundreds of people decided to "ride the storm" for one reason or another. For some, financial hardship made it impossible to temporarily relocate, and all they could afford to do is to huddle down and hope for the best—very sad indeed. But for everybody else who could afford a tank of gas and drive away from the coast, their decision seems less sensible. For some, it was an unconscientious bravado spirit that made them feel invulnerable to the elements, summarized by "no big deal, we've been in hurricanes before," possibly combined with the appeal of financial gains hoping to sell storm videos to the media—the video selfie of a fool drowning in a storm surge may indeed be worth something. For others, it might have been the sheer attachment to material possessions that made them emotionally impossible to abandon everything, or the misplaced hope that the storm would hit somewhere else and that nothing could possibly be as bad as projected (weather forecasts are always wrong after all, aren't they?). When some said after the hurricane, "I don't know why we stayed," they essentially confessed to having made an uneducated bet that they would survive. For some, the bet paid off—there are, after all, always five winners for every loser in Russian Roulette—but others were not so lucky.

Not many of those who decide to "ride the storm" have usually bothered assessing the "seaworthiness" of their home—not unlike those who boarded the *Titanic*—as their home will literally attempt to play the part of a ship during a storm surge. Seawater weighs a little bit less than a ton per cubic yard—1,728 pounds to be exact—which makes it difficult for the flat wall of a house to stop a crashing wave. A better strategy is elevating the house, as if on top of a wharf, to allow waves to travel below it. An alternative approach, applicable when the ground level story is enclosed, is to purposely allow water to flow inside the house, such as to equalize the pressure applied to the walls by the water inside and outside the house—albeit an inconvenient consequence if the first level is anything other than garage space.

When Hurricane Michael made landfall in Florida's panhandle, it chose as its target the coastal community of Mexico Beach, a small seaside town predominantly consisting of 1960s and 1970s bungalows, intertwined with some relatively more recent constructions. Water pushed to shore by the cyclonic winds surged to roughly twenty feet above mean sea level.[43]

In this day and age where everybody has an HD camera in their pocket, with plenty of people dead set on "riding the storm" against better judgment, the entire world (wide web) could see footage of homes collapsing or floating away like boats, and bungalows turned into submarines with water up to (or above) their roofline, among floating debris, crashing waves, and pouring rain.[44] Receding waters revealed roads and lands filled with sand and piled-up debris of wood frame homes, dead bodies that got trapped in the floating debris, concrete slabs where homes used to be, homes standing without a roof and/or missing a few walls, dead vegetation, and pretty much nothing to celebrate.

Likewise, drone videos of "ground zero" taken following the hurricane showed the widespread devastation across town. The fifty-year-old bungalows had been engulfed—their seaworthiness was nil, and they had been for the most part destroyed. However, among the mix of debris from houses, dead trees, boats, and more, the keen eye not distracted by the scenes of destruction but looking for good news can find in these videos some buildings standing up, with little or no visible damage. Data retrieved from real estate websites confirm that these surviving homes were the most recently constructed buildings—in other words, those built in compliance with the latest code requirements that call for elevating the home and designing it for 120 mph wind forces.[45] These were few and far between because, after Hurricane Andrew, while stricter design requirements were enacted in the South Florida Building Code, the legislature did not impose those requirements along the entire state coastline, and certainly not in Florida's panhandle, to minimize the impact on construction costs. This changed in 2007 after Category 3 Hurricane Ivan hit the panhandle in 2004. From that point on, the same building code was applied to all of Florida,[46] but too late for Mexico Beach, as the code update benefited only those that built new homes thereafter, and only 20 percent of the town's population lived in such new homes at the time of Hurricane Michael.

Very little attention was paid to those homes that more successfully "rode the storm" with minimal damage—as the media loves death and destruction more than successes. However, one oceanfront home that survived particularly well—called the Sand Palace—became an overnight celebrity because it conspicuously stood out pristine in a wasteland. Granted that being a survivor in a neighborhood where nearly all surrounding houses have been obliterated and the beach dunes have been washed away is a success of mitigated benefit, it showed that it is possible to design buildings to survive a Category 5 hurricane head-on. This—and maybe the fact that the house had a sexy name to boot and its own Facebook page—caught the media's eye.

It had been designed and built above and beyond the code requirements and splendidly survived the battering by wind and water. The attention to details it received in its construction was lauded by the popular media[47] as well as trade magazines.[48] The popular media was wowed by the fact that the owner had purposely required the house to be designed to resist 250 mph winds, had elevated it on top of piles driven 40 feet deep, and had used sacrificial break-away walls at ground level that could tear-away without damaging the building—although it did rip away the stairs to the elevated house, such that it temporarily had to be accessed by a ladder. Some newspaper articles speculated that the cost to build this house had to be double that of a regular home—but one must never forget that journalists are ready to print any wild guess number they hear as long as it comes from the mouth of an architect they can cite, because citing a source is easier than fact-checking when under a tight deadline.

Beyond newspapers, trade magazines were more factual. Companies that provided products that went into its construction focused on the benefits these products provided, while other professionals discussed the features that helped it survive, such as having a first floor fifteen feet above ground, small roof overhangs, limited number of windows, and reinforced concrete construction[49]—seven-inch-thick insulated concrete forms and reinforcing bars, to be exact.[50] The fact that the owners worked with a structural engineer to design the house to survive "the big one" is significant, as most homes are instead built following the minimum requirements of the building code, using empirical rules deemed to meet intent, without a tailor-made engineering design. Straight from the mouth of the horse, the structural engineer who designed the house indicated that the total cost premium for that house was on the order of 15 to 20 percent more on a per-square-foot basis[51]—a far cry from the more sensational "double the cost" previously reported.

Now, while much attention was paid to that building, and much emphasis was placed on the fact that its wall strength might have been greater than required by codes, it remains that the windows and doors were only certified to meet the code-specified 140 mph winds (as qualified by the projectile-resistant criteria), and that none of those broke during the storm.[52] Keeping the wind outside of the building is winning half the battle, because once windows are broken, the roof is pushed up by the wind pressure entering the building, and pulled up by the wind forces outside uplifting it like the wings of an airplane. That is twice the pressure, and generally too much to ask, turning the roof into a kite. Indeed, small details make a big difference.

So why does all this matter? On one hand, it defines the hazard as it is, and on the other hand it defines the hazard as it is perceived—either magnified or demagnified. Among those residents of Mexico Beach who survived Hurricane Michael, some were determined to move out of the area and never come back, probably to never ever live again near an ocean shore, or even a lake or a river. They could very well end up, unknowingly, in areas prone to earthquakes, tornados, or forest fires, but that is another story—or another chapter. Yet, some will return and rebuild in Mexico Beach, and when they do, new construction will have to contend with significantly more stringent building requirements. For a start, the design wind speed will be 140 mph, as defined by the Florida Building Code.

Paradoxically, at the same time, across the state on the Atlantic shore, in Palm Coast, on the streets where the first floor of some bungalows turned into indoor salt-water pools because of the storm surge created by Hurricane Matthew in 2016, construction also resumed a few years later. In the aftermath of that hurricane, some of these new homes were built with elevated first floors, to prevent flooding in future storm surges, but surprisingly, some were bungalows at grade level and likely to be flooded in future hurricanes. Given that Hurricane Matthew was a "near miss" that did not hit shore, resulting in a storm surge much smaller than it could have been, maybe it will take a hurricane the size of Michael to make a difference there and change the perception of what is a hurricane and what is needed to survive one.

In the meantime, try telling a homeowner in the Florida Keys sitting on a house deck so low that their feet dip in the ocean at high tide that this very house could someday be thirty feet under water. That homeowner will likely reply, with conviction, that waters around the Keys are too shallow for hurricanes and storm surges to happen.

ON THE DISASTER TRAIL

Hello Sandy, We Were Waiting for You
A few years ago, I visited the New York State Office of Emergency Management command center in Albany, New York. It is effectively an underground nuclear bunker, with self-contained water supply, electricity generation, all the needed services, and enough food to keep an emergency response team alive for months. Since nuclear conflicts have been few and far between, the command center has been used for other emergency management purposes and for coordination of response to less radioactive disasters. Rooms full of computers, overhead screens, direct phone lines to key government agencies and private sector stakeholders, and all kinds of other high-tech gadgets filled the command center. This was all impressive, but one thing that caught my eye during the visit was a poster showing expected inundation zones along the coast of Long Island, and New York City, in future Category 1, 2, 3, 4, 5 hurricanes. I eventually found a copy of that map and posted it on my office wall.

When Hurricane Sandy made landfall on October 29, 2012, near Atlantic City, New Jersey, with winds of 80 mph[53] and storm surge inflicting $70 billion worth of damage along the coast of the Northeastern United States,[54] it was sad to read and hear it called "the storm nobody expected,"[55] "the Storm of the Century that no one believed would really happen,"[56] and other similar names. There, on my wall, was the map that said it would happen, and when Hurricane Sandy happened, it flooded pretty much the exact same areas shown on the map.

Surprise is always a relative concept.

Flood

WATER RISING

Water is the most important resource on earth—vital to human life. Yet, it is not advisable to say, "You can never have enough of a good thing" to people who have six feet of it in their living room.

Every little kid discovers early on that water flows down—and how fun it is to jump in puddles. Likewise, the same kids quickly realize that it is futile to push water uphill. The equation is therefore simple: rain, water downhill, and people uphill. Move up to stay dry. End of story.

If only it were that simple.

Worldwide losses due to floods are estimated at $40 billion per year[1]— including $8 billion per year in the United States alone. Most of that is predominantly due to the overflow of rivers due to heavy rains, melting of snow, or both at the same time—and not so much from earthquake-induced tsunamis or storm surges along the coast during hurricanes, although those evidently add up to the bill every now and then. The size of this annual loss due to floods underscores how difficult it is for people to stay away from the edge of water. It is not happening naturally, and will not happen easily. Being close to water is an ingrained attraction—and, often, a necessity. Any real estate agent can tell you that a property that has a bit of a shoreline is worth more than the landlocked one across the street—a lot more.

There is nothing mysterious about the "mechanics" of where flood-water comes from, so the mythological stories are typically not about what creates the rain, but rather about the wickedness of temperamental gods who control the on-off switch for floods or droughts—sometimes on a whim, sometimes out of anger. Indeed, if one is a god, and water is aplenty, what better way to punish people than to flood them. Water was the first weapon of massive destruction in the history of the world—or at least, in the mythology of the world. Zeus—the big honcho of the Greek pantheon—unleashed a massive deluge to destroy all of humanity, but Deucalion and his wife got a hint of his plan ahead of time and managed to float through it all, survive, and rebuild humanity one person at the time.[2] In another region further southeast, Noah and his family did the same, but brought an entire zoo along for the ride.[3] A bit more harshly, nobody survived when an Aztec deity flooded the world with fifty-two years of tears, but thankfully a couple of humans were resurrected from their bones to restart civilization,[4] although in an alternate version of the myth, the couple survived the flood without needing to be resurrected[5]—after all, it was a while ago, so some details understandably get blurry. Likewise, Hindu gods, Mesopotamian gods, Chinese gods, African gods, and many others[6] also flooded their part of the world, showing that gods are usually a well-connected fashionable crowd when it comes to unleashing disasters. Obviously, there were some more timid gods, apparently not part of the in-crowd, who only flooded to oblivion an island, like Poseidon who sank Atlantis—but apparently not for good, as it seems to have resurfaced in the twentieth century as a five-star vacation resort in the Bahamas.

FLOODPLAINS

By definition, the areas that periodically get flooded when rivers get out of their bed are typically defined to be floodplains. Or, as the joke goes, an area is called a floodplain because it is plain obvious that it floods. The area of a floodplain is not of a fixed size because how much land gets flooded pretty much depends on the size of the flood. For Noah's proverbial deluge, the entire planet was apparently the floodplain. For more practical purposes, planners often refer to 100-year and 500-year flood zones, which means that lands within those zones statistically have a 1 percent and a 0.2 percent chance of being flooded each year.

For agricultural purposes, floodplains are desirable because the receding floodwaters leave silt and clay deposits that help make the soil fertile. Ancient agrarian civilizations adapted to the natural flood cycles and thrived because of it.[7] The Nile River in Egypt provided such benefits until 1970 when the Aswan High Dam upriver eliminated the annual flood cycles, but created many other problems—such as forcing farmers to use chemical fertilizers instead, which in turn polluted the river, and long-term soil erosion in absence of the floods that used to bring new silt and clay deposits to replenish the shores.[8]

Modern city dwellers who love (and pay a premium for) their river view are less thrilled by floods. Rare are the riparian residents and businesses who benefit from ten feet of muddy water flowing through the ground floor. When residents of a floodplain wish to take action to keep their land dry, they typically build levees. As a result, all the water that would normally fill a valley many miles wide is channeled between the walls of the levees, rushing through that artificial corridor at greater speed to flood those downstream that are without levees. The cycle of levee building then repeats itself downstream, until the entire river is surrounded by levees. For example, there are currently over 3,500 miles of levees on the Mississippi River and Tributaries system[9] (for comparison, the Great Wall of China is 3,889 miles long,[10]—although China prefers referencing its own State Administration of Cultural Heritage's study which reports it to be 13,170 miles long,[11] or 53 percent of the earth's circumference of 24,901 miles).[12]

Unfortunately, as water levels rise higher and higher, the weakest links are the lowest levees. During the Great Mississippi Flood of 1927, 145 levees were breached along the river, flooding more than twenty-seven thousand square miles ("widening" the river to eighty miles in some locations), leaving five hundred people dead and seven hundred thousand homeless, producing $1 billion in damages (flooding the same area nowadays would produce $1 trillion in damage),[13] and—most importantly—inspiring the song "When the Levee Breaks," composed by Kansas Joe McCoy the same year, and popularized decades later by Led Zeppelin.[14] Yet, contrary to the song's lyrics, the good folks that survived returned to their land once the water receded and the mud dried up.

Many of the levees across the United States at that time were privately owned and originally built to protect farmland from flooding,[15] so maybe it was not surprising that they failed, but even state-of-the-art, top-notch, best-of-the-best, government-built levees can fail when flood levels exceed the levels considered in their design. As unambiguously stated by FEMA: "While levees can help reduce the risk of flooding, it is important to remember that they do not eliminate the risk. Levees can and do deteriorate over time and must be maintained to retain their effectiveness. When levees fail, or are overtopped, the results can be catastrophic. In fact, the flood damage can be greater than if the levee had not been built."[16]

In short, all levees can (and do) fail.

It is estimated that there are one hundred thousand miles of levees in the United States, located in 22 percent of the nation's 3,147 counties (the exact number is unknown). Approximately 43 percent of the US population live in these counties, although, evidently, not everybody in each of those counties is at the same elevation and thus exposed to the same risk.[17] The US Army Corps of Engineers is responsible for roughly 10 percent of the nation's levees, protecting ten million people. Beyond that, maintenance or improvement of levees is the responsibility of their respective public or private owners, which can be federal, state, or local entities.

Breached levees allow water to return to the floodplains in a less than desirable manner. When crops planted behind the levees become inundated, the cost is borne by taxpayers, because the US Department of Agriculture provides a Federal Crop Insurance that subsidizes farmers in the occurrence of such losses.[18] Beyond damage to crops, when residential areas are flooded, the losses are substantially more significant than in rural areas. These losses are covered by the National Flood Insurance Program, a federal government program that carried over $20 billion in debt in 2018, in spite of Congress having previously canceled $16 billion of debt in 2017[19]—therefore, a cost partly borne by taxpayers.

When levees are raised higher to reduce the risk of flooding, that often is done by the Army Corps of Engineers, thus also a cost borne by taxpayers. It is estimated that it would cost $100 billion to repair and rehabilitate US levees to an acceptable level—expensive, but cheaper than doing nothing and waiting for the floods.[20]

Therefore, at its core, the flood protection problem revolves around the issue of what is the desired protection level. In other words, for how rare of a flood should the defenses hold? Should taxpayers' dollars flow before or after the waters flow?

CAN'T STOP THE RAIN

Millions of citizens live in a 100-year floodplain. The boundaries of such a plain are determined for the average size of a flood having a 100-year return period. This implies that there is only a 1 percent chance per year of being underwater. At first glance, it does not look so bad, but this is a statistical delusion. Beware of the real estate agent who confidently tells potential buyers that one hundred years is a long time away. Not only can a 100-year flood happen tomorrow, it can happen multiple years in a row[21]—like getting tails when flipping a coin is not a guarantee that the next outcome will be head,[22] or like getting one divorce is not a guarantee that the next marriage will be divorce free. Those puzzled by the concept that a 100-year flood can happen multiple times within any 100-year period are the same people who are easily fooled by statistics—which is pretty much almost everybody, as will be shown later. In truth, compared to the return period used for many other hazards, a 100-year return period does not provide a very significant protection. This may have been considered acceptable by whoever drafted the flood maps years ago because being wet is generally not a deadly condition—except for drowning—but the financial losses can be significant. Again, like a divorce.

For example, in 1973, the waters of the Mississippi River reached the 100-year flood level; only twenty years later, in 1993, an even larger event occurred, with more than one thousand levees failing or overtopped,[23] flooding fifty thousand homes and seventy-five communities over sixteen thousand square miles (other estimates report thirty thousand square miles),[24] drowning forty-eight people, and producing more than $15 billion in damage[25] over a four-month period.[26] Then, the great Mississippi River Flood of 2011 that followed, again roughly twenty years later, is also considered to be one of the most damaging in history, with tens of thousands of floodplain residents evacuated over seven states.[27] After that, some real estate agents are probably telling potential buyers that since the 100-year flood has happened three times in the past forty years, this has taken care of three centuries of bad weather and that there is now nothing to worry about for the next 260 years (equal to 300 years minus 40).

How to cope with recurring floods of such magnitude? Turning off the tap is not an option, since nobody can stop the rain (besides, doing so would deprive musicians of a steady source of income, since rain has inspired so many Billboard hits). Staying out of floodplains is a logical option, but it can be problematic. For example, prohibiting settlements in the Mississippi River floodplain would be equivalent to creating a "no-man's land" of forty-seven thousand square miles, which is not appealing or practical.[28] Yet not all rivers flood across so many miles like the Mississippi, and even then, people find it difficult to move away from their home.

Take the Chaudière River, for instance. The first dwellers to settle along its banks in 1736 were quick to discover its temperamental seasonal water levels—particularly when its frozen surface melted in the spring and ice floes, hitting obstacles in their downstream travel, created ice dams that blocked the flow of the river and made water upstream rise up fast and high. Those setting up the mail route between Boston and Québec commented in 1773 on how badly the primitive road was flooded, and two years later, flooding in October 1775 slowed down the troops of General Benedict Arnold's problem-plagued expedition on its way from Boston to besiege Québec City.[29] Yet, it appears that nobody thought it worthwhile to redraw property lines to account for this annual event. Instead, in 1778, the locals came up with a brilliant solution to control floods: building a church dedicated to Saint-Anne to implore her to enlist divine protection against floods. A larger one was built in 1828, followed by a third one—in stone this time—in 1890.[30] To no avail, as all three were built in the floodplain and were repeatedly flooded—which suggests that coaxing saints into interfering with the higher powers that control weather is a futile endeavor. A dam completed in 1968 in the hope of improving things did not make a difference either. In some curious mix of resignation and pragmatism, at the first sign of thaw each spring, local residents empty their cellars and make sure that canoes and boats are ready to provide travel across town when the water level will rise.[31]

Flood after flood, the local provincial government provided financial compensation to those hospitable folks that welcomed the Chaudière River into their homes. Then, after the 2019 flood, recognizing that it took more than two years to process the 6,171 claims filed after the 2017 flood, the government eliminated the requirement that invoices be produced to justify expenses; instead, to speed things up and lighten the bureaucratic burden, it issued a list of standard amounts it would refund for specific items: $1,275 for a French window, $800 for a washer, $30 for an ironing board, and so on.[32] Tired of providing financial relief to repeat victims year after year, in 2019, the government also decided to set limits to lifetime indemnification, and offered a $200,000 "buy-out" to residents willing to relocate outside the floodplain. While the government might not have thought through how this plan would work for an entire city, like Beauceville that regularly sees all of its shops, businesses, and institutions visited by flood waters, some owners of residential properties were takers.[33] That proposed Québec policy was essentially modeled on one that has been implemented in parts of the United States for decades.

In fact, returning to the Mississippi for a moment, a *10th-Anniversary Anthology of Stories of Hardship and Triumph* published by FEMA[34] proudly lists a couple of dozen examples of happy community planners and property owners who have benefited from federal and state programs that provide generous financial packages for people to move out of floodplains. However, in spite of all the uplifting stories, this has been a slow and expensive process with limited success. The Harris County Flood Control District, in the Houston area, is believed to have done more such buyouts than any other county nationwide—a coastal flooding area, but a flooding area nonetheless for sake of illustrating the point. Over three decades, from 1985 to 2017, Harris County has spent $342 million to purchase roughly 3,100 properties. This is laudable, but beyond that, the county still had 3,330 more to purchase on its priority buyout list alone—which, in itself, is a relatively small percentage of the 69,000 properties in Harris County that were flooded during Hurricane Harvey.[35] The county's acquisition program budget only allows purchasing about a hundred modest homes per year, so even for homes that previously got flooded up to the roof line, a buyout can still be decades away. Like the little kid in class raising a hand and shouting "me, me, me" but ignored by the teacher, those willing to relocate by taking advantage of this program are simply "not eligible" if they are not on the county's priority list—but even those who are, at the current funding level, might have to wait centuries before being bought-out. Therefore, some of these owners instead elect to sell at a loss, often to private investors who perform quick cosmetic fixes and rent the property to low-income residents, shifting the exposure to hazards to those less fortunate.[36]

At the same time, floodplain or not, some folks simply will not sell. Again, as real estate agents will tell you, people everywhere love their waterfront properties. In fact, part of today's problem stems from the creation of the National Flood Insurance Program in 1968. With the best of intentions, this program's goal was to "reduce the impact of flooding on private and public structures . . . by providing affordable insurance to property owners, renters and businesses and by encouraging communities to adopt and enforce floodplain management regulations."[37] The real estate market was quick to latch on to the fact that this made it possible to build new residential developments in floodplains where no insurance company would have been willing to provide flood coverage before, simply because the federal government was now willing to pick up the tab in exchange for a low/subsidized annual premium. True to the law of unintended consequences, the availability of this flood insurance—from an underwriter that can never go bankrupt, to boot—therefore translated into a population increase of 5 percent in zones where the flood risk is high. More than that, it was found to provide an incentive for people to rebuild at the exact same place after a flood—as long as they were willing to have fishes as roommates every now and then.[38]

Likewise, beyond the federal government, even the incentives of local governments are often counter to moving people out of the floodplain, because that often also means moving them out of the county, which translates into a loss of tax income to the county. To overcome this problem, some counties use federal dollars to buy out and demolish homes damaged by floods, and resell the land for the construction of homes on stilts or walls elevated above projected future floods levels[39]—which often means more expensive homes, and thus larger taxes collected. Isn't it brilliant?

Definitely, not all Little Pigs are born the same.

On the positive side, flooding is a disaster that happens one drop at a time, usually leaving plenty of time to evacuate. Flash floods are the exception. These can be loosely defined as floods that happen so quickly that they catch people off-guard. In arid regions, a dry riverbed on a sunny day can be filled within seconds by a strong current full of debris launched by a downpour that occurred miles away. This is deadly to hikers who cannot escape a canyon when the flash flood arrives, as well as to drivers with the unrealistic expectation that an SUV can cross eighteen inches of raging waters.

LIVING IN A YELLOW SUBMARINE

If one remembers the "water flows down" concept, it would appear odd to build with the intent to live below sea level—unless building a submarine. Yet, 26 percent of the Netherlands is reclaimed land below sea-level, and 21 percent of the 17.5 million Netherlanders live in such negative elevation zones,[40] the lowest point being at minus twenty-two feet. Only 50 percent of the country has an elevation above three feet. It is not uncommon when driving along a road on top of a dike, to see water five feet below on the right, and cows in pasture fifteen feet below on the left. It is up for debate whether it takes a certain temerity, an optimistic faith in the future, a solid dose of risk denial, strong relaxing drugs (legal in the Netherlands), or all of the above, to lounge in one's home sweet home while looking upward to watch boats floating on the other side of the dike.

For over two thousand years, Netherlanders have been building dunes, dikes, and seawalls.[41] While some countries waged war against other nations to expand their territories—and the Netherlanders naturally got caught in a few of those too—the country mostly went at war against the sea. And, as in all wars, they lost some major battles. Tens of thousands died when dunes, dikes, and other sea defenses breached.[42] More than one hundred thousand in one event in 1530; roughly fourteen thousand on Christmas day in 1717; more than twenty-five hundred in 1953. Harsh lessons as they might be, and as much as one would not be surprised if people decided to move out to higher grounds after such disasters, instead, each time, the breaches were repaired and the land was reclaimed anew—in some cases, the dikes were moved even farther seaward.[43] Actually, in spite of the current forecasts of sea-level rises, nobody is packing up and running away. On the contrary, Netherlands is fortifying its sea defenses and its population is still increasing and projected to continue doing so, even in the parts of the country that might someday become octopus's gardens (in sing-along joy, in the shade).

Yet the Netherlanders are not alone in defying a destiny predicated on the laws of gravity. The classic US example that comes to mind is New Orleans. When the French established their *Nouvelle-Orléans* outpost there in 1718, they built it on a couple of slivers of land about ten feet above sea level, surrounded by swamps. These marches were not attractive to urban residents, but, given that an inch and half of rainwater can fall on any given day there (the daily rainfall record being thirteen inches), they served as a convenient outflow destination to all the drainage canals dug throughout the city to help dry the soil after each storm.[44] The advent of steam power helped launch in 1895 an era of massive civil works to build a new drainage system, first to collect and pump out water faster, and then to expand the city by drying up the marshes. By 1926, fifty square miles of marshland had been reclaimed, and within decades, developers turned swamps into attractive suburbs of houses built on concrete slabs at ground level.[45]

However, true to the law of unintended consequences, pumping the water out after each rainfall and building levees to protect against seasonal flooding from the river itself, allows the sedimentary soil of the Mississippi delta (on which the city is built) to dry out and compact itself over time. Therefore, thanks to human activity intended to keep water out of former swamps that were originally more or less at sea level, New Orleans progressively started to sink. By 1960, 321,000 people lived in areas that had sunk by up to seven feet below sea level.[46] Nowadays, measures have been taken to slow down this subsidence, but the lowest point in the city remains at eight feet below sea level.[47] The fact that large parts of the city could be flooded during a hurricane was a predictable outcome well known prior to Hurricane Katrina. In 2004, FEMA started to hold a series of emergency response simulations with state and federal officials using a Category 3 hurricane scenario that assumed that half a million New Orleans residents would be under ten feet of water, but several of those meetings were canceled in early 2005 because FEMA could not find in its budget the $15,000 needed to cover the travel expenses of the people it invited.[48] Sure enough, when Hurricane Katrina made landfall in Louisiana as a Category 3 hurricane a few months later, in August 2005, some of the levees and floodwalls were breached and 80 percent of New Orleans and many neighboring parishes were flooded. Then, the FEMA "floodgates" opened and the agency provided a lot more than $15,000 in post-disaster assistance (public accounting experts can explain how governments can be broke and rich at the same time). The hurricane left behind $161 billion in damages and nearly two thousand deaths.[49]

Beyond the immediate post-Katrina response, Congress approved $14 billion in funding for the US Army Corps of Engineers to strengthen the levees and walls that surrounded New Orleans, such as to provide protection against a 100-year flood—in other words, leaving a 1 percent chance each year that all that work will not be sufficient to prevent flooding.[50] Furthermore, while this massive project was completed in 2018, barely a year later, the Army Corps indicated that, as a result of subsidence and sea-level rise, it projected that the level of protection provided by this $14 billion investment would actually fall below the 100-year flood level by 2023.[51] Technically, once the risk to a property becomes greater than 1 percent per year, it may become impossible for it to be covered by the National Flood Insurance program, but it is fair to bet that there are enough motivated lawyers in the United States to make sure that such a technicality will not "hold water," because it is the government that built these levees in the first place.

Maybe coincidental to the above situation, but most possibly thanks to brilliant statistical and mathematical models that would be fun to hear explained by technocrats, the "base flood elevation level" defined by FEMA as "the regulatory requirement for the elevation or flood-proofing of structures" in some New Orleans neighborhoods was set to be below sea level. For example, in parts of St. Bernard Parish—the very same parish that was completely flooded after levee failures during Katrina, the very same parish that bathed for weeks in five to fifteen feet of stagnant waters contaminated by a large oil spill, the very same parish where only a handful of homes out of the 26,900 were still habitable after the event[52]—the new base flood elevation level is two feet below sea level. Base elevation in some other neighborhoods was set to be as low as six feet below sea level.[53]

Possibly encouraged by these numbers, ten years after Hurricane Katrina, confident that twenty-three miles of floodwalls had been reconstructed, much of the very same St. Bernard parish above had been rebuilt and its population was back to forty-four thousand people—or two-thirds of what it was before Katrina. Most importantly, people were not prevented from rebuilding in the low-lying areas of the parish.[54] The same is happening in many of the city's other neighborhoods.[55] It is a free country after all.

DOUBLE WHAMMY

The California gold rush started when flakes of gold were found in a streambed near Sacramento in August 1848. Of all those lured by the dream of plucking a fortune out of water, a third made the trip by sea. The population of nearby San Francisco boomed from one thousand to twenty-five thousand in 1849 alone. Hundreds of ships that arrived in San Francisco in those days were abandoned by passengers and crews rushing to join the other miners. Some of these boats were turned into hotels or used for storage, but most were sunk in the bay together with landfill as material to expand the city boundaries by reclaiming land from the bay.[56] As a territory ceded to the United States by Mexico in 1848 as part of the treaty signed at the end of the Mexican-American War,[57] California was a patchwork collection of mission towns with a population of 7,300. Two years later, thanks to the gold fever, the first census counted 92,597 people[58]—plenty more than the 60,000 threshold required at the time to request statehood status, which it received in 1850 in a fast-tracked process.[59] By 1853, at the gold rush peak, 250,000 "forty-niners" (as they were called, in reference to the 1849 boom) were at work in the goldfields—hardly any one of them getting rich, even though roughly $2 billion in gold was extracted during the entire ten years that the rush lasted when the easily accessible deposits could be exploited.[60] In contrast, a century later, many Forty-Niners became quite rich and struck gold by winning five Super Bowls—the football franchise alone was estimated to be worth more than $3.5 billion in 2018.[61]

With a population of 379,994 in 1860,[62] and not much future in gold, many became farmers. Turns out the Sacramento Valley where the gold rush started was part of a massive delta of the Sacramento–San Joaquin River. As the delta was mostly a freshwater marsh, it was easy to remodel the landscape. By 1870, enough levees had been built to reclaim 780 square miles of wetland.[63] These early levees were destroyed by the great flood of 1862 that inundated 6,000 square miles of the Sacramento and San Joaquin valley for up to six months, in water up to thirty feet deep at some locations.[64] Undeterred, higher and stronger levees were built, periodically breached, and rebuilt taller and stronger—and so on. Part of the problem lay in the fact that the top layer of soil in the delta is a thick layer of peat, which is essentially partly decayed plant material. In many locations, this layer is thirty feet thick or more. When the process of reclaiming the marshland was started in the 1850s, the land surface was at sea level and thus periodically inundated by floods and tidal movements. Hence, the levees did not need to be very tall. However, once reclaimed and farmed, the land started to "sink" slowly and progressively, year after year—a phenomenon known as subsidence and recognized to occur for a number of reasons when peat is cultivated.[65] As a result, the land progressively sank to lower elevations, but since water remained at the same elevation—it being sea level—the levees built on the sinking ground had to be raised. Nowadays, some of the farmland in the delta is more than twenty feet below sea level.[66]

Today, California's Central Valley has roughly 2,400 miles of levees, with 600 miles within the Sacramento–San Joaquin River Delta.[67] Beyond all the usual matters that must be addressed when assessing whether levees can be breached by floods having various return periods, as mentioned earlier, here is a double whammy: levees can also be vulnerable to earthquakes. In 2011, the State of California Department of Water Resources considered that approximately 291 miles of levees in the San Joaquin River valley alone (effectively 75 percent of them all) were a high seismic hazard,[68] which means that during an earthquake their failure is likely and of great consequences. In 2015, the Department stated, "the likelihood of a major failure of multiple Delta levees in the next 25 years was greater than 50 percent."[69]

A video produced to illustrate the result of a simulation performed in 2009, as part of the Delta Risk Management Strategy, shows the likely flooding that will happen when multiple levees collapse in the delta during an earthquake.[70] Four years earlier, the director of the Department of Water Resources mentioned to the California Senate that this could happen for as small an earthquake as a magnitude 6 having its epicenter near the delta.[71] Beyond the fact that farmland and thousands of homes will be flooded, and the fact that rail transportation would be disturbed and that natural gas and oil pipelines would be ruptured, the truly big problem revealed by the simulation is the penetration of saltwater into the valley because it will corrupt the drinking water for millions of Californians living in the San Francisco Bay Area and in Southern California.[72] It is estimated that it would take over a year to repair the water control network, and several years for the salt leached into the watershed to clear out naturally before its water becomes suitable again for irrigation and drinking. Some communities could find alternative water sources to ride out the disaster while others may become ghost towns for years until water service resumes.[73] All that was in a simulation that was considered a plausible scenario, not a worst-case scenario.

To prevent such a dire outcome, a $15 billion project, called "California Water Fix and Eco Restore," was proposed. It consisted of two forty-foot-diameter water tunnels intended to pipe water through the delta and forward it directly to where it is most desired, but this and multiple other previous similar concepts were defeated by voters suspicious of who will decide where it is most desired.[74] When it comes to the politics of who gets the water, nothing is simple in the desert states.[75]

As always, the simplest approach is to wait for the failures to happen first, and then see. As the saying goes, "If it's not broke, don't fix it."

ON THE DISASTER TRAIL

Non-Volcanic Craters

The Fallingbrook neighborhood, in what used to be called Orleans, in Ontario, Canada (totally unrelated to New Orleans, USA), was established on a flat plateau covered by odd-shaped craters. The thick clay deposits underlying this plateau (and plentiful fossils) testify to the fact that the area was underwater ten thousand years ago when the melting glaciers created the Champlain Sea at the end of the last glacial period.[76] However, nowadays, short of another ice age, Fallingbrook is unlikely to be flooded by the nearby Ottawa River since the plateau is 150 feet higher than the river in elevation.[77]

To top it off, the deep craters there have nothing to do with volcanos. They are massive retention ponds that have been created as part of the city's storm water management plan to collect excess overflow in the case of rare and extreme rainfalls—the type of storms expected to happen (apparently) once in a thousand years. For example, the Apollo Crater (the one in Fallingbrook, not on the moon) is roughly twenty-feet deep[78] and nearly six hundred feet in diameter.[79]

Most years, the craters remain dry. They provide perfect slopes for little kids' snow sliding in winter and serve other recreational use for people (and dogs) in the summer. In the nearly ten years I lived in Fallingbrook, it only happened once that a torrential rain filled enough of the craters to make it possible for excited kids to jump into the temporarily-created pools—not a particularly bright idea considering that storm water run-offs carry oil, dirt, chemicals, lawn fertilizers, bacteria, and other urban pollutants.[80]

All of that to say that I never expected to be flooded when I lived in Fallingbrook—until the day it happened. The sky was blue and the craters were perfectly dry but the sewer pipes at the end of the street became partially clogged in a way that had nothing to do with rainfall. All the houses on our street ended up with a few inches of water in their basement. Not a newsworthy disaster, but an annoying mess, nonetheless.

How somebody managed to clog that sewer pipe is beyond understanding—and something one prefers not to try to imagine—but, if anything, it shows that it is important to carry insurance, no matter what the real estate agent said about floods.

Tornado Alley

DO THE TWIST

Joe and Jane had originally left their home state to attend college in Los Angeles, lured away by the promise of year-round sunshine and beaches. They enjoyed a decade of warm and sunny weather in La-La Land—even though they spent more time parked on freeways, where nobody dances, than on the beach—but when one of those crazy wildfires that spread over tens of thousands of acres was stopped only a few miles from their brand-new residential district, it sparked their decision to move out of California. Sitting in a hotel room, miles away, watching firefighters in the distance struggling through the night to save their neighborhood, they realized how lucky they were to have been able to evacuate to safety. A similar wildfire in Butte County the year before—known as the Camp Fire[1]—had burned for seventeen days, spread over nearly 240 square miles, destroyed more than eighteen thousand structures, and killed eighty-five people who did not learn of the fire or receive evacuation orders soon enough to join the twenty-seven thousand others who managed to barely escape through one of the only three flame-surrounded roads leading out of danger.[2] Joe and Jane could deal with the threat of earthquakes, because "big damaging earthquakes never happened," but after seeing dozens of wildfires in the news every year and the inferno in their own backyard, they felt it was time to return home, to Oklahoma.

Where a bungalow was all they could afford in the crazy Southern California real estate market, now in a state where homes cost four times less per square foot,[3] they were looking forward to building the mansion of their dreams. Pre-approved mortgage in hand, when they sat with the homebuilder to make their final selection of upgrade options, they discussed how to best use the remaining $5,000 in their budget for extras. One option was to build a safe room.

Safe rooms are typically small windowless rooms located in the middle of a house and designed to withstand winds of up to 250 miles per hour, flying debris, and windborne objects.[4] Providing about five square feet of space per person—say, a room four by five feet for four people—it looks more like a safe closet than a safe room, but it is to be occupied for only a few minutes during the passage of a tornado. It is designed to be anchored to the ground slab—or better yet, the basement slab if there is a basement—and to protect the occupants from injury in a sturdy enclosure while the tornado shreds the rest of the house to pieces. The same effect can be achieved by an underground cellar,[5] which is typically more common in the farmlands as they can also serve as true cellars as well. The drawback of cellars is that, depending on whether or not early warning is received, one may need to run outside during the storm to reach the shelter, exposed to lightning, hail, extreme winds and dangerous debris, unless the cellar is accessible from inside the house (for example, some cellars are located in an attached garage).

Builders/contractors and design professionals—and in fact, pretty much anyone else—can get instructions on how to build a safe room using documents made freely available by FEMA.[6] Having grown up in Oklahoma—right in the middle of tornado alley—Joe and Jane were fully familiar with what these shelters are and do. Their city's website recommends that a safe room or underground cellar be considered for every residence, but this is not required by the building code or by any local or state ordinance across the United States. In fact, in Moore, Oklahoma, where four devastating tornadoes struck between 1999 and 2013,[7] only 10 percent of the city's homes have them.[8] A safe room typically costs between $2000 and $10,000, depending on size.[9]

Joe and Jane did not need more than a minute to reflect on that. They looked each other in the eyes and knew exactly what the other one thought. Their priorities had always been clear. Without hesitation, they decided to use the remaining $5,000 to buy granite kitchen counters, with the hope that Joe would eventually, someday, build an underground cellar in the backyard as a part-time project when the family finances would allow it. Besides, what are the odds?

THE DNA OF A TORNADO

Those who fundamentally believe that the world as we know it is at risk of collapsing if items from two different food groups touch each other, that eating beans is tantamount to murder because souls travel through beans while awaiting to be reborn, and that blowing out candles on your birthday cake is a guarantee that your wildest wishes will come true are likely to find this section offensive and can easily skip it without any major indigestion.

Some have joked that given the fact that "tornado alley" is smack in the middle of the Bible Belt, God must have a "twisted" sense of humor. However, while it is true that states in that Belt see more tornadoes than anywhere else, with peaks of more than 10 tornadoes per ten thousand square miles per year,[10] corresponding to 96, 52 and 126 tornadoes per year in Kansas, Oklahoma, and Texas, respectively,[11] it turns out that tornadoes happen quite frequently elsewhere too—that is almost anywhere in the United States, Southern Canada, Europe, South Africa, New Zealand, Australia, and large parts of Asia.[12] In fact, the Freising (Germany) tornado of 788 might be the earliest one documented in Europe's history. The first documented one in the United States dates back to 1671, sighted in Rehoboth, Massachusetts, and making one casualty.[13] It is only later in their westward march to settle the country that European immigrants would meet what some Native American tribes of the prairies called the "Devil Wind."

Tornadoes are violently spinning air funnels that connect the ground and the clouds, making for one of the most visually striking meteorological phenomena. So impressive that storm chasers have started to make a business out of taking passengers with them[14]—a relatively dangerous venture given that the path of tornadoes is wildly unpredictable and that some of these tornado tourists have already been killed.[15] On the positive side, thanks to the chasers, pretty much everybody has already seen dazzling pictures of tornadoes from up close.

The size of tornadoes is measured by the Enhanced Fujita Scale, which ranges from F0 to F5.[16] Like the Modified Mercalli Intensity Scale for earthquakes, it is the damage produced by a tornado that earns it a specific value on the Enhanced Fujita Scale. Wind speed velocities are inferred from this damage. An F0 tornado will rip out some gutters and break a few tree branches, corresponding to winds 65–85 miles per hour—nothing to get excited about since most cities experience days when such wind speeds are reached, without any tornado. An F5 will throw cars, trucks, and train cars a mile away, and will level well-built residential homes, with wind speeds estimated to be in excess of 200 miles per hour—essentially, a wind storm on steroids. In the world of truly horrible Hollywood movies—and in that world only—an F5 can also suck sharks out of the oceans, carry them hundreds of miles over land, and rain them all over the place in an airborne gore-fest of bloody shark attacks, called a Sharknado.[17] Fortunately for both humans and sharks, F5s are rare occurrences. In the United States, there was an eight-year span between the 1999 Oklahoma

F5 and the 2007 Greensburg, Kansas, one. A similar span of more than six years started in 2013 after the Moore Oklahoma tornado.[18]

Given that tornadoes typically form in a few minutes, making it difficult to deploy measuring instruments at the right time and at the right place, and because the high wind speeds tend to destroy the measuring instruments, the science that explains the genesis of tornadoes is still a field in evolution.[19] The ingredients required to generate a tornado (as for instance, a "supercell" thunderstorm) are known,[20] and tornado warnings are issued based on these precursor signs—or based on sightings of actual tornadoes if need be for lack of earlier warning.

However, what matters here is that, on average, a tornado produces damage along a path one or two miles long and over a width of approximately fifty yards.[21] Hence, while a damaging earthquake can affect an entire region, the path of damage from a tornado amounts to a small line drawn on a map (again, on average). As far as record holders are concerned, the longest path ever observed for a pack of tornadoes from the same supercell is 293 miles, while the largest path width is 2.6 miles (but at significantly lower wind speeds).[22] The most deadly one was the Tri-State Tornado of March 18, 1925, which killed 695 on a four-hour rampage that crossed Missouri, Illinois, and Indiana, leaving a signature of debris 219 miles long and three-quarters of a mile wide.[23]

TORNADO-MAGNETS

Considering the fact that only a small percentage of homeowners have found it to be a wise investment to install a safe room or cellar where they can bite their nails, sheltered, while a tornado blows away the rest of the house, there might not be much of a market for completely tornado-resistant houses in the near future. In other words, going beyond small tornado-resistant closets, making an entire dwelling able to resist the 200+ mph winds of an F5 tornado, while possible, carries a bigger price tag. Many new homes along the Florida coast are designed to resist 140 mph hurricane winds, which is equivalent to the wind speeds generated by an F2 tornado, but stepping up the technology to resist an F5 will likely require a bit more entrepreneurial "oomph" down the line.

Then, there are mobile homes. The National Severe Storms Laboratory of NOAA, using data from 1985 to 1995, reported that people were ten to twenty times more likely to die in a mobile home than in other types of dwellings during tornadoes,[24] a trend that apparently had remained the same by 2007.[25] While mobile homes can be designed to resist speeds of up to 110 mph in hurricane areas, winds speeds of only 70 mph have been considered in their design in most parts of the country—including in Tornado Alley.[26] As a result, their damage in tornadoes has been so extensive that some people believe that trailer parks are actually tornado magnets—a severe confusion of cause and effect. It is because of the above statistics that observers get the impression that funnels "sniff around" until they find their favorite food—that is, trailer parks—before touching down. However, tornadoes are "equal opportunity destroyers" and will attempt to equally flatten everything along their path—only the strongest ones will survive, and trailer homes have a limited membership in that club. Tornadoes will more readily flatten a weak mobile home than a sturdier residential construction—so, in that sense, yes, trailer parks are fast food for tornadoes, but not magnets (another urban legend debunked!). However, what is more likely is that trailer parks are news media magnets following the passage of a tornado. News anchors love the sense of drama they project when broadcasting their reports standing in front of massive debris, and what better place to find destruction after a tornado than in a trailer park? A good case of lazy reporting, and nothing more.

Most unfortunately, though, in some states, people who live in mobile homes are less likely to recover from such a disaster because they are the most socioeconomically disadvantaged populations—or, in plain English, the poorest folks in town.[27] It is a sad reality for those who live in trailer parks out of necessity. It is a matter of debatable priorities when people who could afford to do otherwise freely choose to live in a mobile home in tornado alley for sake of being able to buy a fancier car. Double whammy: they are often the ones who believe myths such as "opening windows will help reduce the wind damage"[28]—which should be now recognized to be a bad idea by those who paid attention to an earlier chapter's discussion on how internal and external wind pressures detrimentally add up when windows break.

As for those brilliant minds who suggested that tornadoes could be killed using nuclear bombs to defuse their energy,[29] a sarcastic slow clap is the only possible response.

ON THE DISASTER TRAIL

The Famous 1996 Twisters
No matter how often I repeated that going to see tornadoes would be too scary, my son was determined. All my warnings fell on deaf ears and it was clear that the more I objected, the stronger became his resolve. A tornado is an understandable attraction. "The adventure of a lifetime" as advertised by storm chasers selling tours to tornado tourists willing to pay thousands of dollars for the promise of driving close to one or more twisters—one of the rare natural hazards that will kill everybody in its path and yet can

be approached to within a mile[30] for amazing souvenir pictures, and the occasional near-death experience.[31] As one who has spent his life working with extreme events, I certainly could understand my son's attraction with twisters.

In the end, this was a fight in my power to win, but sometimes parents have to pick their battles. Sometimes, bad decisions can be good ones on the strength of the lessons learned. "OK, you want to see Twisters, let's go see Twisters," I told him. It did not take long to see one, a roaring F5, with cars and cows among the twirling debris, and sure enough, as I had predicted, my son was scared to death. He buried his face in my shoulder, waiting for the noise to abate. Same reaction, tornado after tornado as most ten-year-olds would be expected to have done. After that, he never argued when we refused to allow him to see a PG-13 movie with scary over-the-top special effects and thundering THX sound.[32] "Twisters," released in 1996, grossed $500 million worldwide and scared plenty of little kids—and some adults too.

Sitting on a Volcano

SCHOOL OF LIFE

Like all kids in Japan most early mornings, Sakurajima's first graders and high schoolers walk along narrow sidewalks, in small groups, dressed in a school uniform that generally includes a matched backpack. What differentiates Sakurajima's school kids from those in most other prefectures is that as a standard part of the uniform, either worn on their head or snapped to and hanging from a backpack strap, is a hard hat. What is also different along the sidewalk is that most bus stops are actually little concrete bunkers designed to potentially deflect the flow of lava coming from upstream should the Sakurajima volcano decide to eject more than ash on a given morning—as the helmet alone might not be sufficient protection when that day comes.

Sakurajima is not only a volcano, it is a volcano-made island, built-up as a typical volcanic cone by new lava piling up on top of old lava from prior eruptions. It also has the distinction of being Japan's most active volcano. The 4,500 people who live on the volcano are in a far suburb of the city of Kagoshima that is separated from the volcano by three miles of water. That is not much of a buffer for those across the bay, but it's an even bigger problem for those who live around the base of the volcano.

The Kagoshima city website provides information on the various dangers created by a volcanic eruption:[1]

- Cinders, which are rocks up to three feet in diameter ejected from the crater and that come crashing down as far as a few miles away from the crater;
- Pyroclastic flow, which is a burning hot mix of rocks, ashes, and gases that rushes down the face of the volcano at speeds ranging from 60 to 400 mph (depending on the volcano);

78

- Lava, which is a slower-moving magma (molten rock) that ignites everything it touches and buries it at the same time;
- Ashes ejected into the air (blown up to fifteen miles away in the case of Sakurajima's major eruptions) and falling like gray snow deposits, accumulating multiple feet thick;
- Debris slides of earth and rocks, created when rain loosens ash buildups on the face of the volcano;
- Moderate earthquakes that come along for the ride;
- Rockslides and tsunamis, triggered by these earthquakes; and
- Toxic volcanic gases, such as sulfur dioxide and hydrogen sulfide, emitted by the volcano.

The list is likely educating only visitors and tourists, however, as kids in Sakurajima need not be reminded. They have grown up on a volcano that has been erupting almost nonstop since 1955, with thousands of small explosions each year, some severe activity from 2009 to 2016, and ash clouds often rising a few miles above the tip of the volcano. To boot, the local landmarks are a lava observatory[2] and the stone door of a Shinto shrine partly buried in ashes and pumice,[3] a reminder of the 1911 eruption. And, not to forget, they carry a helmet to school.

In the event that early signs make it possible to warn the island's residents of an impending eruption, Sakurajima's detailed evacuation plan[4] identifies twenty-one designated ferry departure points to which they should converge. They are also encouraged to do so at once, given that nineteen of those are within the delineated danger zone—approximately three miles in radius from the volcano's peak—considered to be within reach of pyroclastic flow and blasts once the eruption starts, and the other three are within reach of volcanic rocks shooting out of the crater at 350 mph.[5] Note that rocks twenty feet in diameter were projected up to two thousand feet from the crater during the 1935 Mount Asama eruption,[6] but smaller ones can reach greater distances, raining down from thousands of feet in the air, which can be equally deadly.

Even if ducking into a concrete bunker is possible, evacuation is definitely the sensible thing to do because nothing stops lava, and trying to predict which way lava will flow on a volcano is a little bit like making wagers on what side of the candle the wax will drip. If some paths have been established and seen to be recurring from eruption to eruption, the odds are good it could flow there again in the future. An aerial view of Mont Vesuvius in Italy clearly reveals where lava last flowed during the 1944 eruption,[7] as vegetation has not yet taken hold on top of the solidified lava—the path of that lava flow has been well documented in scientific publications.[8] Defying the odds, right at the tip of that flow, where the lava solidified and stopped its forward progress, the same aerial view shows a recently constructed mansion with two large swimming pools, across the street from a gourmet restaurant.[9]

Yet, these are odds, not guarantees. In fact, over the long run, volcanoes are cones for a good reason: they are building themselves up because lava flows all around and solidifies, adding to the volume of the cone. The more symmetrical the cone, the more the lava has been uniformly distributed all around. On that basis, if constructing a house on the slope of a volcano, situating it on top of the latest lava flow might be strategic, hoping that the next lava flows will go elsewhere to continue building up a nice symmetrical cone. Unfortunately, recurring flows from multiple eruptions are sometimes needed before a specific valley is filled with lava and future flows are redirected elsewhere, which may take thousands of years to accomplish—geology and humans definitely work on different timeclocks.

To put it mildly, settling on the slopes of a volcano is a dicey proposition, particularly since there is no truly proven way to divert molten rock that has been ejected at 1,300 to 2,000°F[10] and that will flow downhill until it cools and solidifies.

Piling up concrete blocks alone will not work, since concrete (approximately 150 pounds per cubic foot) is typically less dense than lava (approximately 200 pounds per cubic foot) and will therefore "float" on it. With enough advance planning, the concrete could be tied down to counter its buoyancy, but concrete's resistance to high temperature is not infinite. The moisture inside the concrete will turn into steam when its boiling point is reached, and if the thermal shock of hot lava hitting concrete is sudden, the pressure created by that steam could make the concrete explode[11]—unlike in a building fire, in which the slower temperature rise gives time for the steam to escape through the concrete's matrix of materials. In addition, the properties of the Portland cement inside the concrete start to degrade at 1,500°F.[12] Special coatings can be applied to concrete to resist a rise of temperature to 2,200°F for ten minutes, but being in direct contact with lava creates a significantly faster thermal shock. Levees of rock and ash proved more successful in containing lava flow from the Etna eruption in 1983: a first levee 30 feet high, 100 feet wide, and 1,200 feet long was overtaken by the lava, but a second one built parallel to it 300 feet farther away managed to contain the flow.[13]

Given that lava is molten rock, one approach that can be envisioned to block the flow of lava in a specific direction is to cool it down such that the resulting rock creates a barrier to the oncoming lava behind it, thus diverting the flow elsewhere. This strategy has worked to protect the Icelandic port town of Vestmannaeyjar during the 1970 eruption of the Eldfell volcano, where water cannons shot 1.5 billion gallons (12.5 billion pounds) of cold seawater onto a slow moving lava front over a five-month period, eventually redirecting the lava flow to protect the port.[14] Unfortunately, not everybody is so lucky to have the perfect combination of snail-paced lava and an infinite supply of water, so this has been a unique success story so far. Furthermore, it should not be forgotten that the tradeoff to this success is that when 12.5 billion pounds of water hits lava, it turns into 12.5 billion pounds of highly acidic steam—enough to make a jolly bunch of clouds that will eventually drop 1.5 billion gallons of acid rain somewhere.[15]

Another approached tried in the past has been literally to bomb the lava flow, military style, in strategic locations, in an attempt to redirect the flow by using explosion craters and by destroying the tubes of hardened lava through which molten lava flows faster.[16] However, as with many losing wars, much damage was done to no avail, as the lava simply filled the craters and resumed its course—shockingly, turning out to be another world problem that could not be solved by a good bombing campaign.

Of course, in all of the few successful containment cases reported in the literature, it is understood that success was achieved partly because lava speed was only a few feet per hour and—most importantly—the volcanic activity that lasted months eventually subsided, which effectively "turned off the tap" and lava stopped flowing. Unfortunately, that is not always the case. In particular, lava can flow at speeds of up to forty miles per hour.[17] Furthermore, all the diverted lava has to go somewhere else; while some critical location along the flank of the volcano might be spared, it could be at the expense of somebody else's proverbial backyard.

Nonetheless, these are the risks one must be willing to accept when taking up residence on the slope of a volcano. Sure, the view might be second to none, there may be tons of opportunities to cash-up if the volcano itself is a tourist magnet, the soil can be incredibly fertile for farming, and land for residential construction can sometimes be cheaper to buy there. However, it should not be forgotten that after an eruption, one might end up with more land than they paid for—if not in plan, at least in volume, thanks to all that cooled-off volcanic rock piled-up on top of whatever real estate once existed.

Take Vesuvius, for example. Volcanologists agree that Mount Vesuvius is not only active, but also overdue for a major eruption. Only 8.5 miles as the crow flies from its caldera to downtown Naples, the Vesuvius is world famous for its AD 79 eruption that killed many of Pompeii's inhabitants and buried the entire town under more than twenty feet of stones and ash.[18] Located six miles from the caldera, Pompeii's population had prime seats to watch the volcano blow up a plume visible from hundreds of miles away (as recreated in some awesome YouTube videos).[19] Those who recognized the danger had enough time to flee—more than 90 percent of the population apparently escaped in time.[20] A few thousand remained, maybe mesmerized by the awesome show, or maybe diehard optimists reluctant to leave for the same reason that evacuation orders are often ignored nowadays. For them, all was fine until all that had flown up started to rain down. First came ashes, accumulating on the ground like gray snow a few feet deep, then rocks, which sent terrified people running to seek shelter in their homes. To no avail. What followed was a lot more ashes, making it difficult to breathe, and—the killer—the pyroclastic flow of 500°F poisonous gases rushing in at a hundred miles per hour. In a day, Pompeii was erased from the face of the earth, lost and forgotten for centuries, until it was accidentally rediscovered and excavations of its ruins started in 1748.[21] Voids in the ashes left after the organic matter had decomposed served as molds that, when filled with plaster, revealed full bodies in the exact posture they had been in the instant before their asphyxiation—gruesome reminders of death by volcanic attack.

Vesuvius continued to erupt for over a thousand years thereafter, with some other major eruptions, particularly in 203, 472, 512, 685, 787, 968, 991, 993, 999, 1007, 1037, 1068, 1078, and 1138. Then, after a period of quiescence of roughly five hundred years from 1139 to 1631, Vesuvius woke up and has regularly flexed its muscles with eruptions in 1631, 1660, 1682, 1694, 1698, 1707, 1737, 1760, 1767, 1779, 1794, 1822, 1834, 1839, 1850, 1855, 1861, 1868, 1872, 1906, 1926, 1929, and 1944—that last one actually caught on film.[22] On average, one eruption every thirteen years. Since 1944, nothing major—which is not necessarily a good thing, since it is expected that the longer it takes before the next eruption, the more severe it will be.

To make matters worse, no other volcano on earth has as many people living within the danger zone around it.[23] Including the city of Naples, the surrounding urban area is home to more than three million people,[24] including six hundred thousand living within the "red zone" who would have to be evacuated. That task would take roughly a week to accomplish, relying on 500 buses and 220 trains[25]—and, evidently, a truly optimistic outlook on Italian punctuality. Obviously, whether such a mass exodus can be successful is contingent upon scientists being able to predict reliably, and with reasonable advance warning, when a major eruption will occur—something that is still far from an exact science.[26]

Of those living in the red zone, 97 percent are fully aware that the area is within a zone of high volcanic risk, and 61 percent recognize that the presence of the volcano makes it a "hostile place to live" and that they could be displaced by future eruptions.[27] Logical numbers, given that Pompeii is an eloquent reminder, smack dab within that red zone. Yet, in spite of an Italian government program giving 30,000 euros to all those willing to abandon their home and relocate elsewhere, not many have moved out—and some squatters have moved in.[28]

So, why do people live there? Because living in proximity of the volcano seems like a normal condition. Why leave? Nothing has happened since 1944, so life goes on. Maybe divine protection will suffice. Besides, as someone said, living on a volcano makes it perfectly fine to smoke and drink with abandon.[29]

Yet Naples is not alone in hugging a volcano. Many large cities border the fifteen hundred known potentially active volcanoes on earth,[30] including populated areas in Mexico, Chile, Bolivia, Ecuador, Peru, Spain, the Philippines, Indonesia, Papua New Guinea, New Zealand, the United States, and Japan.

For example, Mount Fuji's peak is sixty-two miles from the Tokyo city center, but only thirty-five miles to the western edge of the Tokyo-Yokohama region. This most iconic Japanese landmark has become a symbol of Japan, captured in multiple work of arts over centuries, such as in Hiroshige's famous series of woodblock prints *Thirty-Six Views of Mount Fuji* which nowadays adorn T-shirts, coffee mugs, and everything else that can be bought in the country's countless tourist traps.

In the dreariness of the urban pile up that comprises the country's capital, residents there take pride in being able to see the beautiful, snow-capped Mount Fuji on a perfectly clear winter day—which practically almost never happens. Every year, when all snow has melted (mostly in July and August), three hundred thousand people hike up to the top along one of four well-traveled routes, starting from parking lots at different heights from the base. Most hikers climb at night so as to be at the summit in time for sunrise, which can make for massive pedestrian congestion reminiscent of a Japanese train station. As the Japanese proverb goes: "A wise person will climb Mt. Fuji once; a fool will climb Mt. Fuji twice." The amount of garbage[31] and feces[32] left on the mountain by all trekkers makes some wonder if even a once-in-a-lifetime pilgrimage up Fuji is actually that wise.[33] Nonetheless, climbing or not, the Fuji Five Lakes area around the base of Mount Fuji is packed with resorts, spas, golf courses, ski stations, and even a roller coaster amusement park, all of it attracting nine million visitors every year.[34] This is possible because the volcano has not erupted since 1707.[35] It is considered active, but dormant.

As it turns out, Mount Fuji might be due for a good show soon. Every sleeping giant must wake up and stretch every now and then, and when that happens, things get interesting.

A number of volcanologists over past decades have warned that an eruption is overdue,[36] including some from Mount Fuji's very own Research Institute,[37] who have reported that volcanic activity has been abnormally insignificant in Japan in the past century compared to what has been recorded in geological history. They underscore that Fuji, which used to erupt once every thirty years but has been dormant for the past three hundred, might be about to make some trouble, because an eruption after a long dormancy has the potential to be highly explosive. Predictions of Fuji's impending eruption have been leveraged by recent studies that have recorded an increased level in the volcano's seismic activity; this was blamed on the 2011 magnitude 9 Tōhoku earthquake, which was believed to have increased the pressure building up inside the volcano[38] up to a value estimated to be greater than what caused prior eruptions.[39]

Studies have investigated what would be the consequence of an eruption today. Focusing only on the amount of ash ejected by the volcano, considering over a thousand different scenarios of eruption scale, wind directions, and air pressure to calculate ash dispersion, it was found that the next eruption of Mount Fuji would blanket the entire Tokyo area with somewhere between three inches to three feet of ash. Any of those scenarios would bring the metropolitan area to a standstill. Roads and railway operations would shut down, water drains and ventilation units would be clogged, and exposed machinery would jam. Failures would further cascade from there. Massive power outages would be expected from ash clogging the intake filters of gas turbines or due to short circuits from rain falling on ash accumulated on power lines.[40] There seems to be no estimates of how long it would take to shovel away all that ash, or where it could be dumped. As for lava flows, if any, plans estimate that three-quarters of a million people would have to be evacuated from the vicinity of the volcano—which would obviously have to happen before too much ash piled up in the roads.

When the volcano last erupted in 1707, it killed twenty thousand people,[41] at a time when the entire population in the Kai and Sagami provinces around the volcano was around six hundred thousand.[42] Today, the Tokyo-Yokohama region is a fifty-mile-wide spread of entangled urban development with a population of approximately forty million.[43] Average population density is roughly seven thousand persons per square mile, peaking at forty thousand people per square mile in the downtown areas. It is both the world's most populous metropolitan area and the largest metropolitan economy.[44] It is not known if studies have predicted the expected number of deaths this time around.

On the positive side, this study has apparently prompted Japanese official to start drawing up emergency response plans for such a situation.[45] This sounds like a good idea. With more than 110 active volcanos in Japan,[46] there is plenty of expertise around to tap; some phones will likely ring near Sakurajima.

THE DNA OF A VOLCANO

Those who believe that fruit-flavored candies and soft-drinks are legitimate portions of a healthy diet's daily allowance of fruits, that fireworks on the 4th of July celebrate the United States of America gaining its independence from France, that *Titanic* is a great movie about a mythical boat that never existed, and that laser beams can be cut into three-foot-long lightsabers, might find this section offensive and can easily skip it without disturbing the space-time continuum.

Earthquake duration is counted in seconds. Hurricanes and tornados are only wind—lots of it, but only wind, nonetheless. The pyrotechnics of a volcano though, that is quite something else: the smoke, the rumblings, the eruption, the fire, the lava—and the damn thing grows, from eruption to eruption, asserting itself, and creating new land. No other natural hazard packs as much "shock and awe." No wonder that in many civilizations, it was the same god that controlled both volcanos and fire—like Pele, goddess of volcanoes and fire and creator of the Hawaiian Islands.[47] After all, by definition, a volcano is the chimney of Vulcan's forge, where the gods' weapons of war were created.[48]

Until the development of plate tectonics, nobody understood by what mechanism all that lava from within the earth could explode to the surface. Why would a volcano sleep for centuries and then, out of the blue, violently erupt?

For a long time, volcanoes were thought to reach to the center of the earth. So, logically, in *Journey to the Center of the Earth*, Professor Otto Lidenbrock's 1863 expedition to that destination started in the volcanic tubes of the sleepy Iceland giant, Snæfellsjökull. However, contrary to what is vividly described in Jules Verne's novel,[49] no respectable nineteenth-century volcanologist believed that dinosaurs and giant mushrooms would be encountered on the way to the earth's core. Artistic license prevailed nevertheless, since nobody had ever gotten close enough to the center to contradict good old Jules.

At best, scientist have barely scratched the surface of the earth. The distance to the center of the earth is roughly 3,959 miles, give or take a few, depending on if one stands on top of Everest (5.5 miles above sea level) or at the bottom of Challenger Deep (2.3 miles under the sea). By comparison, the Kola Superdeep Borehole, which is nine inches in diameter and took twenty years to drill, reached down 7.5 miles and stands as the deepest manmade hole to date.[50] For the record, a couple of recent oil wells in Qatar and on Sakalin Island (offshore Russia) are a few hundred feet longer, but do not go as deep in terms of true vertical depth,[51] because oil drilling can be done at different angles.[52] Looking forward, an ongoing billion-dollar scientific project, using a specially rigged boat to drill into the earth's crust from the bottom of the sea, is expected to go a little bit deeper—which is literally a sophisticated way to throw money into a hole.[53]

While poking a few holes here and there over the next centuries will greatly enrich knowledge on the composition of the earth, in the meantime, plate tectonics can be used to explain the existence of volcanos. Given that lava is magma that reaches the surface and that magma is molten rock, what better place to melt rock than at the interface between two tectonic plates exerting great pressure when rubbing on each other. Subduction zones, which are where one tectonic plate is pushed under another,[54] are perfect settings to create magma, particularly at greater depths, where temperatures are higher. That remains the best explanation so far, and a compellingly logical one. At least while awaiting a robotic descendant of Professor Lidenbrock that could dive into the lava and navigate down the volcano's chimney, which probably will never be possible. So far, one wild soul wrapped in special protective gear made it down into the crater of the active Marum volcano in the South Pacific, so close to the edge of the gurgling lava that micro-splashes of it melted bits of his GoPro cameras, but he did not dare jump all in.[55]

Beyond the fact that humans can travel to visit a volcano, it is also possible for a volcano to come and visit humans—sort of. In Paricutín, Mexico, on February 20, 1943, a farmer saw a half-meter-deep crack in the ground of his cornfield suddenly swell into a five-foot-tall mound blowing sulfur-smelling smoke, ash, and semi-molten rocks. Within a day, the mound grew into a 150-foot-tall mini-volcano. Within a week, it was 300 feet.[56] After a year of activity, the cornfield had become a 1,475-foot-tall cone that buried the village of Paricutín.[57] If any corn was left, it was probably popcorn.

Similarly, a geyser suddenly popped out of a backyard garden in Rotorua, New Zealand. In due time, the geyser gazers who came for the show saw it turn into a bubbling mud crater that grew and engulfed the garage next to the bungalow.[58] That is not quite a volcano, but another one of the potential hazards that cannot be ignored when living in a geothermal area.

VOLCANIC DOUBLE WHAMMY

Historians may debate whether it is because cunning Vikings wanted to trick enemies with an eye on those lands that they called the country covered by ice over 80 percent of its area Greenland and the other one nearby with only 11 percent of its surface frozen Iceland.[59] In any case, a more fitting name for the latter would be Fire-and-Ice-land. Beyond being famous for its geothermal energy and world-renowned hot springs, which attract flocks of tourists, Iceland is also home to thirty active volcanos, and one hundred more believed to be inactive.[60]

The magic of mixing fire and ice is that it makes it possible for volcanoes to erupt from under a glacier, thereby melting a significant part of the glacier. Eruptions too small to break the surface of a glacier can still melt lots of ice from its base. That water seeps under the glacier and eventually bursts out somewhere as a surprise flood—many Icelandic villages have been wiped out that way.[61] Scale-up the eruption, scale-up the flood. In pure Icelandic tradition, these unpredictable glacial floods are given an unpronounceable name: *jökulhlaups.*

When eruptions punch through the glacier, Iceland gets the full double whammy of a full-blown volcanic eruption—with ash, lava, bombs, and so on—and a menacing jökulhlaups.

During the April 2010 eruption of the Eyjafjallajokull volcano (yep, another tongue twister), punching through the glacier and melting much ice around it, the biggest fear in far distances from the volcano was that major roads and bridges would be washed away; also, people in the upcoming path of the flood were evacuated.[62] Destroyed roads and bridges were exactly the consequences during the September 1996 Vatnajökull eruption,[63] which punched through a 1,500-foot-thick glacier and sent water downhill at a flow rate of 45,000 cubic meters per second;[64] or twenty times that of Niagara Falls.[65] Luckily, nobody lived near Vatnajökull. As for Eyjafjallajokull, whose massive jökulhlaup was caught on video,[66] the engineers had time to dig trenches through the causeway of a threatened national highway to divert the upcoming floodwaters through those channels to protect a bridge—as it is cheaper to rebuild segments of a causeway road leading to a bridge than the bridge itself—and it worked.

THE END OF THE WORLD

In a special class of their own are the supervolcanoes. They are a sort of "end-of-the-world" type of event, best explained by making some comparisons.

Much like the magnitude scale for earthquakes, the Volcanic Eruption Index (VEI) measures the severity of eruptions on a logarithmic scale ranging from 1 to 8, considering how much and to what height material is ejected, and how long the eruption lasts. At mid-range of the scale, a VEI 5 eruption projects between one and ten cubic kilometers of volcanic material,[67] from ash particles that will eventually mix with clouds and come back down as acid rain to large molten rocks that crash to the earth like bombs. One cubic kilometer is 0.24 cubic miles, or 1.3 billion cubic yards. This means a one-inch-thick coat of debris over a surface of 15,000 square miles—or one foot thick over 1,250 square miles. The VEI 5 Mount St. Helens eruption in 1980 scattered trace amounts of ashes over eleven states and five Canadian provinces. By comparison, the 2010 Eyjafjalla-jökull eruptions that shut down the European airspace for five days was only a VEI 4. On a logarithmic scale, that is ten times smaller than a VEI 5. VEI 6 eruptions eject so much ash into the atmosphere that it typically affects world climate for a few years. For example, because of the 1883 VEI 6 Krakatoa eruption, summer temperatures in the Northern Hemisphere dropped by 2.2°F and it took three years for things to return to normal.[68]

The domain of supervolcanoes is in the VEI 7s and 8s range. There have been ten VEI 7 eruptions in the past ten thousand years, and four in the past two thousand years. The more recent one is the 1815 Mount Tambora eruption in Indonesia, which disturbed the world climate for a few years— to the extent that it snowed in Albany, New York, on June 6. Crops were ruined in parts of North America and Europe, leading to famine in a few countries.[69] And that's only VEI 7.

VEI 8s are the truly dreadful mega-monsters. One thousand times more powerful than the Mount St. Helens eruption. The devastating impact of a supervolcano can be as significant as that of a fifty-mile wide asteroid colliding with earth—keeping in mind that the fifty-mile wide rock that hit the Yucatan peninsula 66 million years ago projected so much dust in the air that it has been blamed for the extinction of dinosaurs. In fact, two theories[70] have been proposed to explain this massive extinction: asteroid collision or a massive burst in volcanic activity. Both phenomena have the ability to fill the skies with gigatons of particles that end up blocking the sunlight that plants need and cooling the planet, a double whammy that— much like with the Krakatoa eruption but on a grander scale—starves the planet through years of darkness.

The occurrence of supervolcanos is not unusual from the perspective of the earth's geological time scale, but from a human perspective, the odds are good that one thousand generations could squeeze between two events. The last VEI 8 occurred 26,500 years ago, at the Taupo Volcano in New Zealand—obviously, there are no photos or paintings of that event. Known as the Oruanui eruption, it ejected 1,170 cubic kilometers of rock, ash, and lava, with deposits found more than five hundred miles away. The one before that, seventy-four thousand years ago, in Indonesia, ejected twice as much material, standing as second place record holder in that category.[71] The United States has many competitors in that category. Although not a medal holder, the Yellowstone caldron has exploded three times in the past 2.1 million years,[72] with the last eruption 640,000 years ago. Statistically, if the Yellowstone eruptions were taken to occur like clockwork, as Old Faithful almost does, that would leave another seventy thousand years before the country is significantly remodeled by such geological forces. The gold medal apparently goes to the eruption of La Garita Caldera, which is believed to be the largest volcanic eruption ever, estimated to have been so large that it has prompted volcanologists to add a level to the VEI scale, just for it—and they assigned it a VEI of 9.2.[73] Erupting twenty-seven million years ago in southwest Colorado, it ejected five thousand cubic kilometers of materials. If that amount of material could be evenly spread over a geographic area, it would be enough to cover the entire contiguous United States with two feet of debris—not to forget the impact on climate from all those ashes obscuring the sky.

It is hard to find a worse way for an entire civilization to die. Short of a zombie invasion, there is probably not a more painful way to reach the end of the world than a supervolcano eruption.

ON THE DISASTER TRAIL

Dust in the Wind

Dormant volcanoes can be sleeping giants, picturesque and great for hiking, but for someone like me who thinks that visiting volcanoes is an absolutely essential thing to do, active volcanoes are more fun. Most of them are national parks or official tourist attractions, each with their own special character that makes them uniquely enjoyable.

The Vesuvius observation area allows a peek into a steaming cauldron that is relatively peaceful and, all things considered, immensely less dangerous than the drive to it from Naples—the road is safe, but the Italians drivers are "*completamente ignorante delle regole del traffico.*" The drive to Mt. Aso Nakadake Crater, in Kyushu, Japan, is more rewarding, as far as "volcano tasting" is concerned. The large parking lot located along the rim can be filled with school buses, but the kids run away from the crater observation area at the first whiff of sulfur. Multiple signs warn that those with asthma and other breathing problems might be at risk, and the observation platform is closed when unfavorable winds engulf it in the toxic sulfur dioxide gas emitted from the crater lake two hundred feet below. Over the years, a number of people have died from volcanic gas inhalation, and many more have been severely injured.[74] Every now and then, the volcano also erupts.

White Island in New Zealand provides an even more intimate encounter with the lethal chemistry of volcanos. A ninety-minute boat ride away from the coast, White Island is a desolate landscape in a geothermal frenzy. After having signed a waiver acknowledging that eruptions are unpredictable and that we assumed the risk, our tour guide provided us with a gas mask along with a stern warning that anyone stepping outside the marked trail risked punching through an unstable crust and landing in an acid bath. The trail meandered between bubbling mud pools, steam vents roaring more furiously than locomotive chimneys, hissing hot volcanic streams, and an acidic crater lake known to instantly dissolve anyone falling in (this fun fact was probably discovered the hard way). Every now and then, White Island erupts too, as it did on December 9, 2019, killing twenty-one tourists and injuring twenty-five more with severe burns.[75]

Yet, of all my volcano visits over the years, the "quaintest" one—if volcano visits can ever be described as such—was an overnight stay on Sakurajima Island. Given that Sakurajima is the most active volcano in Japan, there is no public road to its crater, but there are roads and villages all around its base. We had elected to stay overnight in a Minshuku, which is the Japanese equivalent of a bed and breakfast. It was a traditional Japanese home, the hosts were charming, and conversation was limited to clumsy sign language, but sleeping at the base of an active volcano was an eerie experience—no waiver of responsibility was necessary to sign in this case, as this was normal everyday life as far as our hosts were concerned. The "icing on the cake" of this volcanic experience came in the morning, as we left and discovered that our car had been covered by a sheet of ash during the night—courtesy of an active volcano. Many times before, I have had to "dust away" snow from my car in the morning, but this time, the "dusting away" was as dusty as it gets.

Technological Disasters

NO SAFE PLACE ON EARTH

Homer Simpson works at the fictitious Springfield Nuclear Power Plant, a facility that has received hundreds of safety violations and has averted multiple total meltdowns by sheer luck alone.[1] Homer is an absolute, unabashed, incompetent safety inspector who would not be woken by flashing red alerts and would not see anything wrong in using fuel rods as paperweights. However, while Homer is a cartoon character, there had to be a real Homer Simpsonov operating the Chernobyl nuclear power plant when one of its reactors exploded in 1986 during a safety test gone awfully wrong. In fact, there had to be an entire cohort of Simpsonovs along the entire chain of command, including the engineers who designed the flawed nuclear reactor,[2] the poorly trained plant operators who had turned off the automatic shutdown mechanism the day before the explosion,[3] the officials who waited a day before issuing an order to evacuate the fifty thousand people in the neighboring town of Prypiat twelve miles away, and all the top-level comrades who downplayed the event for as long as possible.[4] All of them contributed to the resulting mess. How else to explain why the USSR kept quiet about the incident, if not for fear of utter embarrassment?

While the communist apparatchiks acted as if nothing had happened on the day the plant blew up, and the state-owned media devoted a full twenty seconds during the evening news to mention that a minor incident had happened (giving it less attention than the weather forecast),[5] the world discovered that something had gone terribly wrong behind the iron curtain when radioactive rain came down in Sweden two days later.[6] The steam from the explosion and the fire that burned for a week before being contained, spewing radioactive clouds into the atmosphere, not only made Chernobyl the largest nuclear disaster in history—until the Fukushima Daiichi nuclear disaster twenty-five years later[7]—but it also arguably played a significant role in bringing down the USSR six years later.[8] At least, it was perceived as such by Mikhail Gorbachev,[9] general secretary of the Communist Party of the Soviet Union at the time of the disaster. To its people, raised on the propaganda that the USSR was the top world superpower, best at everything, with truckloads of Olympic medals to prove it, the widespread incompetency at managing a public health crisis exposed the fallibility of their government and rapidly eroded whatever confidence they might have had in its institutions.[10]

Three decades later, for a mere 3,300 Ukrainian Hryvnia (roughly US$130), tourists were able to visit Chernobyl by booking a tour with the state enterprise in charge of decommissioning the nuclear power plant or with other organizations that marketed Chernobyl tours as an "eye opening experience of post-apocalyptic world."[11] One needed to sign a waiver releasing the tour organizers from "any damages caused to the health, property of the visitor" from a list of possible causes, including "the effects of radiation and other harmful factors,"[12] and then, off to the tour—not exactly Disney World, but there was apparently a market for it.

Whether or not the Chernobyl incident has had an impact on the nuclear power industry has been a subject of debate,[13] but there is global agreement that the 2011 Fukushima nuclear incident did. In the aftermath of the Japanese disaster, the industry faced significant project delays and—in some cases—roadblocks. The difference is that, in 2011, it was more a case of nature exceeding design assumptions than problems created by a bunch of Homerima Simpsonakis asleep at the switch.

When the March 11, 2011, magnitude 9 Tohoku earthquake struck 110 miles away from the Fukushima Daiichi Nuclear Power Plant, the ground accelerations recorded by instruments at two of the six nuclear reactors on site exceeded the values considered in the design of the plants by about 20 percent. The plant was subjected to two minutes of strong ground motions,[14] but this did not create an issue because, unlike buildings where damage is expected if the earthquake considered in their design occurs (as will be described later), nuclear power plants are designed to remain undamaged for the considered design forces, for obvious reasons, as damage to them can have severe consequences.

As soon as seismic activity was detected by Fukushima's instruments, the reactors were shut down. Emergency generators powered by diesel fuel kicked in when power from the grid was lost due to damage to nearby transmission lines. So far, it all played out by the textbook, and most of the world would never have heard the word "Fukushima" if the impact of the earthquake had been limited to that. Unfortunately, the tsunami wave arrived, and all hell broke loose.

When the power plant was designed in the 1970s, close to the ocean, it was designed to be protected from possible future tsunamis. As for all things, it all boils down to what is considered "possible." At the time, a ten-foot tsunami had been recorded from the 1960 Chile earthquake, so that was taken as the design basis in planning for future tsunamis.[15] Therefore, the power plant was built thirty feet above sea level and the seawater pumps that circulate the water necessary to cool down the reactors—which remain hot for quite a while even after a shutdown—were located twelve feet above sea level.

Then—oops—the Tohoku earthquake generated a massive tsunami wave that was forty-five feet tall when it reached Fukushima. Flooded by fifteen feet of water (and more for the pumps), the plant lost its emergency diesel generators. Electrical switchgear and batteries located in the inundated basements also became useless. This delayed cooling and was the beginning of a long process in the attempt to control the reactor. It did not help that many access roads had also been taken out by the tsunami. Without going into the mechanics of how a nuclear reactor works, suffice to say that out of this mess, as dedicated personnel worked to prevent a bigger catastrophe,[16] three nuclear reactors suffered meltdowns, three hydrogen explosions occurred, and radioactive contamination was continuously released over a three-day period.[17]

Post-disaster investigations revealed that studies conducted in the years prior to the Tohoku earthquake had brought to the attention of the government and the power plant owners that tsunamis of up to fifty feet were possible and pointed to the need to implement protective measures in anticipation of such events,[18] but those warnings were not acted upon, awaiting further expert review.[19] Some post-earthquake critics denounced the collusion that existed between regulators and industry in Japan, stating that many of the regulators were former employees of the power companies.[20] Given the consequences of a nuclear reactor failure, one would indeed expect due diligence when a red flag is raised. As a result, beyond all the damage to the power plant itself, there was also damage to the people's trust in their decision makers. The faith that public safety would be upheld by all stakeholders as the upmost priority was shattered. The Big Bad Wolf had blown off the house that was supposed to be safe; in the aftermath, baffled, many wondered how they had ended up with a straw hut.

In some countries, Fukushima eroded whatever trust remained in nuclear energy. For example, following Fukushima, Germany shut down eight of its seventeen reactors and promised to close the rest by 2022.[21] Siemens, the German multinational giant, terminated its nuclear engineering industry, announcing it would no longer build nuclear plants across the world. Similarly, South Korea, Switzerland, Taiwan, and others enacted laws to phase down nuclear power in their respective countries.[22] Arguably, support for nuclear energy in some of these countries was already shaky, and Fukushima only gave the needed incentive to trigger or accelerate plans to move away from nuclear energy. Likewise, a few countries that intended to open the door and adopt nuclear energy decided to keep that door closed for a while longer.[23] Some countries—like India—have elected to still move ahead with the building of many nuclear power plants but have slowed their construction schedule to ensure greater safety (or for political reasons).[24] Not surprisingly—it takes all sorts of people to make a world—in some countries, Fukushima did not make a difference, and plans for new nuclear plants are moving ahead undeterred, or faster than before, out of need or for opportunistic reasons.

Part of the shockwave of the Fukushima failure on the global nuclear industry is attributable to the fact that, since the 1980s at least, Japan had been recognized to be a producer of top-quality, engineered products, with a track record of reliability, performance, and world-class quality controls. Japan is not one of those emerging or corrupt economies where crappy products and lack of safety are the norm—and where, incidentally, nuclear power plants are also being constructed.

In Japan itself, some fancy footwork was in order. On one hand, large companies like Toshiba, Hitachi, and Mitsubishi, were in the profitable business of selling nuclear power plants worldwide; on the other hand, public trust had collapsed. Mega-conglomerates cozy with politicians may have more money, but the public has more votes. As such, following the Fukushima incident, the Japanese government ordered all of the country's thirty-eight nuclear reactors to shut down within a year and required compliance with new safety standards so stringent that only nine reactors had resumed service by 2019.[25] Elected officials also insisted that there would be no new construction of nuclear power plants in the country—fourteen new ones had been planned to open by 2030, but Fukushima put a halt to those plans. Paradoxically, Japan's hope to continue building reactors abroad was undermined by the construction freeze at home. Who would buy nuclear power plants from a country that prevents them from being built at home—from the very country where the largest nuclear disaster of the millennium happened? This would be like buying steroids from suppliers in a country that has stopped winning Olympic medals. In 2017, when Westinghouse filed for Chapter 11, facing billion-dollar losses on power plant projects, its parent company—Toshiba—could not keep it afloat, as it was itself sinking.[26]

The downfall of German and Japanese nuclear plant builders following Fukushima opened up the world market to reactors manufactured by China and Russia. In 2018, two-thirds of the nuclear reactor projects worldwide were being constructed by Rosatom, a Russian state-owned energy company[27]—presumably on the expectation that the new Russia is not as corrupt and contemptuous of safety as the USSR of the bygone Chernobyl era. As for China, on the basis that some Chinese decision makers aim to build five hundred nuclear power plants across the country by 2050 (which is possible if extrapolating the country's $57 billion 2010–2020 investment with twenty nuclear plant projects under construction, together with the fact that it is planning to spend an additional $850 billion doing so until 2050),[28] some scientists are already calculating the probabilities of a major Chinese nuclear disaster.[29]

One of the lessons of the Fukushima disaster is that designing for probabilistically based scenarios is—well—just that, namely setting up an acceptable risk and living with the consequences if an extremely rare event exceeds the design basis. Regardless how rare the event—be it a one in a thousand, one in ten thousand, or one in a million years event—eventually technology will fail. All technologies will. Instead of relying on probabilistic calculations to bury a risk in mathematics, a better solution is to plan fail-safe approaches—that is, defense mechanisms, embedded in the infrastructure itself, that assume things will fail but that will be able to control and minimize the consequences. As a case in point, various types of elevators have existed for over a millennium—chairs and platforms pulled up by rope for the most part.[30] Yet rare were those who wanted to buy elevators until Mr. Otis pulled a stunt at the 1854 New York Exposition, cutting the rope of his elevator, sending it into freefall (gasps!) until it was stopped by an automatically released mechanism that gripped the elevator guide rails and brought it safely to a stop (applause).[31] Fail-safe. Sales boomed.

EVEN THE SKY IS FALLING

Nothing ever fails because of an error in the laws of science. Bridges do not collapse because of a sudden glitch in the law of gravity, airplanes do not crash because molecules in the atmosphere misinterpret the rules of fluid dynamics, and nuclear power plants do not melt down because of an unexpected change in the universal laws of thermodynamics and nuclear fission.[32] The laws of the universe rule everything, from subatomic particles to ten-billion-light-years-wide superclusters of galaxies. They have created the world as it is, and as chaotic as this world might seem at times, the physics of it all is constant. Plainly said, the universe does not make "mistakes." Therefore, when technology does not perform as intended by humans, it can only be because of human factors.

One of the primary laws of physics is the principle of minimum energy, which states that everything in the universe has a natural tendency toward the state where the least amount of energy is needed to be in equilibrium. Any parent with a teenager slouching on a couch can immediately relate to the purest expression of this law of physics.

Therefore, constructing a road hundreds of feet up in the air to connect the two sides of a canyon requires carrying material up there and taking appropriate measures to keep it up there forever. The point of least energy for all that material is at the bottom of the canyon. Whether fighting gravity that wishes to bring it down in big chucks, or corrosion that wishes to bring it down one atom at the time, it is a constant battle—and the same goes for any created technology. A number of human factors conspire to make the battle arduous—factors that could lead to failure. Pressure to reduce costs and deliver on time may lead to "cutting corners" in design, fabrication/construction, quality control, and safety checks. Incompetence, negligence, or outright corruption (or even sabotage) can lead to a product that will eventually show its deficiencies either in the long or short term. Incomplete or erroneous data, ambiguous specifications, inaccurate models, and inability to account properly for the complex nonlinear interaction of interdependent systems, combined with lack of experience, can lead to improper design decisions. Lack of sleep, emotional or health problems, internal politics, personal biases, and other similar factors that affect performance can also allow errors to sneak through the process. And then there are errors that happen simply because people sometimes make mistakes without any reason.

In short, to err is human. The only way to achieve a successful outcome is to put in place enough mechanisms and processes to counteract all of the human factors that might lead to failure. This does not mean getting rid of every engineer who has spousal problems, but rather enforcing the application of accepted practices; using conservative and fail-safe designs; having realistic budgets, resources, and timetables; questioning existing practices and perceptions; continuously updating the state of practice to account for new knowledge; and developing effective and practical plans for continued monitoring during and after construction. That is a lot of work—just saying it, is exhausting. Yet this is most important for infrastructure projects because, contrary to other industries where manufacturers can stress-test a product for hours—or even crash-test it on a wall—before releasing it to the market, it is quite difficult to "crash-test" a nuclear power plant.

On the positive side, infrastructure projects are engineered today with more care and due diligence than most manufactured products. This is most reassuring given how manufacturing has evolved in recent decades, particularly in the high-tech sector where quality control has sometimes been thrown out the window under pressure to beat the competition and/or make a quick billion. Some decision makers have apparently concluded that it is less expensive to replace a small percentage of products shipped defective than to invest in quality control to identify them in the plant and prevent their release. In less politically correct terms, this means that too many companies do not mind selling crap as long as there are suckers buying it. This mentality is a direct extension of the despicable software engineering philosophy that consists of releasing flawed products early to cash in as soon as possible and then fixing bugs on the fly as problems are encountered by users—essentially being inconsiderate to these aggravated users stuck with the problems. In essence, this is shifting the responsibility for a part of product testing from the manufacturers to users.

When a bug in poorly written software prevents a computer-controlled enemy to activate as intended on the thirtieth level of a video game,[33] or freezes half the screen at the 256th level of Pacman,[34] who cares? The software developer who wrote the "spaghetti code" that led to the error is insignificant in the grand scheme of things—just like the video game in itself. However, when buggy software impacts the lives of real people (who usually have only one life, contrary to pixelated characters in games), that is more consequential. If code writers wish to promote themselves as software *engineers*, then they must be held to the same standards of responsibility and ethics that are required in other engineering disciplines—something many argue is far from the case.[35] Software bugs are widespread, and defective updates are released via the internet for everybody's pleasure and enjoyment. Some even make the news. For example, many people purchased the Nest smart digital home thermostat that promised to learn from the habits of its users, by identifying their preferred temperature at various times of day and programing itself to save and reduce the energy bill accordingly. Noble goal. However, as smart as the thermostat might be, in a less smart move, those who maintain its software released an update in December 2015 that uploaded itself to all these internet-connected devices and caused them to drain their batteries and stop functioning by mid-January 2016.[36] With relatively mild temperatures at the Nest headquarters in sunny Palo Alto, California, those venting their frustration on social media were not impressed that Nest's programmers did not seem to have considered the possible consequences of a thermostat failure in less balmy mid-winter climates when they released their update.[37]

They were equally unimpressed when Nest stated it was aware of the issue and expected to have a solution ready to "roll out in the coming weeks," together with instructions on how to manually reboot the thermostat in the meantime as a temporary fix. This compassionate message must have provided great solace to all, especially those from northern states who were traveling when the "glitch" happened and had nightmares of returning to a water-damaged home buried in ice due to burst pipes.[38]

At the other end of the fire and ice spectrum, it was even more news-worthy when it was discovered that the battery in the new Samsung Galaxy Note 7 phone/tablet tended to overheat, release smoke, catch fire and sometimes explode.[39] Not exactly an insignificant bug. Devices were recalled and reissued with different batteries supplied by a Chinese company instead of Samsung itself, but—adding insult to injury—a second recall had to be issued as some of the replacement units suffered the same problem of failure and combustion.[40] Given that Samsung's mobile business chief, in response to the crisis, stated, "I am working to straighten out our quality control process,"[41] one can only wonder what quality-control processes he was referring to that could have allowed a million devices to be produced and shipped without anyone catching this problem.

The scary thought is that this computer science mentality of quick-to-market and imperfect quality control could be spreading. When a retired quality manager who had spent decades working for Boeing filed a whistle-blower complaint with the Federal Aviation Administration alleging that the flagship 787 planes had been shipped with problems that could lead to electrical shorts and cause fires,[42] and when all the 737 MAX planes were grounded after two crashes while Boeing was trying to fix—surprise!—a software problem,[43] regulators and lawmakers launched investigations, issued subpoenas, held hearings, and so on. Yet, given that not all corporations have the engineering skills and wherewithal of Boeing, or its visibility, that leaves many citizens wondering what might be the norm in various industries nowadays? Are existing practices, regulations, and assorted corporate cultures adequate to ensure that the lure of quick profit does not take precedence over safety?

Note that software errors having tragic consequences is not a new phenomenon; modern history is replete with episodes where fatal bugs were found, often where one would least expect it. For example, in 1985, many patients were killed or seriously injured before somebody realized that this was caused by a flawed software upgrade to the Therac-25 radiation therapy device[44] that made the machine bombard people with one hundred times the intended dose of radiation.[45] During the investigations launched by the US Food and Drug Administration after a few cases of a lethal dose of radiation had been reported, multiple software design errors were found.[46] In particular, it was discovered that the software had not been independently reviewed and that testing of the device had not been redone after a software upgrade.[47]

Saying, "it's not a bug, it's a feature" is not as funny as the marketing department may think. To make it worse, from teenagers of the generation born with smart phones USB-connected to their pacifiers to senior nerds raised on punched cards and stacks of folded printouts, too many keyboard wizards have a condescending view of those they regard as being "technologically challenged." Such egotism, lack of care, and, in some cases, lack of life experience, sadly can make the computer geek "tech-not-logical," or "tech-illogical," resulting in products whose newest version is lacking some of the best features of its previous one, and device operations that defy logic.

It takes time to do a good job. It takes time and dedication to make sure everything will work as intended—safely and reliably. That slows down the release to market and does not generate more profits. For example, how many engineers get big bonuses or promotions for doing all the due diligence to prevent disasters? Probably not that many—if any—because nobody wakes up in the morning grateful for all the disasters that did not happen in the previous twenty-four hours but that would have happened if not for the duty of care upheld by those who keep things safe: the true pillars of society. Instead, it is those who spring to action after a disaster to pick up the pieces and set things on some sort of a course to recovery that are called heroes and get all the medals and accolades—sometimes because they are genuinely good at fixing up big messes, and sometimes only because they are blabber mouths who are quick to jump in front of cameras to claim credit—even when those disasters might have been prevented in the first place if these same heroes had displayed foresight, adopted different priorities, and taken different actions prior to the events.

Fortunately, on the positive side, as stated earlier, comparing the state of the built infrastructure with that of manufacturing—or, God forbid, that of the high-tech industry—should be avoided, because these operate in different worlds when it comes to accountability.

Yet, on the negative side, things are still far from rosy when it comes to infrastructure. Technological disasters constitute a subset of all possible disasters that can be caused by human action or inaction (a category known as anthropogenic disasters). The truth of the matter is that there are millions of buildings, bridges, and other infrastructures worldwide that have been designed according to obsolete requirements, waiting for hazards to bring them down.

MASTER BUILDERS AND ENGINEERS

Theodore Cooper had been hired as a consultant because he was one of the foremost bridge engineers at the time, recognized for his pioneering work on the analysis and construction of steel structures, and particularly long-span bridges.[48] Cooper had even twice won the prestigious Norman Medal of the American Society of Civil Engineers, awarded in recognition of the impact of a specific technical contribution on engineering practice— a reward received by only a select few, with repeat award recipients being even more rare.[49] The bridge was to become the world's record holder for longest span, but as Cooper was in his sixties, in poor health, and unable to travel away from his New York City office, he hired Norman McLure—a young Princeton graduate—to be his eyes on the construction site.

At one point, mid-construction, McLure traveled to meet Cooper in his office to review in person, and in more details, worrisome measurements that had been taken on some of the bridge members over the previous weeks (a member being a component or element of a structure, such as a beam, column, or truss diagonal). Problems had first been reported on June 15, 1907, when some members of the bridge could not be connected because the pre-drilled holes on adjacent members did not line up. Errors of up to a quarter inch were noted. At the time, these were assumed to be fabrication errors, presumably in the pre-cambering of the member that had been performed prior to shipping—yet, this was odd, considering that pre-cambering is something that was calculated and done on purpose to make members line up perfectly when subjected to loading. However, subsequent inspections in early August revealed out-of-straightness of as much as three-quarters of an inch in some members, which caused the field engineers to wire a message to Cooper. This triggered much discussions as to whether these members had been banged up and bent prior to shipping or not. In spite of McLure's conviction that the bend in the members occurred after they were installed in the bridge, Cooper argued that the members had to have been hit by other members before being installed. Over a two-week period, the lateral deformation increased from .75 inches to 2.25 inches. That is enormous for a member that is supposed to remain straight, so on August 27, the contractor stopped all work asking for this issue to be reviewed. McLure traveled to New York to meet with Cooper on the 29th.

While McLure and Cooper met, unknown to them, work had resumed on-site because the local chief engineer employed by the entity that had issued the constructions bonds—essentially, a bunch of businessmen and politicians incorporated for the sole purpose of this project—declared that the stoppage was bad for the morale of the construction workers. This decision was backed up by the top brass at the Phoenixville (Pennsylvania) headquarters of the construction company, who called the site to assure them that it was safe to proceed. Meanwhile in New York, McLure and Cooper had concluded that no more load should be added to the bridge until they had thoroughly reviewed all the facts. Cooper wired this information to the Phoenixville construction company and traveled there at once for a 5 p.m. meeting the same day. Unfortunately, in his haste, McLure forgot to wire the same information to the construction site.[50] The meeting at the Phoenixville headquarters was brief and concluded with a decision to reflect on this overnight and resume discussions the next morning.

At 5:30 p.m., on August 29, 1907, the steel members having the large out-of-straightness suddenly buckled. The Québec Bridge collapsed, killing seventy-five of the eighty-six workers on the structure when the nineteen thousand tons of steel plunged into the Saint Lawrence River.

Why did it happen? Could it have been prevented? How to prevent it from ever happening again? These are some of the questions asked after any failure with fatalities, be it a bridge collapse or a plane crash. The truth of the matter is that pushing the boundaries of technology means entering uncharted territories. As stated multiple times throughout history—by French Revolution decrees, Churchill, Roosevelt and many more,[51] and popularized as the Peter Parker principle by Spider-Man fans—with great power comes great responsibility, which means here that extensive testing on components and systems must be the norm when developing new technology. Yet testing implies that one knows under what conditions something is to be tested—that is, what phenomena are to be replicated. Knowledge might simply be lacking, or the scale of the system might be so large that only small components can be tested, from which behavior of the entire system must be inferred.

For example, transmission lines have been weaved across continents for decades, but in building those, nobody anticipated that solar flares could be a problem. It turns out that the wild burning gas ball that is the sun, every now and then, has sudden outbursts of energy equivalent to thousands of nuclear bombs exploding simultaneously. Each of these powerful explosions, called solar flares, happens to shoot a magnetic wave straight toward the Earth at a million miles an hour. An unusually intense solar flare occurred on March 13, 1989.[52] When it hit home, it disabled the entire Hydro Québec power grid, putting six million people in the dark for nine hours. The entire northeastern region of the United States barely escaped its own blackout, thanks to the fact that the loss of power from up north (which feeds a lot of the US grid) happened at 2:44 a.m. when demand for electricity was low. When stringing electrical wires over thousands of miles, who would have thought of the sun as a hazard—other than construction workers getting sun burns. Learning from experience, the power grid has been upgraded to prevent future recurrences of this problem,[53] but this illustrates how learning by failure is an important aspect of technology, and how foreseeing every possible problem and testing every system ahead of time for every possible technological failure can be difficult—in this case, it would have "simply" required blowing up, somewhere in space, thousands of nuclear bombs at the same time.

Coming back to the Québec Bridge, the Royal Commission that investigated the collapse revealed a combination of shortcomings in knowledge, engineering judgment errors, and organizational failures. In the 1900s, steel bridges had been built for over thirty years, but it was still a relatively new technology—somewhat like rocket science in the 1960s. Problems were many. First, the existing knowledge on the buckling strength of members in compression was far from perfect, and the entire Québec Bridge strength relied on the strength of large compression members. Then, in May 1900, the bridge's clear span length was increased from 1,600 feet to 1,800 feet. This relocated the piers outside of deep water and thus eliminated the need to design them to resist the strong current at that site and large ice forces on the piers what would have come with it—and at the same time, it made Theodore Cooper the chief engineer of what was to become the longest cantilever span bridge in the world. In principle, this should not have been a problem in itself, but relying on his engineering judgment and because of budget constraints, Cooper did not deem it necessary to revise the design to account for the extra weight added to the bridge due to its lengthening. Instead, he increased the value of the maximum stress allowed on members, essentially shaving away the margin of safety. Finally, Cooper was a prominent consulting engineer, apparently "self-confident to the point of arrogance,"[54] without a counterpart on the construction site who would have had the confidence to challenge and contradict him. Therefore, as many failures often do, this collapse happened due to a combination of unfortunate circumstances. It ended Cooper's career, who retired from public life.[55] Defeated politicians may sign multimillion-dollar book deals to brag about their pretended accomplishments and greatness, but defeated engineers simply vanish, quietly.

Cooper was not the first nor the last bridge builder to see his work destroyed by forces of nature, be it gravity, wind, ice jams, or other hazards. Most of the time, phenomena unknown to exist—or unknown to have an impact on bridge performance—were discovered or rediscovered because of failures.

Collapses of large bridges due to lack of engineering knowledge have been well documented throughout the nineteenth and early twentieth centuries. For instance:[56]

- 79 people died in 1845 because the chains of the Yarmouth suspension bridge in England fractured when too many people assembled on the bridge to view a clown floating down the river in a barrel—hard to think of a more ridiculous reason to die. Oddly enough, the bridge had been designed by an architect with little relevant prior experience, who did not bother to perform the tests necessary to ensure adequate strength of the construction materials, and who showed up on the construction site for the first time on the bridge's opening day.[57] It can be speculated that this architect's specialization in designing churches throughout his career[58] possibly provided him with the kind of holy contacts needed to avoid jail after the bridge collapsed, in spite of harsh criticisms by the British Institution for Civil Engineers.

- 226 soldiers died in 1850 when troops marched in step across the Angers suspension bridge in France and created resonance—a dynamic amplification of vibrations due to repeated impulses at the same fundamental vibration frequency as the bridge. The Broughton Suspension Bridge in England had similarly failed in 1831,[59] so soldiers on the Angers Bridge knew of the dangers of walking in lockstep across a bridge—something armies going as far back as the Romans also knew. But the Angers Bridge was crossed during a wind storm that made it oscillate, so even though the soldiers had been told to space out and break step, the severe swaying of the bridge made them involuntarily take steps in sync with the bridge motions to be able to keep their balance. In other words, because the soldiers merely attempted not to fall down, the bridge did. Nowadays, soldiers are still often cautioned to break stride when crossing bridges; for many large modern bridges, this might not be necessary anymore, but old habits die hard.

- Many died when a number of bridge spans in the United States collapsed due to train overloads in the 1850s, and in many other countries in the subsequent decades, including the Gasconade Bridge in Missouri in 1855,[60] the Desjardins Canal Bridge in Ontario in 1857,[61] and the Sauquoit Creek Bridge in New York State in 1858,[62] to name a few. What makes the Gasconade Bridge uniquely interesting is that it occurred during its inaugural train run—what better way to impress six hundred dignitaries and guests on board than to have the train plunge into a valley along with bridge debris. In the purest tradition of US trains arriving late, this one simply never arrived to destination. Around forty passengers died—or, from the point of view of a die-hard optimist, more than 90 percent of those on board survived. Who would have thought that train travel could be so hazardous?

- Many more died due to lack of knowledge on the various conditions that could lead to fractures in metals, a shortcoming that led to collapses of the Ashtabula River Bridge in Ohio in 1876,[63] the King Street Bridge in Melbourne, Australia, in 1962,[64] the Mianus River Bridge in Connecticut in 2003,[65] and many more—although some more recent failures can be attributed not to ignorance but rather to deferred maintenance practices.
- Many bridges have been washed away by floods, or by flash floods such as the Dry Creek Bridge, in Colorado in 1907, for which the sudden surge in water level unfortunately happened at the same time as when a train was crossing, resulting in a hundred deaths.[66]
- Ice jams also did the trick every now and then, such as for the Upper Niagara Falls Honeymoon arch bridge which failed in 1938 when the ice in the gorge pushed the arch off its abutments.[67]
- The world-famous Tacoma Narrows bridge failure, in 1940, whose deck vibrated in a twisting motion when subjected to a steady 40 mph wind before breaking apart,[68] brought to the attention of bridge engineers the critical need to consider aerodynamic forces in the design of slender suspended spans. It also brought to the attention of high school students the fact that science teachers love to show videos of the Tacoma Narrows bridge collapse to explain the importance of resonance in physics—although, strangely enough, it is not resonance that killed the bridge but aerodynamic fluttering, a topic way beyond the high school physics curriculum.[69]
- The Duplessis Bridge in Québec collapsed in January 1951, not because of sabotage by communist terrorists (as alleged by the prime minister at the time), but because the specific kind of steel used in those spans had brittle properties at cold temperatures and it was minus 34 degrees Celsius when it collapsed at 3 a.m. in the middle of a harsh winter night—definitely crisp weather, even by Canadian standards.[70]

And so on—the list is long.[71] And that is not even considering bridges that are blown to pieces in war zones, or those doomed from the start because they are built in corrupt countries where infrastructure projects are, in large part, a way to divert public funds to line the pockets of politicians and other crooks.

Nowadays, bridge failures still occur for a wide number of reasons (some crazier than one can imagine), but mostly due to boat collisions, construction errors, or extreme events. For example, in 1980, a freighter boat suddenly caught in a wind burst and torrential rain that reduced visibility to zero did not have time to drop anchor before hitting a pier supporting the main span of the Florida skyway, causing 1,200 feet of the bridge and thirty-five people to plunge 150 feet to their death.[72]

Failures due to extreme events more commonly occur to bridges designed at a time when the magnitude of the possible extreme demands was not well known, or when knowledge on how to design these bridges to enable them to survive such events did not exist. For example, a mile-long segment[73] of the double-decked elevated highway in Oakland known as the Cypress Freeway collapsed during the Loma Prieta earthquake—killing forty-two people in the process—because it was built in the 1950s, at a time when engineers did not know how to detail reinforced concrete structures to prevent their collapse during earthquakes.[74] After the earthquake, the road was rebuilt at ground level rather than elevated, at the mere cost of $1.2 billion dollars and nine years of construction.[75] Although the vulnerability of the viaduct had been brought to attention and measures could have been taken to strengthen the structure prior to the earthquake and thus avoid its collapse,[76] waiting for the earthquake to wipe it out and reconstructing it anew afterward had the distinct advantage that the federal government footed 90 percent of the bill.[77] Maybe money flows more easily after a disaster for the same reason that nobody is rewarded for preventing a disaster. The more visible the pain, the more opportunities for well-intentioned politicians to shine (apologies for the oxymoron), and the greater the largesse.

Going forward, with millions of smart phones and surveillance cameras on the ready, bridge failures have become YouTube events. Some relatively recent collapses caught on video include an overloaded new pedestrian bridge in China in 2013,[78] a century-old stone bridge in Greece in 2018,[79] and a modern 450-foot-long arch bridge in Taiwan in 2019.[80]

"Wait a minute!" would say the astute reader at this point. "Are there not supposed to be design codes that engineers are supposed to follow and that exist for the sole purpose of preventing problems—and most importantly, failures and collapses? After all, there are codes for plumbing, electricity, roofing, and just about anything that can be constructed." The answer to that excellent observation is a categorical "Yes, . . . but."

This requires a few explanations that will come much later in this book.

ON THE DISASTER TRAIL

What Does This Guy Know Anyhow?

In 2007, the post-apocalyptic movie *I Am Legend* starred Will Smith as the only living man in New York City, having survived a virus that decimated most of the world by either killing humans or turning them into vampire mutants that only came out at night.[81] Capitalizing on the popularity of the blockbuster that raked in nearly $600 million in revenues, the magazine *Popular Mechanics* published an article titled "*I Am Legend*'s Junk Science: Hollywood Sci-Fi vs. Reality"[82] that sought to determine if a number of aspects of the movie were scientifically plausible. It asked whether an abandoned Manhattan infrastructure could really be covered in tall weeds after only three years, whether a retrovirus could spread and wipe out humanity that fast and that thoroughly—recognizing that there is yet no known virus that turns humans into vampires—and whether the Brooklyn Bridge (and the Manhattan Bridge next to it)[83] could be destroyed in the manner depicted in the movie. This last question arose because the movie showed both bridges missing about three-quarters of their center span, the remaining quarter projecting outward from one of their towers but not supported by cables, while their east span all the way to the Brooklyn shore was still supported by their suspension cable.

When contacted by the magazine to comment on whether this kind of damage was realistic, I mentioned that it was odd, to say the least. This is because the way suspension bridges are designed, the main cables are run from the shore to the first tower, to the second tower, and then to the opposite shore, like a giant clothesline—one that sags a lot. Smaller vertical cables hanging from the main cable are attached to the bridge deck at various intervals to support it. Therefore, as would happen if one would cut a clothesline in its middle, taking out the cables in the middle of a suspension bridge would cause all the spans to collapse, not just the middle one; only the towers would remain standing. And even if the cables were somehow fixed to the top of the towers—they are typically not—and the mid-span of a suspension bridge was bombed, the forces from the shore-spans would pull the towers significantly laterally, and it is unlikely that the towers could withstand those sorts of forces. I suppose a writer could imagine a scenario where the cable was welded at the top of the towers, or got jammed or stuck there by debris, but that is science fiction. In truth, to complicate matters (and something I did not mention to *Popular Mechanics*, to keep things simple), contrary to the Manhattan Bridge, the Brooklyn Bridge is a hybrid design: it is both a suspension bridge and a cable-stayed bridge, which means that the bridge deck is supported two different ways (much like a person can wear both a belt and suspenders at the same time to make pants fight gravity).[84] In cable-stayed bridges, the tension force in each cable is resisted by the deck in compression between the cable and the tower, such that the middle span could be removed without collapsing the rest of the bridge. In fact, cable-stayed bridges are constructed by adding pieces of deck (and cables supporting them) one piece at the time, projecting outward from each tower, until everything meets at mid-span. However, if the suspension cable were cut and cable-stayed action became the mechanism to carry the Brooklyn Bridge's deck, the suspension cables could not be left standing up in the air as shown on the *I Am Legend* movie posters.

It was nice that *Popular Mechanics* went through all that trouble with that article, to debunk "fake-science" and "fake-engineering," given the nonsense that Hollywood usually produces in attempts to jazz up stories with visual effects (see later section on earthquake movies). However, the thing I found most interesting when the article was first published online, was the section at the end of the article where others could post comments. The first comment was, "What does this idiot know about bridges anyhow?" That eloquent commentator did not expand further, making it unclear as to whether he was frustrated that I did not expand on the hybrid concept described above, or he if was simply frustrated about everything in life. Apparently, trolls come from a different engineering school where the phrase "everybody can be an anonymous expert when online" is engraved on their diploma. (To *Popular Mechanics'* credit, the practice of allowing any wannabe to add comments at the bottom of their online articles has been discontinued, and all comments to previous articles have since been deleted.)

Terrorist Attacks

BOOM

It is a gorgeous day. Not a cloud in sight along the entire eastern seaboard. It is September and fall colors are soon to adorn the northern states. It still feels like summer in the southern ones. Noisy kids fill the school cafeteria—at least in those schools that still allow them to have a decent lunch break.[1] Then, at noon sharp, in one hundred schools across the entire eastern time zone, with the fury of a firework finale, the usual cafeteria boisterousness is drowned by the sound of banging metal parts and hypersonic bullets leaving the barrels of AK-47s. A butchery—two orders of magnitude worse than Columbine[2]—by commandoes of armed fanatics that orchestrated a synchronized mass murder to hurt innocent people. A horrible deed intended to strike Americans where it would hurt the most—their kids

Fortunately, this has not happened. If it had, it would have shocked the nation, spreading the message that nobody is safe anywhere at ground level, unleashing chaos in every neighborhood. One can only speculate whether unprecedented national outrage would escalate to the point of launching nuclear warheads in retaliation.

There is no shortage of opportunities for anybody who wishes to be a terrorist. Banking on the promised reward of (maybe) seventy-two virgins awaiting in paradise,[3] some have been eager to turn themselves into a human-bomb, but others have done it for free.[4] And some have even used children for this purpose.[5] Nobody would mind suicidal believers exploding in private, without taking hostages in their delusion, but unfortunately, that is not how they operate.

Sadly, there is an infinite number of ways to disrupt the workings of a modern society. One highly effective way to throw a wrench into the gears is to take out one large piece of the infrastructure. Another way is to scar the nation's psyche. Driving commercial airplanes into the World Trade

Center Towers and the Pentagon managed to do both, as it shut down the entire airspace for days and left the population in shock at the sight of the towers collapsing. It was the violation of a way of life.

It has often been said that, for many Americans, the terrorist attacks that occurred on September 11, 2001 (a.k.a., 9/11) were a wakeup call. Everybody alive then remembers where they were on that day. To make things worse, in the weeks that followed, five people were killed and seventeen sickened when they received letters laced with anthrax—a toxic bacteria.[6] A lot of stuff and people suddenly became suspicious.

Behind the scenes, all government security and intelligence agencies worked around the clock. More visible to the general public, much energy focused on screening the traffic driving across bridges and through tunnels to Manhattan and on confiscating nail clippers and other similarly dangerous weapons from the luggage of airport passengers. On these fronts, things ramped up over time. A gigantic Department of Homeland Security was created, and the government provided airports with equipment making it possible to scan though clothes and see passengers effectively naked—but not in color, for a modicum of modesty.

Yet, only three years later, experts were already commenting that the country had returned to complacency.[7] For example, it was reported that a more serious anthrax attack could produce a hundred thousand deaths and that the health care system did not have the capacity to handle the flux of victims arriving at once with respiratory problems due to anthrax or other bio-terrorism agents—and that it is still the case today.[8] Note that the same is potentially true for any other rapidly ramping-up contagious diseases of natural origin, as highlighted by the recent COVID-19 pandemic in many countries.

This complacency was also highlighted for critical infrastructures. While massive resources were invested to make airports safer, many targets other than airports remained unprotected, including some that could be highly attractive to terrorists seeking to send the nation's economy into disarray. Many scenarios are possible. A sample of some of those, well known to the government and enemies, have been exposed in details by experts in many books that can be shipped overnight from anyone's favorite bookstore—so these are not state secrets. Simplest to execute includes driving a tanker truck through the flimsy security gate of a chemical plant to explode it next to tanks holding deadly chemicals, spreading toxic fumes for miles around. Among the fifteen thousand chemical plants and refineries that store hazardous material across the United States, the Environmental Protection Agency (EPA) had identified hundreds where such an attack could kill up to a million people.[9] If targeting maximum deadly impact, it does not take a PhD in decision-making to judiciously pick a tank close to residential areas or transportation hubs. More elaborate schemes consist of running speedboats next to large transatlantic tankers carrying either oil or—for a bigger bang—Liquid Natural Gas (LNG) as they enter the ports of Long Beach and Boston, and using "shaped" charges to penetrate the hull of the ships.[10] A shaped charge is a piece of steel that can be turned into a sharp projectile by using a little bit of explosive—a simple and inexpensive trick that has been used quite effectively to penetrate the tanks of US forces during recent wars in the Middle East. The resulting oil spills or LNG fires would be disastrous and deadly on their own, polluting waters in urban centers and incinerating buildings within a certain radius of the ignited LNG burning at 3,000°F. Sinking the vessels would be a bonus. However, either way, the actual objective of the terrorists would be to create major economic losses to the nation by shutting down these ports for an extended period. Half of the LNG coming into the United States transits through Boston, and all main pipelines to the oil refineries in California are located in Long Beach, which also happens to be the largest port on the West Coast, where $10 billion of goods transit every year.[11]

As far as anthropogenic disasters go, acts of terrorism are the most vexatious because they are disasters inflicted on purpose, unpredictably. The question is not will they happen, but where? More specifically, where are the vulnerabilities? Where are the straw huts and where are the brick houses?

SELF-REGULATING BODIES

A nuclear power plant may be subjected to massive regulatory scrutiny and be designed to resist some of the most extreme load conditions—including earthquakes, tornadoes, and a direct hit from a crashing airplane, to name a few—but what is reasonable to expect for the rest of the infrastructure?

Governments in most capitalistic countries, including the United States, do not necessarily wish to get involved in regulating every industry, because regulations require allocating resources for inspection and enforcement, which can quickly get out of hand. Nobody has endless resources. And why harass responsible organizations when no visible harm is being done? The assumption of responsible behavior is the key idea here, bolstered by the belief that irresponsible acts by an enterprise would effectively be tantamount to brand suicide—thus automatically cleansing the market of dubious players. This frees the government to focus on regulating only those industries where errors and failures produce great dangers, typically in response to public perception and thus political pressure. If Tiny Tim breaks a tooth on a one-inch nail that should not have been in his brand-new cereal box and posts a video of this discovery on YouTube, a nationwide recall will follow together with possible incommensurable damage to the brand (on top of a lawsuit by Tiny Tim's newly found personal injury attorneys). The problem almost takes care of itself. If a nuclear power plant emits massive amounts of radioactive material into the air for days, the problems do not stop with bankruptcy of the shamed power company.

With that mindset, many governments worldwide have typically adopted a "hands-off" approach when it comes to a large part of the private sector and industrial infrastructure—and particularly so in North America.[12] Successful self-regulation is a good way for an industry to keep the government off its back, because the consequence of government oversight is a mountain of paperwork that adds nothing to productivity and profits. By analogy, except for accountants who can earn a living out of the process, nobody enjoys working on income tax forms. It is a drudgery that requires plowing through a maze of forms and nonsensical rules, takes an inordinate amount of time, and brings no profit—a tax refund is not a profit, it is a recovered loss. Being under the thumb of government regulations can be like filling tax income forms every single day of one's life.

The best way to keep bureaucrats at bay is to have a good record of accomplishment when it comes to preventing accidents and failures. Take the chemical industry for instance. It produces thousands of products that are needed across the world by businesses and individuals, and this requires producing and handling some of the most hazardous and toxic materials in existence. Some would argue that maybe it should not handle such dangerous products, but that would be a fundamental misunderstanding of chemistry. It may take dangerous products to produce safe ones. To illustrate the point, take the extremely poisonous chlorine gas—first used as a lethal weapon in World War I.[13] Then take sodium, which is so reactive that it explodes when it comes in contact with other products—particularly water. Put chlorine and sodium together, and, boom, out of the explosion comes regular salt—a perfectly safe staple of life. Mother Nature has a pretty dark sense of humor indeed. Of course, nobody would produce salt that way, given that it is much easier, safer, and cheaper to simply let water evaporate out of salt-water fields and collect the remaining salt, but some mundane staples of life have no such safer alternative ways to be produced and must be assembled by combining some nasty, toxic, hazardous gases, liquids, and solids.

For a long time, as far as operations were concerned, because the chemical industry had a relatively good safety track record in the United States (maybe less shiningly so in some less developed countries), it had been for the most part self-regulating, with some oversight from the EPA and the Department of Transportation[14]—understanding that regulations and number of regulating agencies have tended to increase over time,[15] for example including the Toxic Substances Control Act updated in 2021[16] to better monitor the more than seven hundred new chemicals introduced into the marketplace each year.[17] However, safety under normal operations does not imply security against extreme events. Post 9/11, some experts expressed concerns that little had been done to secure hundreds of chemical plants in the United States to prevent a terrorist attack that could injure or kill up to a million people.[18] The same was said about the trains that ship hazardous materials through neighborhoods and the barges that do the same through unmonitored inland waterways.[19] Realistic scenarios that could lead to such catastrophic outcomes (such as the ones outlined above) are in the public domain.

Therefore, in hope of keeping terrorists away from hazardous chemicals that could be weaponized, in 2007 Congress authorized the Department of Homeland Security to establish (at the same time as many other Security Acts[20]) the Chemical Facility Anti-Terrorism Standards (CFATS) program.[21] This spawned the "Protecting and Securing Chemical Facilities from Terrorist Attacks Act" that became law in 2014 and that regulates high-risk facilities by imposing some security standards.[22] This likely helped plug some security holes (using plenty of paperwork to fill them, some might be tempted to say), but are the security measures in place effective? Hard to know with certainty because there has been no such attack yet—other than an ill-conceived one that failed in France.[23]

The alternative approach though, when everybody is self-regulating, requires answering the question: Who is in charge of the common good?

THE TRAGEDY OF THE COMMONS

The "tragedy of the commons" is a theory elegantly postulating that when rational individuals share a common resource, they will inevitably deplete the shared resource. Politely said, each individual will attempt to get a greater share of the common resource to receive more benefits out of it than the others do.[24] Since each member in the group is naturally inclined to act that way, the resource will eventually be exhausted. The name "tragedy of the commons" was coined in the 1960s by an ecologist concerned about human overpopulation, who co-opted an 1833 story about cattle herders bringing their cows to graze in a common park.[25] With each herder trying to bring as many cows as possible to feed in the free public park, the combined effect of this selfishness forever spoiled the resource. Apparently, the take-away message from the story was that even in the nineteenth century, there were people acting like modern-day CEOs. In the 1960s, with common resources of clean air and water being polluted, fish stocks being depleted, and the earth being stressed by accelerating population growth, it was a timely topic to repackage.

To call that a theory is maybe a stretch, because it is not a profound discovery; most everybody knows freeloaders who empty the bowl of chips before others have had a chance to get any. However, it underscores the fact that if a resource is shared by a collective, some will always try to benefit by taking more than their proportional share. Turning the problem around, if there is a common resource—or a shared market—from which benefits ensue, if an investment is required to make the resource more sustainable or secure and there is no immediate return on the investment, the theory implies that nobody will be motivated to be the first one to incur costs in improving the resource. In other words, nobody wants to be the sucker that foots the bill for the benefit of all the others who wait on the sideline.

Thus, when it comes to making an investment to enhance the security of a private sector infrastructure—such as, hypothetically, a chemical plant— who will do it first if the cost for producing the same chemical product as competitors will be higher for the one making that investment? As a result, short of a concerted effort, very little—if anything—beyond what is legally mandated gets done to enhance security against terrorist attacks. That is a first "tragedy of the commons."[26]

To make things worse, if one CEO decides to "bite the bullet" and spend on some measures to increase security of its chemical plant (to be "a good corporate citizen" if for no other reason), at the potential cost of some loss in market-share, it may not necessarily be of any value down the line. When a terrorist blows up the plant of a more vulnerable competitor, the government, forced to act decisively for political reasons, could step in and indiscriminately require all plants to be temporarily shut down—with every company losing money at the same rate, irrespective of whether or not they had invested to make their plant more secure. Then, subsequently, since there are always different technologies that can be used to achieve specific objectives, the government may decide to impose on each plant a standardized and completely different set of security measures than those initially implemented by the pro-active CEO, rendering useless the initial investments. Using a very simplistic analogy—security measures in chemical plants obviously being a lot more complex than that—imagine a CEO spending hundreds of thousands of dollars installing metal detectors everywhere, when no other competitor does, only to be told after a terrorist attack months later that metal detectors are not acceptable and that body scanners are now required. To add insult to injury, imagine that to make the adoption of its policy more expedient, the government offers to subsidize the purchase of body scanners for every plant, so that it costs hardly anything to those who never bothered with security. That is the pernicious effect of the tragedy of the commons.

Arguments against enhancing the safety and security of industrial facilities and infrastructure are easy to find. Suffice to say that more than a hundred people die every day in the United States in motor vehicle accidents—that is more than thirty-five thousand per year, and more than 70 percent of them were not drunk.[27] By comparison, how many died of terrorist attacks on an average year? Of course, that is a moot point, since one could likewise argue that comparatively few people die in mass shootings—only twenty-six per year—but that does not make it a reason to not care.[28]

Clearly, safety is a relative concept, since nobody is proposing a driving ban to save thousands of lives every year.

CASCADES

Near closure of a business day, the CEO of a large company learned that he lost $100 million in a deal that went sour during the day. Mad as hell, he convened an emergency meeting of his top executives and screamed in their ears that they were nothing more than a bunch of idiots. Rushing out at the end of the meeting, the executives returned to their respective departments to yell at all their staff, to threaten to fire them all, and to tell them to "scram!" The army of disgruntled employees left the office fuming, drove home, slammed the door, and shouted at their spouses. Minutes later, the frustrated spouses snapped at their kids for not having picked up their toys, washed their hands, or whatever other thing that had been repeated millions of times but not done yet at that very moment. The upset kids got out of their house, heads down, hands in pockets, and kicked their dogs. By the end of the day, thousands of dogs in town were left puzzled, wondering what they had done to be kicked in the ass, out of the blue.

A cascading failure is exactly like that: namely, a series of events that cascade through a chain of related systems, one after the other as a domino effect, often in unexpected ways, because of a single triggering event, leaving those unaware of the linkages along that chain wondering what just happened—like the poor dogs above. In today's modern society, multiple complex systems operate in entangled ways that are not fully understood.

A regional power outage triggered by an earthquake makes the first domino fall. Whoever did not invest in an on-site backup power generation system is vulnerable. Water utilities that rely on massive amounts of power to pump water across geographical obstacles or up into water towers will suffer. The processes to treat drinking water may also fail, leaving the water supply exposed to contamination and the spread of diseases. Sick people will not show up to work and ambulances trying to reach affected citizens will need to navigate a city gridlocked by nonfunctioning traffic lights. And so on—all disrupting the economy.

When Hurricane Sandy made landfall in New Jersey close to Atlantic City on October 29, 2012, it had become a post-tropical cyclone with hurricane-force wind gusts exceeding 75 mph.[29] More importantly, it caused a significant storm surge along the New Jersey and New York coastlines. Beyond the facts that the hurricane reportedly damaged or destroyed six hundred fifty thousand houses,[30] that power outages were extensive with six hundred thousand people without electricity for nearly two weeks, and that the New York Stock Exchange was closed for two days,[31] it is also significant that many cascading failures also occurred. In particular, a large tank ruptured at the refining storage facility in Sewaren and spilled 335,000 gallons of fuel into the adjacent waterway.[32] Pollution also occurred from wastewater treatment plant failures, with nearly one billion gallons of raw sewage spilling into the adjacent bay at one plant alone.[33] And most importantly, 8 percent of the total US oil refining capacity was located in the affected area and damaged, severely disrupting the nation's production and requiring intervention by the federal government and the temporary suspension of a number of rules—such as the EPA's clean gasoline requirements[34]—to ensure overall fuel availability through the country.

The cascading effects from any hazard can be far reaching and hard to imagine. In 1998, an ice storm dumped freezing rain and drizzle for more than eighty hours in many parts of eastern Canada and the northeastern United States, but predominantly in southern Québec near Montréal. Accumulation of ice reached three to four inches in many locations. The weight of the ice coating power lines became excessive and the transmission towers that carried these lines collapsed. Hydro-Québec reported fifteen hundred damaged towers, with nearly 1.4 million households without power. The outage affected more than three million people for several days—more than 40 percent of the province's population of 7.3 million at the time.[35] In large parts of Montréal's south shore, 150,000 people were without power for up to three weeks following the storm.[36]

The list of infrastructure systems, organizations, companies, and individuals that were severely impacted by that ice storm and power outage goes on and on, and most people would be able to name a lot of the problems that occurred if asked today. Closed shopping malls and businesses, roof collapses under the weight of ice, treacherous travel conditions, carbon monoxide poisoning in stranded vehicles or when propane grills were run indoors, electrocution from downed power lines, and many more, easily come to mind. Yet most people would not think to mention losses to the agricultural sector.

Indeed, many barn roofs collapsed under the weight of ice, crushing livestock to death, with great losses to farmers—a problem exacerbated when some insurance companies played hardball and shrewdly classified the ice storm as an Act of God.[37] More broadly, though, when power failed, heating and ventilation systems stopped. Many farm animals literally froze to death or died of asphyxiation—and that kind of frozen meat is not allowed to end up in supermarket freezers.[38] Pigs, poultry, and cows died by the thousands, and farmers had to wait weeks to get the carcasses carted away by companies that dispose of dead animals.

Dairy farmers were also hard hit, first because this is a highly mechanized large-scale industry and, without power, only a small percentage of the cows could be milked the old-fashioned way, and second because whatever milk could be collected (using emergency generators or otherwise) could not be cooled, picked up by cistern trucks, and treated at the processing plants.[39] Country roads were icy and sometimes blocked by fallen trees, but, more importantly, the processing plants were not operating. A total of 13.5 million liters of milk[40] from 5,500 Ontario and Québec dairy farmers had to be dumped.[41] The problem was further compounded by the fact that, in a production environment, dairy cows that are not milked regularly become vulnerable to mastitis, an infection of the udder that is extremely painful and can lead to death.[42] When mastitis is not fatal, problems ensue with the quality of the subsequent milk production.[43] Some of the farmers had to kill or sell their cows.[44]

The agricultural sector that depends on production from trees also suffered as the massive ice accumulations on trees bent and snapped many in half. For example, roughly 20 percent of the trees used for maple syrup production in Québec were lost.[45]

This example highlights that all industry sectors—even some that do not intuitively come to most people's minds, such as farming—depend on the existing infrastructure to a point of maximum optimization. While optimization is good when it comes to operations in normal conditions, it also means operating without redundancy, thus relying on a fragile equilibrium. Optimized systems are slim, efficient, cost-effective—all good things—but not resilient.

ON THE DISASTER TRAIL

The Ultimate Tragedy of the Commons

In a Communist country—in theory—everything is part of the commons (both words come from the same Latin root, *communis*).[46] In 1990, the European Conference in Earthquake Engineering was held in Moscow, which gave me a chance to visit the USSR in the years of Gorbachev, Glasnost, and Perestroika. The country talked about changes, openness, and transparency, but hard-core Communism still ruled. It felt as if I had time-traveled to the 1940s. Phone receivers weighed as much as a barbell, subway stations were decorated like renaissance museums, and folks drove banged-up Ladas that embodied the art of frugality in automobile manufacturing. Changes: Thousands of people lined up at the first McDonalds ever to open in Russia, to get a taste of Western decadence—with longer wait times there than for Avatar Flight of Passage in Disney World nowadays.[47] Hard-core Communism: After sitting for one hour at a table in a restaurant waiting for a menu, I was forced to conclude that the word "service" was absent from the Russian language—but I found out later that the word "bribe" was not.

At the conference, the Russian engineers seemed obsessed with complex mathematical approaches for problems that the rest of the world typically solved using simple computer algorithms (I later learned that there were no personal computers in the Soviet Union in 1990). Strangely, nobody bothered to close the curtains during the keynote lecture, so the screen remained white—drenched in sunlight—throughout the entire presentation as the entire audience could distinctly hear the slides change in the projector. Equally strange, most toilets in the brand-new convention center were also clogged.

As part of my visit, one of the lead engineers of the USSR National Laboratory where earthquake engineering research and experiments were being conducted gave me a private tour of their facilities. Walking through their enormous laboratory, I noticed that the place was filled with a large number of technicians—which is usually a good thing—but that they all seemed to be sitting around idle, talking and smoking. I asked my host if this was a break time. He responded, "The government pretends to pay us and we pretend to work."

Then, after the tour, he invited me to meet his family for dinner at his apartment. It was in a complex of multistory residential towers. It reminded me of the infamous subsidized housing development projects that had been constructed in some big US cities. There was garbage on the ground, dirty walls everywhere, even dirtier staircases, and plenty of things falling apart for lack of maintenance. However, as soon as we entered the apartment, the place was warm, bright, clean, nicely decorated, and welcoming. The whole family was charming and delighted to have a visitor. The contrast between the inside and the outside was striking. I asked my host why was it that the apartment was so nice while the exterior of the building looked so drab. He responded that people had pride and took care of what was inside their own apartments. Outside, however, that was the government's responsibility. The divide was clear in their mind. Why would anyone take the initiative of being the stupid one working outside to make the grounds look better while others would not contribute? It was either "all in" or nothing, and the "all in" part fell within the domain of the Communist government's responsibility. The ultimate tragedy of the commons.

Annoying Doomsday Scenarios

ZOMBIES, VAMPIRES, AND THE PANDEMIC OF THE DAY

Those who believe that the outcome of World Cup games can be predicted by an octopus, that other future events are best predicted by throwing spears of asparagus in the air (asparamancing),[1] that cancer can be removed from the body by touch alone or by reaching for it after making an incision on the stomach with two fingers (psychic surgery), that the electrons your body needs to balance its overabundance of free radicals must be absorbed by walking barefoot in dirt (body earthing), or that the ultimate remedy against every known disease that can at the same time make your skin softer and enlighten your third eye is to drink urine (urine therapy), might find this section offensive and can easily skip it without breaking out in hives.

Earthquakes, hurricanes, volcanoes, and other natural hazards unleash powerful forces of gigantic proportions. Yet, the elephant in the room—one of the biggest disasters ever—is actually smaller than an elephant, smaller than an ant, and smaller than a single cell. Technically, it is not even alive. It is a microscopic, self-replicating, invader. Not satisfied with creating a local disaster, it seeks to wreak havoc globally. Meet the pandemic.

If there is one hazard that needs little introduction by now, it is the pandemic. After months of social distancing, home confinement, travel bans, rationed toilet paper, and scrutiny of data, riding the waves of new cases, intubations, deaths, and upturn in the stock market, it is fair to say that little explanation is needed. Since COVID-19, everybody worldwide can now viscerally relate to what a pandemic is. Previously—like most disasters described above—it was a vague concept; something that happened in underdeveloped worlds; something that happened in medieval times; something that happened when medicine did not know any better; something that the average Little Pig did not plan for because that kind of Big Bad Wolf would never come. Unfortunately, as for all the other hazards above, this false sense of security was unwarranted.

In the aftermath of the 2014–2016 Ebola epidemic, a Global Preparedness Monitoring Board was tasked to provide a frank opinion as to whether or not the world was ready for a pandemic; it published its findings in September 2019, four months before the existence of COVID-19 was first reported.[2] The Board concluded that the world was not prepared at all to deal with a rapidly spreading lethal virus that had the potential to kill up to eighty million people worldwide and wipe-out 5 percent of the world's economy—a catastrophe that it considered to be a real possibility given the fact that the World Health Organization had identified 1,483 epidemic events in 172 countries between 2011 and 2018.[3] A prescient warning if any.

This false sense of security and lack of preparedness is not surprising. In many ways, a pandemic is an earthquake in slow motion. It is something everybody knows is possible at any time but is more convenient to ignore. Then, like every other ignored threat, when it happens, it is a mess.

Nobody is pleased to see disasters unfold—be they pandemics, earthquakes, hurricanes, or whatever else derails life from its peaceful course. Except for those who get spiritual or supernatural titillation out of it. In that category, there are groups who sincerely believe that a raging pandemic is the long-awaited "end of the world" that will, at last, allow them to meet their creator, or fuse with the universe, or whatever else is supposed to happen on the big doomsday, and who are driven to congregate to fulfill this destiny. Orders that prohibit large gatherings for public health reasons are pointless to fanatics eager to meet their end. Nothing prevents such folks from exercising their freedom of religion, but during a world pandemic, they should do so packed in a stadium, all doors welded shut (with all the utilities working and periodic food drops for humanitarian reasons), to live their dream secluded from those who do not wish to be infected by their contagious joy. Cynics would add that apocalypticists[4] of all denominations should be locked together on a remote island and left to count their dead to resolve which creed or cult held the truth in the end—if they do not kill each other in a religious war before the virus does.

For the rest of the population that, given a choice, prefers to survive pandemics, the best course of action is evidently to prepare. This means planning on how to contain the spread of the disease by preventing transmission from infected humans. In this case, coming back to the Three Little Pigs, whether the huts are built of straw, wood, or brick does not matter. None will stop a virus—although stronger huts could prevent the spread of "diseases" transmitted by the bite of vampires or zombies by keeping them out.

Previous chapters have dealt with earthquakes, hurricanes, floods, and some anthropogenic disasters, because these shared a common attribute: each of these hazards could have devastating effects on the built infrastructure, and consequently on the population. If releasing the straitjacket of that commonality, the list of possible causes of disasters becomes much longer. It can include droughts, snowstorms, avalanches and landslides, animal plagues, locust invasions and other insect-related problems, food chain poisoning, famine, concentration of the world's wealth in the top 0.1 percent of the population, illiteracy, political instability, mass riots, dictatorships, war, massive solar flares and geomagnetic storms, asteroid impact, extraterrestrial invasion, the in-laws moving in, and more. Each of these by themselves can be—and have been—the topic of many books. Here, the primary focus is on hazards that affect everybody as a consequence of damage to the infrastructure, but without too much of a stretch, it remains that many of the matters addressed in the subsequent chapters are generally applicable to other hazards that do not affect infrastructure. All these hazards are "earthquakes" that can produce disasters of their own.

For some people, the earthquake will actually be an earthquake.

For others, the earthquake may be losing something/someone precious as a consequence of their negligence.

For alcoholics, the earthquake may be killing somebody when drunk driving.

For drug addicts, the earthquake may be hitting the absolute "bottom of the barrel."

For chain smokers, the earthquake may be a heart attack.

The way humans prepare (or not) and react to rare high-consequence events remains relatively the same, irrespective of the nature of the disaster.

ON THE DISASTER TRAIL

Mopology Wisdom

One of my first summer jobs was in a general hospital, working as one of the many "mopologists" hired to help keep the maintenance operations running while members of the regular mop squad took summer vacations. In the process, I learned two important lessons.

The first one was taught by the regular employee I was assigned to shadow during my first week of employment, to learn the ropes. After punching the time card, I followed him to the closet where all the mops, buckets, towels, rotary floor polishers, and other miscellaneous cleaning products were stored. He sat on a bench, looked me straight in the eyes, and said, "Listen carefully, kid. When the end of August comes, you kids are going back to school, but we are going nowhere. So don't make us look bad," which was a stern order to respect the regular pace of work and to not even think of being zealous. He then sat on a bucket and read his newspaper for thirty minutes. Thereafter, we worked for about ninety minutes, spent a good forty-five minutes in the cafeteria for the morning fifteen-minute union break, worked another hour, went to lunch fifteen minutes early and repeated the pattern in the afternoon. Amazingly, we managed to squeeze a solid five hours of work into an eight-hour workday.

The second lesson was learned progressively over the summer. As maintenance people from various wards of the hospital took vacations somewhat in succession, the "kids" hired to fill-in for the summer were rotated from wing to wing. So, one week, I would be mopping the maternity ward, the next week the oncology department, then the burn center, and so on. Most of the mopping and waxing of floors happened around and under occupied hospital beds, but maintenance employees were strictly forbidden to talk to the patients—presumably because the private life of patients was none of our business. That rule, however, was most absolutely disregarded by all, first because talking is more fun than mopping, but more importantly because many patients are bored stiff and lonely—they can barely squeeze more than a few words out of the nurses and doctors, and visitors are in many instances rare and few.

What struck me was the striking contrasts in patient behavior from ward to ward. In the department of terminal pulmonary diseases, those with last-stage lung cancer or throat cancer, final-stage emphysema, end-stage cystic fibrosis, and the like knew that they had reached the end of the line. It was mindboggling to see some patients smoke cigarettes through the tracheostomy tube sticking out from their throat, in some desperate attempt to lighten the gloomy mood. These patients were not the talkative type, particularly those who had to rely on the "robot voice" of an electro-larynx speaker to communicate because their vocal cords were taken away by larynx cancer.

In contrast, the cardiology department was a blast. Patients arrived by ambulance nearly dead, but those that survived and made it out of the intensive care unit were as giddy as if they had miraculously found a spare life in a Cracker Jack box (in those days, each box contained a prize).[5] They had been given a second chance. They should have been dead, but because of advancements in research, knowledge, and technology, instead they were alive, waiting for the green light to walk out on their own two legs, inspired to adopt a healthy lifestyle of exercise without junk food—best intentions at the very least, successful in many cases. The "big earthquake" had hit hard and had been good for them, as they had survived.

Interlude

THE GALLERY OF HORRIBLE EARTHQUAKE MOVIES

After all the grim facts of the previous chapters, it is time to relax a bit—before more grim facts in the next chapters. Toward that goal, reviews of disaster movies are presented below, focusing on earthquakes only, to keep the list to a manageable size. The deeply curious who may wish to test their pain tolerance can find information on where to find these flicks from the Internet Movie Database (IMDb.com).

Note that the purpose here is not to comment on the quality of the scripts (because most storylines stick to formula for the genre) or acting (because some movies showcased A-list actors on the marquee, while others hired those in the studio parking lot that held highest their "Will work for food" sign). Rather, the purpose here is to comment on the credibility of the whole thing, or sometimes of the earthquake damage as depicted by a film industry that must embellish for box-office benefits. Remember that, as with all art forms, beauty is in the eye of the beholder; there will always be a critic to call the worst film of all time a masterpiece, and vice-versa.

Warning: Do not Trust Hollywood! It loves disaster movies but is clueless on how disasters actually happen. Any teacher who believes that asking kids to watch a two-hour flick is worth some points as part of a science project should be covered in tar and feathers and thrown out of town.

- *Earthquake* (1974)
  For unknown sociological reasons, the 1970s were sort of the heydays of disaster movies.[1] Topics covered ranged from sinking ocean liner (*The Poseidon Adventure*), air disaster (*Airport*), high-rise fire (*The Towering Inferno*), and many more, including, of course, earthquake. *Earthquake* is the granddaddy of earthquake movies. Beyond being a "classic," it is the film that launched "Sensurround" in movie theaters, a technology that pumped sub-audio waves at 120 decibels to

immerse the audience in earthquake waves. Without that rumble, it is just another cookie-cutter disaster movie, with an all-star cast—although it is a "classic." As for the earthquake damage: B for effort—we are talking about mid-70s special effects here, after all. But it is a "classic"—in case that was not made clear.

- *Aftershock: Earthquake in New York* (1999)
  Surprising. Bravo for looking at an earthquake outside of California (although, arguably, Hollywood went overboard, for effect). Notwithstanding the exaggeration, one of the most credible set of earthquake damage of the bunch. A-.

- *Aftershock* (Again, due to severe shortage of imagination for original titles) (2010)
  Spoiler alert: This Chinese production cleverly used the 1976 Tangshan and 2008 Sichuan earthquakes as bookmarks to human drama, weaving thirty-two years of China's history into a tearjerker that is surprisingly watchable (and probably even more so in its IMAX version). Although predictable (by those who know their seismic history), it is a testimonial to the 250,000 who died during the 1976 event. B+.

- *The Impossible* (2012)
  Not surprisingly, in the aftermath of the 2004 Indian Ocean earthquake and tsunami, and the 2011 Tōhoku earthquake and tsunami, it was just a matter of time before Hollywood tried to cash in. Although, in fairness, this one is actually watchable. Once upon a time on the web, a film critic called Mr. Cranky[2] reviewed movies from the perspective that all of them are terrible, inflicting various level of suffering. Instead of using stars, Mr. Cranky's ratings used bombs, and ratings ranged from 1 bomb ("almost tolerable") to 4 bombs ("as good as a poke in the eye with a sharp stick"). Uniquely bad movies were rated either a bundle of dynamite sticks ("so godawful that it ruptured the very fabric of space and time"), or an Atomic Explosion ("Proof that Jesus died in vain"), that latter one awarded to masterpieces like the 2011 remake of *Conan the Barbarian*. If one were to use Mr. Cranky's rating scale, *The Impossible* would deserve only "1 bomb"—which is a compliment.

- *The Great Los Angeles Earthquake* (1990)
  Nothing memorable here. Made-for-TV movie, produced after the 1989 Loma Prieta earthquake to capitalize on the public's sudden awareness of the risk in California, it preceded the Northridge earthquake (Los Angeles) by about three years (both the movie and real earthquake were about hidden faults, but the similarities stop there). A generous C grade, because there have been worse.

- *10.0 Earthquake* (2014)

  The largest earthquake ever recorded was a magnitude 9.5.[3] The USGS has stated that a magnitude 10 is impossible because the fault rupture that would be needed for that to happen would have to extend around the entire planet (no such fault exists).[4] But, who cares? In Hollywood, everything is possible. Fun geological fact: As well demonstrated in multiple dramatic sequences of this masterpiece, when a fault opens up and propagates, swallowing everything in its path, it does so at the speed of a pickup truck trying to escape, following it as it turns left and right, as in a wild a car chase.

- *10.5* (2004)

  Why stop at 10.5, instead of going all out to 11 and becoming the Spinal Tap of disaster movies? Also, why stop at literally separating Southern California from the United States instead of splitting the entire continent along the Rockies? Obviously, two huge lost opportunities, by a timid director. Beyond that, a supreme package for those ignoramuses who like three-hour doses of nonsense. A well-deserved F across the board. Mind-boggling fun-fact: A sequel was produced as a TV mini-series.

- *The Day the Earth Moved* (1974)

  As the movie's budget apparently could only afford to destroy five homes and a gas station in a hellhole in the middle of the desert, they were trashed beyond artistic license (with the means available in 1974). A few minutes might be of interest to those interested in nonstructural damage. Another non-negotiable F.

- *Aftershock* (Again. Evidently, there are always lots of aftershocks) (2012)

  The genius who got the brilliant idea of mashing an earthquake movie with a slasher horror flick succeeded in plunging the seventh art to new lows (and burning $2M in the process). Watching the film's trashy characters flub their lines, soaked in gallons of fake blood, will make *your* eyes bleed. Films like this one perfectly highlight the failure of the letter grade system, because F covers a far too wide numerical range (from 0% to 60%). This one earns a solid F—of the 0% kind.

- *Disaster Wars: Earthquake versus Tsunami* (2013)

  It is a rare event to find all critics agreeing on the rating for a movie. Exceptionally, in this case,[5] they concurred that this one deserves a 0/10 grade (only because negative grades cannot be given). Bears no resemblance to actual earthquakes or tsunamis, but shows that movies can technically be done by randomly recruiting the cast in a Walmart. Forcing detainees at Guantanamo to watch this might be a violation of human rights.

- *San Andreas* (2015)

  Sure, chunks of concrete fall out of nowhere with no rhyme or reason, sure the ground moves in ways that would baffle any respectable seismologist, but . . . IMAX 3D!!! How not to love nonsense when it is projected on a screen measuring 70' × 50' with 30,000 Watts of sound? All done with big-name actors (a first since 1974). Just think of it as the *Fast and Furious* of earthquake movies, and enjoy the ride. Kudos for recognizing that shattered glass falling from buildings is a hazard (not all buildings have tempered glass). No kudos for destroying the Golden Gate Bridge . . . again, after Godzilla (2014), Kaiju (*Pacific Rim* 2013), and just about everything else (a video compilation of Hollywood attacks on the Golden Gate Bridge can be seen on YouTube).[6] San Andreas is as much about earthquakes as *Star Wars* is about rocket science, but lots of bonus points for its sheer entertainment value, for the link on earthquake preparedness in the lower left corner of the movie's official website,[7] and for Sia's eerie slowed-down version (first minute only)[8] of "California Dreaming" from the Mamas and the Papas. All this raises it up to an A-.

- *San Andreas Quake* (2015)

  Apparently purposely named[9] (and released at about the same time) as the big-budget *San Andreas* movie—with a slightly longer title but a massively smaller budget. The targeted audience is those same folks who, driving to Orlando, would get off the highway and follow signs to DazeneyWorld, pay admission and wonder if the drunk clown and the ramshackle House of Mirrors are the Goofy and Space Mountain that everybody talked about. To avoid at all cost, unless one is fond of special effects done with paper and scissors. A solid F.

- *Mega Fault* (Director's cut!) (2009)

  Claim to fame: Features the only earthquake in the world capable of spontaneously igniting heads (with lame CGI flames). This could be seen at twenty-six seconds into a thirty-second clip once posted on YouTube.[10] That says it all. If the grades scale was not truncated at F, this one would deserve a Z.

- *Pandora* (2016)

  Straight from the high-tech country that offered the world the exploding Android tablet batteries, and applying the same quality-control standards to movie scripts, this Korean dud dramatizes a nuclear power plant meltdown following a small magnitude 6 earthquake. It is deserving of a Razzie award (these awards have been created to honor the worst of cinematic under-achievements).[11] Computer animations of the meltdown, reactor explosion, and panicked mass evacuation must have chewed-up the budget, because the earthquake damage itself was limited to less than one minute of fallen suspended ceiling panels, toppled file cabinets, overturned fruit stands, ripped awnings, one fallen telephone pole, and rocks rolling down the hill (hardly exciting stuff, except for a specialized kind of engineer fascinated by damage to nonstructural components). Not a single crack in the village's old buildings, which is magical for a quake that brought down a nuclear power plant. The movie also offers undeniable proof to those not fluent in Korean that bad acting cannot hide behind subtitles. Will be greatly enjoyed by those who consider nonstop screaming to be legitimate dialogue. C+ on the strength of professional camera work.

- *The Quake (Skjelvet) (2018)*

  If Hollywood has no shame in stretching a credible magnitude 8 earthquake threat for California into a magnitude 10, why should Norwegians hesitate to inflate the tiny magnitude 5.4 Oslo earthquake of 1904 into an 8.5 balloon? ACT 1: The hero is moping for seventy-two minutes, depressed because nobody believes him. ACT 2: A thirty-foot tall earthquake wave ripples across town like a surfer wave in Hawaii, a clear indication that there were illegal substances hidden in the CGI team's office. ACT III: The hero suddenly becomes an accomplished acrobat and attempts saving his wife and kid from a building that defies the laws of gravity. Total US box office gross: $6,235 (not a typo).[12] Highly recommended for devoted cinephiles with a fondness for dejected, apathetic, miserable characters. For everybody else, it is a definite C- (for the special effects that incidentally won the Norwegian equivalent of an Oscar—a trophy that may be worth something on Ebay).

- *The Wave (2015)*

  An enormous chunk of mountain falls from two thousand feet up into a fjord and the resulting 250-foot tall splash wave ripples upstream toward a small village nested at the end of the fjord. OK, more tsunami-like than earthquake, but that kind of wave is something that actually happens every now and then in Norway, with seven hundred people killed in a 1934 event (among many examples). Beyond that, the whole plot unfolds per formula, complete with predictions by a scientist nobody believes who becomes the hero saving (almost) everybody in the end. Branded as Norway's first disaster movie, one wishes it had been the last, because—sadly—its sequel was "*The Quake (Skjelvet)*." *The Wave* is surprisingly watchable: B-.

- *San Andreas Megaquake* (2019)

  Guess what! Scientists are predicting that in two days an earthquake will sink California in the ocean. What an original, never-heard-before, unique plot! The world needed that movie like the producer needed a kick in the groin. It is so unique and original that Meryl Streep and Tom Cruise probably would have signed up for the lead roles—if there had been a script. This masterpiece is to cinema what forks scratching plates are to music. A solid F (for FUBAR).

- *Quake* (1992)

  Has the dubious merit of: (1) being the only movie with someone who believes that the best way to survive an earthquake while driving on a highway is to accelerate and bang all the surrounding cars for over a minute; (2) trying to convince the audience that a San Francisco highway can have only four cars on it; and (3) offering the worst-ever genre-bending combination by blending a bad earthquake movie with an even worse psycho thriller. All while "borrowing" newsreel images of damage in San Francisco from the 1989 Loma Prieta earthquake, which is either artistic appropriation or pure cinematographic laziness. If ever offered the choice between watching these seventy-eight excruciating minutes of nonsense while sitting in a comfortable sofa, or wearing a winter coat in a sauna while listening to two hours of screaming by Yoko Ono, definitely go for the second option without any hesitation.

- *The Earthquake* (2016)

  This one is not a laughing matter. On December 7, 1988, a devastating earthquake struck Armenia (then part of the USSR). It killed more than 25,000 people and wiped out many cities, leaving more than half a million homeless in a freezing, soviet winter. This was a grim earthquake and this is a grim movie—almost a memorial. No scientist that knows it all but that everybody ignores, no hero saving the day against all odds, no clichés, no nonsense. Meet a few folks, witness earthquake destruction right away, and, from there, the rest is coping and suffering amid rubble in ways strikingly similar to what was seen thirty years ago. Professionally done. Not a fun watch, but a rare A!

- *Earthquake* (the band) (1971–79)

  Gotcha! This is not fiction but rather a poorly edited concert movie by the rock band called Earthquake. The very same band that will be inducted in the Rock 'n' Roll Hall of Fame hundreds of years from now when they will be running out of ideas for nominees—but still more enjoyable than some of the movies mentioned above. Giving it a major F-C-F-A-C-F rating (fingers on six guitar strings).

- *Geo-Disaster* (2017)

  Everything at the same time: a super volcano, a mega earthquake, and a massive twister. Why not? A cinematographic achievement can launch the career of an actor, but if any Hollywood doors were left open after this disaster orgy, they might have been only those leading to the porn industry. Grade inflation is the new norm, so this one gets a D-.

- *San Francisco* (1936)

  What better way to close this list than with another classic. There are movies where earthquakes are not the topic but rather an accessory, to offer comic relief (such as for The Three Stooges to escape a prison damaged by a three-second tremor)[13] or, better yet, to provide a significant plot twist, as in the 1936 award-winning *San Francisco*. Looking through the lens of the 1930s, *San Francisco* got its quake right, offering an earth-shattering rendition of the 1906 "big one," with showers of bricks, stones, statues, and even a piano, killing those at the receiving end by closely (albeit not perfectly) replicating the manner in which unreinforced masonry buildings of the era collapsed. Starring Oscar-winning actors Spencer Tracy and Clark Gable (the King of Hollywood of the 1930s best remembered as the dude who, frankly, did not give a damn for his dear in *Gone with the Wind*), the 1936 blockbuster also gave the city of San Francisco one of its official anthems.[14] The film is watchable (and broadcast pretty much each year close to the anniversary of the 1906 earthquake), including the earthquake scenes, which is why it gets an A (for its time).

Unfortunately, this demolition derby is far from over. Now that everybody has an HD video camera and free movie editing software on their smartphone, the barrier of having to buy expensive film stock to produce a disaster has been removed. If that barrier has not prevented the above in the past, it will be amazing to see how bad things will become with future releases. While music recording software can fix bum notes and put everything in tune, there is still no software that can convert a bad script into a good one.

MEET THE LITTLE PIGS

The Wonderful Ability to Forget

WHO CARES?

In 2019, a 5,631-square-foot Colonial brick home built in 2000, with five bedrooms, four full and two half bathrooms, an additional 1,000 square feet of finished basement, a two-story foyer, 12-foot ceilings elsewhere, hardwood floors, granite countertops, pro appliances, adjacent butler's pantry, masonry fireplace, built-in bookcases, central vacuum system, whirlpool tub, wet bar, security system, and three-car garage; located on a lot of 0.64 acres, 108 feet by 250 feet with perimeter gardens, stone walls, lawn sprinkler system, and tennis court, located in the city's best school district and only twenty-eight minutes from downtown Buffalo, New York, was offered for $995,000. It had been listed for over eight months, because those who do not wish to vacuum 5,631 square feet of livable floor space could find many other homes nearby, not too shabby either but with only 4,000+ square feet, for hundreds of thousand dollars less, and 3,000-square-foot ones for half that price.

The same year, for those who break out in hives at the sight of a snow flake, a more modest three story, 2,981-square-foot Mediterranean style home, built in 2001, with three bedrooms, three bathrooms (twelve rooms total), tile floor, French doors, 10-foot ceilings, Palladian windows, hurricane shutters, security system and cameras, two-car garage, with second- and third-floor oceanfront balconies, on an 0.51-acre oceanfront lot with private walkover to a quiet beach, a heated saltwater pool with spa, professionally landscaped gardens and sprinkler system, thirty-seven minutes from Saint Augustine, Florida, sold for $925,000. Many similar homes nearby were less expensive, by as much as 40 percent when not oceanfront.

Meanwhile, on the San Francisco Peninsula, hard-earned dollars did not quite stretch that far, but detached homes in the million-dollar range were common. A 990-square-foot single-story ranch-style stucco home built in 1948, with two bedrooms, one bathroom, hardwood floors, and a one-car garage, on a 0.1-acre, 50' × 90' lot, with a narrow patch of lawn in front of the house and another one in the backyard, in the pristine and sunny Brentwood neighborhood, only twenty minutes from downtown San Francisco when driving at 3 a.m., but an hour away on a good day when traffic is bumper-to-bumper, was offered at $868,000. A similar home, but with an extra bedroom, could be found for an extra hundred thousand dollars. At the time, a couple of dozen of million-dollar homes were listed for sale on websites such as Zillow.com or Realestate.com. They typically stayed listed for only a few days, snatched by those quick to make an offer, the exception being homes having backyards abutting an eight-lane freeway, as it generally took a few weeks to find connoisseur buyers appreciative of such prime views.

This reality of the San Francisco market also applies to those homes located in the southwestern-most corner of Daly City, nudged between a loop of Highway 1 and the Mussel Rock open space, which also happens to have the special privilege of being inside what is known as an Alquist-Priolo zone. Repeat visits to Zillow or Realestate.com show that it is as difficult to find a home for sale there as in any other specific neighborhood elsewhere across the peninsula. This implies that—most interestingly for the topic at hand here—as far as real estate value is concerned, the price tag for 1,100-square-foot homes near San Francisco stays roughly the same, irrespective of whether it is inside or outside of an Alquist-Priolo zone.

To reemphasize, if Joe and Jane were shopping for a home in San Francisco—for a measly million dollars, maybe not the mansion of their dream, but a home nonetheless—whether that home is within or outside that zone will not be a factor.

OK, so what is this Alquist-Priolo zone anyway?

Simple.

In 1972, California passed the "Alquist-Priolo Earthquake Fault Zoning Act," which effectively prevents new construction from being built within 750 to 1,000 feet of a seismically active fault.[1] The locations where such faults are known to exist are documented in geological maps produced by the State.[2] The law grandfathered in existing construction, but required real-estate transactions to disclose if properties transacted lie within that zone. Making the zone where construction is prohibited a wide band rather that a slim line was wise, as it certainly helped prevent protracted legal debates to argue whether the fault slices through one's property or is "actually" on the neighbor's lot. Most notably, single-family wood-frame homes are exempted from the law,[3] but earthquakes do not care about human laws: no exemption can make the fault disappear. In fact, on the University of California Berkeley campus, the football stadium was built right on top of the Hayward fault in the 1930s, slicing it in half—so it was built as two separate half-stadiums, to the extent possible (and retrofitted in 2011 to allow similar movement).[4] The west and east halves have since moved by five inches with respect to each other due to creep along the fault, making it undeniably clear where the fault is located.[5] Incidentally, if the Hayward fault were to rupture during a game, the NCAA football rulebook is silent as to whether or not it will be an automatic first down if the ball is on the ground when the west half of the stadium moves south by ten yards.

Anyone with a pen, a ruler, a map of the San Francisco Bay Area, and the skills to draw a straight line, could easily trace on that map the path followed by the San Andreas Fault west of San Francisco. Going north to south, that segment follows the long and narrow Tomales Bay in Point Reyes National Seashore, takes Highway 1 for a few miles, plunges into the Pacific Ocean at Bolinas, travels underwater six miles west offshore from the Golden Gate Bridge, and hits shore in Daly City from where it does a beeline to Southern California.

More precisely, the San Andreas Fault hits the southwestern-most corner of Daly City, nudged between a loop of Highway 1 and the Mussel Rock open space, which is why parts of that neighborhood are inside an Alquist-Priolo zone. That area was developed in the 1960s, before the Alquist-Priolo Earthquake Fault Zoning Act, which is why roughly three hundred homes, lined up along a handful of city blocks there, fall within the designated fault zone. If fact, some even have the distinction and privilege of sitting right on top of one of the world's most famous seismically active faults—the very same one whose rupture destroyed San Francisco in 1906. It is possible nowadays to use web-based real estate software and maps from the California geological survey to check the price of homes sitting on top of the fault, or left or right of it, and notice that there is no difference whatsoever in assessed values. The house whose kitchen will be sheared-off twenty feet away from its living room by the next major earthquake is worth as much as any other not sitting on the fault. Given the narrow width of some lots—typically 33' × 105' [6]—many will get quite a change of scenery after the next significant rupture of the San Andreas Fault. When it comes to the Big Bad Wolf, houses straddling a fault might be expensive straw huts, but straw huts nonetheless.

At first, this might seem stunning—but not really.

It would be hard to argue that the good citizens of Daly City did not know about the whereabouts of the San Andreas Fault when urban sprawl spilled over that fault zone in the 1960s—but they could be given the benefit of the doubt. That Joe and Jane, moving from Nebraska to California a few years ago, ended up buying a house on top of a fault, might also be excusable; maybe they were ignorant of the forces unleashed by an earthquake, since they likely never experienced any—and at 6 percent commission on a million-dollar sale, some real estate agents might excel at reassuring customers when doubts arise. However, what is harder to explain is why a different pair of Joe and Jane, moving from elsewhere in California, would not know a thing or two about earthquakes. Either they are blissfully unaware of faults and Alquist-Priolo Zones, or something in their subconscious fogs it all up and makes them rationally accept the risk, because nothing beats the view from the house or from the park nearby, or because the fresh salty air is so rejuvenating, or simply because the real estate market is so hot that one grabs whatever is available (incidentally, not only hot, but literally on fire: the charred skeleton of a house heavily damaged during a two-alarm fire, listed at $850,000 and twenty-five miles from San Francisco, received multiple offers and sold for $1 million,[7] greatly broadening the meaning of *fixer-upper*).

Or maybe they expect FEMA to bail them out when their home will be sheared into two halves by the San Andreas Fault, and that they are planning to use that money to rebuild. Maybe they hope to be allowed to locate their brand-new house on the very same lot, but if not, they can always build on top of a fault in another state where there are no Alquist-Priolo zone restrictions. Or maybe they will move away to never have to live though another such traumatic experience. Like those who have suffered massive losses during a hurricane and relocate inland—anywhere away from a coast—as if they suddenly discovered that hurricanes are more powerful in real life than they appeared to be on television. It is always after it devastates the local community that it hits home and becomes real.

However, unless a damaging earthquake actually happens, nobody leaves in fear of earthquakes. In fact, more than half a million Californians move away from the state every year,[8] and few mention earthquakes as the cause. The leading reasons for this exodus include the high cost of housing, high taxes, and being fed up with the political culture.[9] Pollution, crime, poverty, uncontrolled immigration, traffic, cost of living, are other reasons mentioned.[10] Earthquakes? Not on the list. Until the next big one, of course.

DENIAL OF DISASTER

To some degree, denial of risk when it comes to rare and extreme events may be rooted in human nature and urban legends, irrespective of the type of hazard. For example, prior to 2016, many residents of Florida living along the coast from St. Augustine to Melbourne were convinced that this part of Florida would never be in the path of a hurricane—in spite of the fact that maps showing the path of all known past hurricanes provide clear evidence to the contrary. As if enticed by the songs of mermaids, blind to reality, they would confidently explain to anyone asking that this protection from hurricanes was due to either the shape of the ocean floor, the shape of the coast where Cape Canaveral projects into the ocean, or both—although no oceanography expert had ever uttered such nonsense. Many real estate agents and homebuilders bought into that fantasy, believing it firmly, as it was particularly convenient to reassure potential buyers moving to the coast. Fortunately, in 2016, when Hurricane Matthew hugged the Florida coastline south to north, about thirty miles offshore,[11] short of landfall but still producing significant wind and storm-surge damage along the shores north of Cape Canaveral, this urban legend was put to rest (at least, for a while, until people forget again).

When it comes to explaining why some believe they are immune to disasters, it is not that everybody's brain has turned to mush, but rather that everybody is predisposed to hear the story they wish to hear. To some, urban legends are sufficient; to others, perceived notions get in the way and can lead to the strangest, highly nonscientific, explanations—like the one occasionally heard that earthquakes in California are divine punishment in retaliation for the gay lifestyle that the state has condoned.

Denial of risk is also the indirect expression of the fact that immediate concerns always take precedence over long-term ones. Concerns about whether today's weather will affect the planned family picnic trump concerns about global warming of the planet over future decades (that may prevent many more future picnics).

Even when a threat is acknowledged as a reality, denial of risk implies transferring the consequence of risk to others. Those living inland will rationalize that global warming may be real, but are convinced that it will mostly flood those on the coast, while those living on the coast will rationalize that it will flood other parts of the coast than where they live. Other deniers of risk will argue that if the threat were real, everybody would be doing something about it now—which never seems to be the case. Or they may concede that the threat is real but that it pertains to an event so far off in the future that it is not sensible to make sacrifices today for something that might happen down the line—or that may actually never occur in one's lifetime.[12]

Whether such denial of risk is a healthy and defensible mechanism to be able to live in an uncertain world is a topic best left to psychologists, anthropologists, and other such specialists. However, because of the natural inclination to deny risk, promoting disaster resilience is not something that people will readily embrace, which is a challenge, particularly to engineers and other professionals interested in enhancing the resilience of communities.

NOT WANTING TO KNOW

Some forms of denial are more damning than others. Like shades of gray, they range from honest, unconscious denial reflexes, all the way to the deliberate deception of calculating minds. In the latter case, when strong interests are at stake and it is clear that recognition of the risk posed by a hazard can deeply hurt the wallet, skepticism is expected. If Big Tobacco can still deny that smoking is harmful,[13] in spite of more than fifty years of reports by the US Surgeon General and unequivocal statements by the Centers for Disease Control and Prevention identifying smoking as the leading cause of preventable disease, disability, and death in the country,[14] it should not be surprising when a single earthquake does not necessarily "convince" those whose bottom line is at stake. Following the 1906 earthquake that devastated San Francisco, with the ensuing conflagration engulfing two-thirds of the city, many folks with something at stake downplayed the earthquake damage. Interestingly, most fire insurance policies in effect at the time did not provide coverage if a building had suffered damage before a fire. Maybe not surprisingly, as the city was still burning, the California governor said that the destruction was mostly due to fire and, likewise, many of the city residents insisted that their property had not suffered any earthquake damage before the fire. Even though photographs taken before the conflagration revealed extensive earthquake damage, within a few weeks of the earthquake, the San Francisco Real Estate Board passed a resolution to only refer to the disaster as "the great fire" instead of "the great earthquake"[15]—which shows that fake news existed before the term was coined.

Public relations "spins" occur not only following earthquakes. When a major hurricane struck Miami in September 1926, killing 372 people, damaging 8,600 homes, and leaving 43,000 homeless, at a time when the population of Miami was approximately 100,000—estimated to be more than double the 42,753 from the 1920 census,[16] as it was in the midst of a construction boom—the mayor issued a public statement to reassure visitors that the hospitable, enjoyable, comfortable city of Miami would be there for their winter vacation.[17] Looking at pictures of the devastation in Miami and along the shoreline,[18] it would appear that the mayor forgot to remind the visitors to bring their camping gear for the trip, as the sunshine and warm weather might still be there, but maybe not the hotel rooms.

In a similar mindset, after the moderate magnitude 6.4 Long Beach earthquake of 1933 that killed 120 people and produced roughly $40 million in damage—in the midst of the great depression when a gallon of gas cost ten cents and a loaf of bread seven—the *Los Angeles Times* and many other newspapers spent much of their coverage to downplay the event and urged their readers to spread the news nationwide that that there was no greater place on earth than California if one wanted to be free from the dangers of natural elements.[19]

In spite of the denial of some, because the Long Beach earthquake destroyed nearly half of the city's schools and several more in nearby Los Angeles—fortunately, at 5:54 p.m., long after classes, as these collapses would have killed thousands of children otherwise—those who strongly believed that buildings should not collapse during an earthquake seized the occasion. State legislation was enacted, banning unreinforced masonry buildings because that type of construction had shown repeatedly its vulnerability to earthquakes, and most schools in the state had been built using this construction material. The legislation also required schools to be designed to resist a small horizontal earthquake force—a tiny force by today's standards, but it was a start. Some local school officials complained that this requirement was too stringent and expensive, but their denial was put to rest when the federal government supplemented the funding from bond issues.[20] Then, in the late 1960s and early 1970s, after many more damaging earthquakes had demonstrated the need to fix the deficiencies of many old schools, many voters blocked the passage of bond issues intended to fund this work. Some of the opponents stated that no kid had ever been hurt by an earthquake while in school—which was true, since none of the past destructive earthquakes in California to date at this time had occurred during school hours.[21] That is equivalent to telling a flock of turkeys, weeks before Thanksgiving, that there is no reason to worry about the future because their experience so far proves that nothing bad ever happens to turkeys.

Many systemic denials are rooted in concerns over financial hardship. In other words, dollars trump facts. Yet massive cost increases routinely occur for a number of reasons in all things that exist, and the extra cost incurred by these increases often by far overwhelm the cost of other measures that can have major long-term benefits when it comes to resisting hazards.

For example, since 1956, the Federal Highway Trust Fund has been collecting a tax on each gallon of gas sold, as a way to fund ground transportation programs across the nation, and most notably highway projects. Great idea, except that the framers of the bill did not—or did not want to—think in percentages. By the end of 1993, the tax was 18.4 cents per gallon. Twenty-six years later, in 2019, it was still 18.4 cents per gallon.[22] In 1993, on average nationwide, a gallon of gas sold for $1.17. In 2019, it was $2.47 (and $3.58 during the peak in oil prices that occurred in 2013).[23] In other words, the 15.7 percent tax in 1993 became a 7.4 percent tax in 2019 (and a 5.1 percent tax in 2013)—a great discount, thanks to the framers of the bill. Every attempt at changing the rate has been a major political fight. Yet, as basic transportation expenses are unescapable, the Highway Trust Fund regularly runs a deficit of tens of billions of dollars and special transfers from the Treasury general fund[24] are needed to fill the budget hole—without necessarily fixing all the other real potholes that the fund was intended to fill in the first place. Strangely, campaigning politicians often promise to fix the nation's decrepit infrastructure via massive infrastructure programs, but, after the election, they instead fight endlessly among themselves, arguing that there is no money for such wasteful programs. All in possible denial that changing the Highway Trust Fund formula to a percentage of sale price rather than a fixed decimal number of cents per gallon would be a financially sound approach, compatible with widely accepted payer-user concepts, and more sensible than rolling out the red carpet for foreign-owned-and-operated toll roads as is happening now.[25] Incidentally, contrasting with a swift bipartisan vote that went smoothly for $8 trillion over ten years in military spending,[26] it took more than eight months of haggling, drama, in-fighting, political posturing, and debate to agree on a 2021 infrastructure bill[27] that budgeted half-a-trillion dollars of new spending over ten years (2022–2032), as a small but positive step toward the additional $2 trillion needed to meet national infrastructure needs, as consistently reported by the American Society of Civil Engineers over past decades.[28]

The fear of greater costs has often been used as the "deal killer" in multiple initiatives proposed to enhance the safety of construction against multiple hazards. Special interest groups have often argued that any new requirement that makes homes better in this regard is prohibitively expensive, alleging that implementing such measures could go as far as killing the housing market. This is an argument of dubious merit, given the fact that the cost of residential homes in many locations across the country has increased wildly in the past decades (except for a few years due to the bursting of the "housing bubble"), generally without any improvements made to their ability to resist any hazard, and new homes have continued to be built and sold nonstop.[29] In many parts of the country, cost does not even bear any relationship to the quality of construction, let alone protection measures against various hazards; it is rather a matter of offer and demand, overarching political and economic factors, and, of course, location, location, location.[30] This makes it ill-advised to use the fear of prohibitive costs as an excuse against enacting logical hazard-protection measures in construction practices.

On account of ignorance and a good dose of California's pioneering spirit, it might be possible to forgive the Chambers of Commerce and other promoters who wrote newspaper editorials and telegraphed messages to all of America and Europe that diminished the extent of damage after each large earthquake of the nineteenth century and early twentieth century, in an attempt to reassure the oncoming masses. It is obvious that statements like "earthquakes are trifles as compared with runaway horses, apothecaries' mistakes, accidents with firearms"[31] (made after the devastating 1868 earthquake),[32] or "one Western cyclone will do more damage than all the earthquakes California has ever known"[33] (in newspapers after the milder 1892 one),[34] or "no place on earth offers greater security to life, and greater freedom from the dangers of natural elements, than Southern California"[35] (affirmed after the 1933 Long Beach earthquake), were dictated by the self-interest of developers.[36] In light of today's knowledge, it is much harder to forgive people who make similar comparisons nowadays. For example, stating that earthquake damage over the past twenty years has been insignificant compared to that from many other hazards is disingenuous; it compares the relative impact of various hazards during a few decades when no major earthquake has happened. Like the turkeys before Thanksgiving mentioned previously, some people fall for that stuff—Paul Joseph Goebbels (Nazi Reich Minister of Propaganda, 1933–1945) could not have done better himself. Truth is, a single major earthquake by itself can produce enough mayhem to turn these damage statistics on their head.

NOT INVENTED HERE AND PROFESSIONAL DENIAL

Bridges are designed to carry truck traffic on an everyday basis, and engineers are compelled to ensure that collapse will not occur under a truck overload. As such, maximum truck weights and maximum load per axle are specified, and bridges are designed to resist these loads—and, in fact, to resist greater loads than that because "safety factors" are built into the design process to account (to some extent) for the probability of overloads. In addition, to protect against renegade members of the trucking industry that may be tempted to disregard the law of the land, weighing stations are located at multiple points along highways. There, trucks must slowly proceed over a scale to confirm compliance with the legal load limit. Violators are subject to fines, and in most states, the driver of an overloaded truck will be charged with a misdemeanor,[37] which, technically, is a crime punishable by up to twelve months in jail—or maybe more if facing a judge who had a truck back up into his brand-new BMW earlier in the day.

In short, tons of trucks drive on roads every day, everybody is familiar with the fact that they are heavy, and engineers use safety factors when designing bridges because nobody wants to see bridges collapse under the weight of the trucks they carry.

Bridges have collapsed in the past, as shown earlier, for a number of reasons, but extremely rarely due to truck overloads. There could be room to revisit the magnitude of the safety factors applied to truckloads when designing bridges, but what would be the point? There is broad acceptance of the existing load factors by the engineering community, and no outcry on the extra cost incurred in each bridge designed to protect against collapse due to truck overloads. These safety factors have been "calibrated" partly on past practice and partly on observed truck traffic, and no pressure exists to reduce them. Trucks are everywhere, every day, noticed by everybody, and nobody is compelled to debate whether the failure of a bridge under truck traffic every 2,500 years would be a more acceptable outcome than once every 500 years. Arguably, there may also be no incentive to have such a discussion because, when a bridge failure happens, the engineer of record for that bridge is held accountable—which makes it an individual responsibility/liability.

In contrast, when it comes to earthquakes, they are not part of everybody's daily lives. Hardly anybody has lived through one or expects to. As such, they are intangible concepts—they do happen around the world, here and there, but they are essentially something "not invented here." The consequences of earthquakes elsewhere are not perceived as highlighting possible similar vulnerabilities at home—it is somebody else's problem.

Now, consider the following scenario. Imagine a small army of rogue drivers, with an axe to grind, who would pack-up their trucks solid, way above the legal limit, and start to drive together with the intent of collapsing big bridges. Imagine that the trucks themselves can carry that load without their axles snapping, and imagine that the roads themselves would not crumble under the heavy axles. Imagine that the heavy axle load of the first truck would not punch through the bridge deck and get jammed, thereby stopping the entire convoy right at the entrance of the bridge due to failure of a single small structural member. Imagine that this criminal convoy of overweight trucks is actually capable of collapsing an entire bridge by grossly overloading it. Then, at best, it will travel until it encounters the first bridge unable to sustain this unusual burden, which will then collapse, bringing down all the trucks with it. That would effectively be the demise of the convoy. The consequence of that overload would be the collapse of a single bridge.

By analogy with the overloaded truck, imagine an earthquake bigger than what has been considered in the design of bridges—which is not a big stretch of imagination, as many bridges are expected to suffer significant damage when a severe earthquake will strike, irrespective of where in the world that happens. Whereas the convoy of super-heavy trucks managed to damage a grand total of one bridge, the "overload earthquake" will damage all at once an overwhelming number of bridges within miles of its epicenter—hundreds of miles in some cases. This can disable the entire road transportation network of a region for weeks—or months. In such a case, given that the state of practice will result in thousands of damaged structures simultaneously, responsibility/liability generally becomes a social problem rather an individual one. The pressing need to immediately respond to the emergency pushes to an undetermined future any blame and accusations. Furthermore, when scrutiny does occur, the engineering community will collectively explain that severe damage is the expected outcome given existing seismic-design specifications (notwithstanding the subtle nuances between the definitions of expected damage and collapse). Such a "forgiving" situation hence provides little incentive to change practice.

Furthermore, engineers are people too. In spite of the inescapable evidence, the fact that seismic maps are consensus-based products of the best scientific minds, and the fact that seismic design provisions in design codes and standards are similarly consensus-based requirements developed by top engineers, there will still be professionals convinced that either earthquakes do not occur, or that if they occur, they will most certainly be less severe than expected. This is expressed incidentally by the fact that some engineers consider the design rules to protect against earthquakes or other extreme events to be "too conservative," or in some cases not warranted—which is effectively "denial," or at least "skepticism." While denial, as indicated earlier, is a deliberate or subconscious belief that catastrophic outcomes will never occur, or only occur to others, skepticism is a judgment, generally made in presence or absence of a deep understanding of all facts and data, which is based on the conviction that "bad science" has been used. Coupled to those two is often the "gambler" attitude that while earthquakes will occur, the rare "design earthquake" will not occur in the gambler's lifetime. Not here.

In most universities, the baccalaureate civil engineering degree provides general courses with basic exposure to design codes. Emphasis is generally on gravity loads and generic codified information on earthquake or wind loading requirements in a way by which they can be applied without special knowledge on the nature of these loads and the philosophy underlying these design requirements. This is also true of the graduate civil engineering curriculum in many universities. With some notable exceptions, the places where this seems to not be the case is where disasters have struck before. Without such a precedent, there is apparently little incentive to do otherwise.

REMEMBERING THE PAST IS NOT A HARBOR OF THE FUTURE

On September 19, every year, Mexico City holds an earthquake drill. Partly in commemoration of the 1985 earthquake that killed thousands on that date, partly as an earthquake preparedness measure, at 11 a.m. on that day, all buildings are evacuated. Everybody then goes down the stairs of their building—calmly, because everybody knows that it is a drill, that everybody will reach the street safely, and that it will be back to work after that bit of morning exercise.[38] The older residents remember the traumatic devastation of 1985, with its more than four hundred collapsed buildings and ten thousand casualties.[39] Yet, to those in their thirties or younger, this is only a historical event—much like Pearl Harbor is to nearly all Americans today: something to read about in textbooks. Understandably, without the benefit of having lived through a disaster, without any sensorial imprint to sharpen the senses, the younger Mexicans cannot approach the drill as seriously as their seniors can, and might go down the stairs while texting their friends. Textbook memory is worthless compared to lived memory.

That is not a Hispanic cultural trait but rather a universal one. How many people will admit to being so fed up with fire drills that they close their office door and continue working when an alarm is triggered? Those with an office on the ground floor or near an exit can argue that they will smell the smoke and have time to run out before roasting when it will be real, but what about those who cannot be bothered with going down two, five, ten, twenty, or thirty flights of stairs? Some people argue in favor of unannounced fire drills, to include the element of surprise as part of the drill, to avoid it simply being an exercise in herding to the exits blasé participants whose behavior bears no resemblance to what would happen in an actual fire. Others argue that unannounced drills have the undesirable consequence of many people ignoring the alarms during an actual fire, thinking that it is another bloody, annoying drill—until they realize that it is not the case and it is too late to escape.[40]

Anyway, two hours after the September 19, 2017 earthquake drill, thirty-two years to the day after the commemorated 1985 disaster, a real earthquake hit Mexico City. Apparently, for those who trust bogus statistics, September 19, like January 17 (for Northridge and Kobe) is a very propitious date for earthquakes. Indisputably, people who survive an earthquake learn a few things. That does not necessarily mean that they learn everything that could potentially be learned. After the 1985 earthquake, the Mexican building code was tightened, reconstruction happened, major efforts were made to improve earthquake preparedness, and some corruption problems were resolved.[41] Still, during the 2017 earthquake, 220 people died and forty-four buildings collapsed in Mexico City—smaller numbers than 1985, but a far cry from zero, and disappointing given that it was a much smaller earthquake this time around (magnitude 7.1 instead of 8.0, although striking much closer).[42]

Most of Mexico City is built on an old lakebed, which means that most buildings in town are sitting on top of more than 150 feet of soft clay.[43] Such a thick deposit of soft soil has its problems. One can think of Mexico City as if sitting on a gigantic bowl of Jell-O: small shaking at the bottom of the bowl is enough to madly wiggle the top. Likewise, earthquake vibrations at the bottom of the lakebed are greatly amplified by the time they reach the surface where everything bounces around more wildly—so to speak. This makes it possible for distant earthquakes that normally would have no impact hundreds of miles away from the epicenter, to become destructive events. The devastating magnitude 8 earthquake that killed so many in 1985 had its epicenter three hundred miles away; the 2017 one was about one hundred miles away. Considering that seismic waves travel at eighteen thousand miles per hour,[44] it takes twenty seconds for these waves to travel one hundred miles. This inspired Mexico, in response to the 1985 disaster, to develop an early-warning system. As one of Mexico's points of pride, it was launched in 1993.[45] As a result, thanks to the early-warning system, hundreds of thousands of Mexicans received an alarm on their cell phone, warning them of the incoming seismic waves racing toward town like a runaway train. Whether anyone can do anything that makes a difference in so few seconds is debatable, and depends on a number of factors, but the warning system was credited for the fact that fewer casualties occurred than would otherwise have been the case without it.[46]

However, another reason why fewer people died—other than the fact that the 2017 earthquake was much smaller than the 1985 one—is because Mother Nature can be an ally by weeding out bad buildings with each earthquake. If the source of the problem (that is, bad buildings) is eradicated by the earthquake (because they collapse) and no new problematic buildings are added to the existing inventory, this should help achieve a better survival rate in future earthquakes. That is the approach taken by

Chile when it comes to unreinforced masonry buildings. Recognizing that this kind of construction is known to be at high risk of collapsing during an earthquake, laws were enacted years ago to stop construction of such buildings there, but it is most difficult to force people to abandon the existing ones constructed prior to the ban, and fixing them to make them safer during earthquakes is more expensive than most owners can afford. As a result, the Chilean strategy is simply to wait for them to be eventually "wiped out" of the building inventory, one earthquake at a time, one collapse after the other, and to outlaw their reconstruction.

However, be it cigarette smoking or shoddy construction practices, bad habits not only die hard, but also tend to return after some time. As such, surviving an earthquake is good in that it brings better things forward, but time is an enemy that can erode gains in awareness. Similar to the often-heard truth that "institutions have no memory"—and are thus prone to repeat the mistakes of the past—societies equally lose memory because hard-earned wisdom can be forgotten within a generation, if not sooner.

ON THE DISASTER TRAIL

Not a Big Deal

When my wife, son, and I arrived in Berkeley in the early 1980s, the area was already suffering from a severe housing shortage. We found ourselves lucky midway through the first academic year to be able to move into one of the University apartments in the Smyth-Fernwald complex. The hillside residence complex had been built in 1945 as dormitories for women to ease the housing crisis created by the large number of GIs enrolling at the university when World War II ended, and they were converted into apartments for married students in 1970.[47] To put it mildly, these were Spartan accommodations, but conveniently close to campus. As a bonus, we had a great view of the Bay Area and—as a bonus to a structural engineer—of the Bay Bridge and the Golden Gate Bridge (fourteen miles away as the crow flies).

During our time there, we were surprised to hear many Californians comment that earthquakes were "not a big deal." When learning that I was studying earthquake engineering, some even went as far as adding that this was all very interesting but that there was no reason to worry about earthquakes—they were problems in the old days, maybe, but not nowadays. These were obviously not words from engineers, but still mindboggling statements considering that the 1906 San Francisco earthquake had happened right across the bay.

A few years after we left California, during the 1989 Loma Prieta earthquake, one of the Smyth-Fernwald buildings moved off its foundation, but since the epicenter of the earthquake had been quite far away, structures on the Berkeley hills did not suffer any major damage. However, this wake-up call reminded the university that it was pretty much sitting on top of the Hayward fault (mea culpa: our apartment was 550 feet from it), so it embarked on a massive program to identify and retrofit its seismically vulnerable buildings. The Smyth-Fernwald housing complex was flagged as problematic, because not only did the Hayward Fault run through it,[48] but also the hill slope on which the complex was built was not considered to be seismically stable. Therefore, the apartments were demolished in 2013.[49]

A sad outcome for something that is "not a big deal."

Airport Proctologists

POST 9/11 AIRPORT SECURITY

Once upon a time, it was common for airline pilots to invite little kids—accompanied by at least one parent—to visit the cockpit during long-haul flights. Kids sitting on their knees, some pilots would even temporarily turn-off the automatic pilot system, to veer the plane a tad bit left, a tad bit right, just for fun.[1] In some rare cases, a lucky passenger would even be invited to sit in the cockpit's jump seat through landing.[2] That was eons ago, when the fun was in the entire flying experience, rather than limited to the goofy arrival and departure announcements by flight attendants of discount airlines.

Every now and then, concerns were raised that easy access to the cockpit was an invitation to hijackers. There certainly was no real barrier preventing a passenger from reaching and threatening the pilot—and many did.[3] According to the Guinness World Book of Records, the first recorded hijacking took place in 1931 in Peru,[4] when an eight-passenger tri-motor airplane—the kind of rust bucket having a square fuselage of corrugated sheet-metal that Charles Lindbergh and Amelia Earhart flew, but slightly larger—was surrounded on the ground by members of a revolutionary army. In-flight hijacking started in the 1940s, more so with commercial airline flights,[5] but certainly not exclusively.[6] Then, through the 1950s and 1960s, it became more common, as all kinds of attention-seeking psychopaths and rebels with a political or social grievance saw it trendy to seize control of airplanes. Between 1968 and 1972, alone, there was an attempted plane hijacking every 5.6 days.[7]

In response, by the end of 1972, the US Federal Aviation Administration (FAA) required screening of all passengers and carry-on luggage, which led to the broad implementation of metal detectors and X-ray machines in airports. It helped a bit, as the number of hijackings dropped, to a handful on a good year, to a few dozen on a bad year.[8] Still a significant number, including many cases where hard lessons were learned too late—when learned.

For instance, in 1999, a Japanese man wrote to All Nippon Airways, multiple Japanese government agencies, and newspapers, that he had discovered multiple security flaws at the Haneda airport. It may not have helped the credibility of his claims that he also asked in the same letter to be hired as security guard, so he was ignored by all. Yet the flaws were real, and to prove his point, he took a flight from Osaka to Haneda with an eight-inch-long kitchen knife in his checked luggage. He collected his bag on arrival and used the unguarded passageways that he had identified as a security flaw to reach the departure gates without going through any security checks. Technically, he could have made his point very clear by pulling the big knife out of his luggage in the middle of the departure gates area—even running around while swinging it wildly and shouting "Bonzai!" if needed, for more spectacular effect. Instead, he quietly boarded an All Nippon Airways Boeing 747. After take-off, he pulled the knife out of his bag, entered the cockpit, forced the co-pilot out, stabbed the pilot, and took command of the plane for a ride—a frightening idea given that he had never piloted before. The airplane dived by tens of thousands of feet, down to an altitude of one thousand feet, before he was neutralized.[9] *In response*, after the hijack, the Japanese authorities reviewed security procedures in all airports across the country and apparently eliminated all such security flaws.

Typically, hijackers took passengers as hostage, seeking either ransom money, a free ride to some political asylum, or concessions such as the release of political prisoners. Some governments were compliant to such demands, some not. For example, in 1985, Trans World Airlines Flight 847 to Rome was hijacked after taking off from Athens. It was diverted to Beirut where nineteen passengers were traded for fuel; then to Algiers where twenty more were dropped; back to Beirut where they killed a passenger, dumped his body on the tarmac, and sent seven that had Jewish names to a Beirut prison; back to Algiers where seventy more were released; back to Beirut where they managed to get the release of an Islamist accomplice from a Greek prison in exchange for eight Greek passengers.[10] All that bouncing around like a ping-pong ball happening while the United States refused to negotiate with the same hijackers—who had requested the release of 766 prisoners held captive in Israel in exchange for the forty US passengers on board—on the basis that it would only encourage further hijacking.[11] After it appeared that no more juice could be squeezed from various governments, all hostages were eventually released and the hijackers walked away unpunished, since Beirut was the capital of a rogue country (Lebanon) in the middle of a fifteen-year civil war.[12]

Things changed though, in 1994, when the Armed Islamic Group of Algeria hijacked an Airbus 300 and its more than two hundred passengers, with the goal of crashing it on the Eiffel tower, or at the very least blowing it up above Paris.[13] The mujahidin had boarded Air France flight 8969 with machine guns and explosives, but thanks to a lack of subtlety, these armed criminals were identified by ground security before the plane took off. To make it worse, it never occurred to the hijackers that the majority of the passengers on board would be Algerians compatriots—more problematic than a planeload full of infidels. The standoff lasted three days and ended-up with a twenty-minute raid of the plane by the French counterterror unit and snipers that, in the end, killed all the hijackers.[14]

In response, Air France stopped flying to Algeria for eight years.[15] Beyond that, the French gave medals to the airplane crewmembers that survived the ordeal, posthumous honors to those that died, accolades to members of the counterterror units, and considered the entire operation a great success. Somehow, overlooked by everybody worldwide was the fact that, for some players, the hijacking game had changed; the objective was not anymore to use passengers as hostages in exchange for various concessions, but rather to use commercial planes as weapons—like missiles or bombs. [16]

Parenthesis: In the eyes of the author, who is a structural engineer, anyone who wishes to destroy a landmark structure—regardless of the cause of action or legal, theological, social, or political theory pled or asserted—shall be deemed nothing more than a complete moron. Irrespective of the directives, legal opinions, and pedagogical wisdom of Piaget, the Summerhill School, Child Protection Services, UNICEF, and the Global Initiative to End All Corporal Punishment of Children, bullies who laugh while stomping on the sand castles of other kids are on the path to become adult morons and deserve a damn good slap on the back of the head before it is too late. Do not touch the Eiffel Tower!

In spite of all this, cockpit doors in aircrafts prior to 2001 remained a rather thin divide between the pilots and their passengers. In fact, some international airlines did not require the doors to be locked during flights,[17] which makes one wonder how many passengers looking for the toilet mistakenly found themselves entering the cockpit—and how many were too drunk to realize their mistake? US airlines required that the cockpit be locked, which was undoubtedly of great assistance to drunkards who randomly pushed all visible doors, but both pilots and flight attendants had keys and flimsy cockpit doors were not exactly breach-resistant—and had been punched through in multiple occasions by deranged passengers.[18]

Then, 9/11 happened, which was a "big earthquake" for the airline industry. Except for those who have been without contact with world history since 9/11—either for having lived in a monastery or for having spent every hour of their waking life playing video games while flunking high school—everybody knows that on September 11, 2001, nineteen fanatics hijacked four airplanes. Three were flown into US landmarks, namely the two World Trade Center towers in New York City and the Pentagon in Washington, while the fourth one crashed in a field on its way to the White House when the passengers learned what happened to the other planes and fought to regain control of the plane.[19] The entire US airspace was closed for three days while government agencies tried to figure out how it happened and how to tighten security in case more of the same was waiting in the wings.

In the months that followed, many heads of US agencies, and even the president of the United States, stated that nobody had foreseen that terrorists could possibly ram hijacked airplanes into buildings.[20] Hard to believe given statements to the contrary by many other former US officials, such as the former FAA security chief, the former head of the CIA's counterterrorism operations, and even officials at the World Trade Center and Pentagon, who all had expressed concerns that commercial airplanes could be used as weapons. The FAA's very own self-published history recognized in 1972 that the hijackers of Southern Airways Flight 49 threatened to crash the DC-9 into the Oak Ridge National Laboratory's nuclear reactor if their demand for a $10 million ransom was not met.[21] In 1974, the perpetrator of another failed hijacking had documented on an audio tape that his intention was to crash the plane into the White House. Again, terrorists captured by the FBI in the mid-1990s confessed that they had been planning to dive a hijacked airplane into the CIA headquarters. Even the North American Aerospace Defense Command (NORAD) ran simulations from 1991 to 2001 that considered a commercial aircraft hijacked for the purpose of crashing it into a landmark building in the United States.[22] Not to forget that "not having foreseen" such a scenario would imply that after the Air France flight 8969 experience, everybody assumed an ostrich position and buried their head in the sand to make the problem go away.

Irrespectively, on 9/11, the big earthquake had now happened and action was called for. *In response*, the FAA recommended the hardening of existing cockpit doors and all major US airlines implemented these measures within six months (it only became a mandatory requirement by November 2003). The Transportation Security Administration (TSA) was also created to put the federal government in charge of airport security, taking over this responsibility from miscellaneous private security companies. That is when nail clippers and forks became potential terrorist weapons and hour-long line-ups at security screening became common occurrences.

Then, on December 22, 2001, on American Airlines Flight 63 from Paris to Miami, another fanatic attempted to detonate plastic explosives stashed into his shoes—unsuccessfully, as passengers noticed his repeated attempts to light the damp fuse sticking out from his shoe and ganged up to restrain him.[23] *In response*, TSA started requiring that all shoes be removed and screened separately.

On August 10, 2006, British Intelligence arrested twenty-five fanatics who had been plotting to use liquid bombs to blow up multiple airplanes over the Atlantic Ocean. *In response*, TSA started prohibiting the transportation of liquids, aerosols, and gels in carry-on luggage, except in minute amounts. This also resulted in countless water bottles being thrown into garbage cans near airport security screening, and a boom in the sale of water bottles past security screening.

On December 25, 2009, on Northwest Airlines Flight 253 from Amsterdam to Detroit, another fanatic failed to detonate plastic explosives hidden in his underwear and set his pants on fire instead, inflicting second degree burns on his thigh and genitalia[24]—maybe with the side benefit of making it harder for this "underwear bomber" to reproduce. *In response*, the TSA used the event to accelerate the implementation of full body scanners across US airports—the wonderful technology that reveals the surface of the skin under clothes, which many consider to be a kind of indecent strip search and an invasion of privacy.[25]

In response, in response, in response. Always *in response*. That is at the crux of the problem. It is fortunate that explosives have never been found in certain body cavities of terrorists; otherwise, *in response*, passengers would now also have to be screened by a TSA proctologist when going through security.

SECURITY MEASURES: FOOLPROOF OR FOR FOOLS

In spite of best efforts and best intentions, it remains that no security measure is foolproof, because there is no such thing as absolute safety. For example, although it has long been recognized that even if a vault door as thick and secure as those used in banks were used to secure access to the cockpit—which would be prohibitive in airplanes due to their excessive weight—that by itself would not be a guarantee of ultimate safety. Simulations have showed that a pack of terrorists would only need three seconds to rush into the cockpit when the pilot opens the door to go to the toilet.[26] Therefore, after all airplanes cockpit doors were replaced by the more secure ones mandated by TSA, a new procedure was developed that required flight attendants to block the aisle with the service trolley to keep passengers at bay when the pilot leaves the cockpit to use the front toilet. Another security hole plugged!

Does all of the above now make airplanes impossible to hijack? Having grandma and grandpa "strip searched" by a machine after throwing away their water bottle might make everybody feel that amazing measures are now in place to keep terrorists at bay. But, arguably, an unstated objective of creating TSA and tightening up the screening measures in airports may have been, in large part, to give the public the *perception* of safety and thus save the commercial airline industry from bankruptcy. Surely, the bar has been raised for anyone planning to hijack an airplane, but it has not made it impossible. In fact, in 2017, inspectors from the Department of Homeland Security whose job is to travel incognito from airport to airport with guns and explosives in their carry-on luggage to test the effectiveness of the TSA screening found that they were able to smuggle them through security checkpoints 80 percent of the time. This was seen as a major improvement over 2015—fourteen years after 9/11—when the TSA failure rate at detecting these weapons was reported to be 95 percent.[27] When attempting to create a security barrier by building on job experience and skills, it certainly does not help when one in four TSA screeners quits the job within six months;[28] in part, because screeners are among the lowest-paid federal government employees, with a starting salary close to minimum wage— less than if they worked at the airport sandwich shop.[29]

Yet, even if airport security screening were watertight, that would not make hijacking impossible. Anyone who has flown business class has observed that rules in that cabin are more relaxed. While the masses in economy class get plastic forks and knives—because a butter knife might be too much temptation to a budding hijacker—the elite in business class on international flights sometimes enjoys real cutlery. While the proletarians in economy class will be yelled at by the flight attendant if they attempt going to the toilet while the seatbelt sign is on—and has been on for three hours—the rule does not always apply to the privileged few who shelled out four times more money on the airfare. Nothing prevents a malicious group of individuals from buying all the seats in the business cabin, to have an easier time storming a cockpit defended by a 120-pound flight attendant and her service trolley—or from buying all the seats in an airplane for that matter to eliminate the presence of heroic passengers that might interfere. Where is it said that someone on a suicide mission cannot splurge on a business class ticket and the comforts it will bring for an hour or so before dying?

However, it may not even be worth the logistical trouble for terror-ists to develop such a plan when so many other options exist. Bombs can be detonated in trains, as was done in 2004 when ten near-simultaneous explosions in four trains in Madrid left 193 people dead and roughly 2,000 injured,[30] or in subways, as was done in 2005 when bombs exploded on three London Underground trains as part of a coordinated operation that killed 52 and left more than 700 injured. Even easier to achieve, trucks can be used as weapons, as was done in 2016 when a nineteen-ton cargo truck was purposely driven into a crowd celebrating the national holiday in Nice, killing 86 and injuring 458—comparably as deadly as the London Underground event, without even needing explosives.

Yet, while hundreds of billions of dollars are being spent in airport security, nothing is done to control access to trains and subways[31]—except maybe for the addition of security cameras that can provide after-the-fact information. The truth of the matter is that nothing happens for a number of reasons. First, trucks and trains cannot be broadly "weaponized" eas-ily—meaning that they cannot be crashed into the White House, as could be done with an airplane. People who took a train assumed the risk that came with it, while innocent people not in the train will not be victims. Second, there is truth in the "official line" that providing rail stations, sub-way stations, and bus stations with the type of screening done at airports is impractical and would bring the transportation network to a halt.[32] The normal activities of life cannot stop because there are ill-intentioned people and nutcases in this world. Nobody wishes to see massive concrete barriers installed along the edge of every sidewalk in crowded cities just in case someone goes on a truck rampage.

However, there are many instances—universally, not just on terrorism-related issues—when things can be done and should be done to prevent events with massive undesirable consequences, but are not done. Instances when, after the fact, the population will insist that these preventive mea-sures should have been implemented before. That is not only true for ter-rorism, but also for every other type of low-probability large-consequences event, symbolically called "earthquakes" for the current purpose.

"WHY WAS IT NOT DONE BEFORE?"
ASKED THE OUTRAGED

Like it or not, as illustrated by the above example, it is part of human nature to react after the fact, rather than plan beforehand. While all would agree in principle with the wisdom of expressions like "an ounce of prevention is worth a pound of cure," all that sagacity tends to be thrown out the window when the said ounce of prevention requires an effort or entails some costs.

That is when the idiom "we'll cross that bridge when we get there" takes over because there is always something more important, urgent, enjoyable, or rewarding that will take precedence in the grand list of things to be done and that will monopolize all available time and resources. Everybody can think of things that should be done at home or at work "when time allows." In fact, the excuse "I'll do it when I get around to it" is so often heard that those tired of waiting can purchase wooden tokens with a large "TUIT" engraved on both faces, to give friends or relatives the round "tuit" that they so desperately need.

As a result, when things get done, they are often done "in response" to the disaster that could have been prevented or mitigated if the actions had been taken before the event. All of a sudden, the extensive damage, the body bags, the tragic stories of loss and pain, provide inescapable evidence that something should have been done, and push the topic to the top of the list. "Something needs to be done, now!" is trumpeted. As if patching potholes in a poorly designed road could compensate for failure to have designed a good road in the first place. As if asking a professor, after having flunked the final exam, if there is any extra work that could be done to pass the class, instead of having done the assignments to learn the material before the exam (yes, that question gets asked). As if trying to shove some of the toothpaste back into the tube makes any sense.

It obviously cannot. Nonetheless, the post-event flurry of activities to clean up and rebuild—be it well intentioned or simply the product of weasel politics—gives the impression that those in charge are all doing what they are supposed to do, in a capable and responsible way. If Penn and Teller (the illusionists renowned for revealing how magic tricks are done) had to describe this, they might call it an effective misdirection to hide the fact that the very same people failed to do what should have been done years earlier. It redirects the "do not fix it until it is broken" dunces before the event, to dress-up as "let's fix the damage" men and women of action after the event—conveniently blurring the fact that the whole thing might have been set up to fail by the inaction of the same men and women.

"Even a fool may be wise after the event," as stated in *The Iliad*.[33]

AHEAD OF THE GAME: WHAT WAS DONE BEFORE, NOT AFTER!

There are obviously examples that will come to mind of individuals and organizations that are preparing ahead of time for disasters. Particularly when it comes to agencies and corporations, it may appear today that they are "ahead of the game" and truly getting ready to prevent disasters. Indeed, some of them truly are. However, in some cases, these outstanding activities to prepare against future damage have happened after massive losses or near misses, or because the sight of damage to others prompted them to assess their vulnerability.

For example, after the Loma Prieta earthquake of 1989—known as the World Series Earthquake because it happened thirty minutes before the start of Game 3 between the Oakland Athletics and the San Francisco Giants, making it the first major US earthquake to happen live on television[34]—when the Goodyear blimp originally intended to provide overhead views of the baseball game flew around town and broadcasted views of the extensive damage to the Bay Bridge and elevated freeways, as well as of the burning fires in San Francisco, the message was loud and clear. Most of the civil infrastructure built in Northern California had been built over decades during which no major damaging earthquakes occurred, and for the most part in an era when knowledge was seriously lacking on how to design it so that it does not sustain fatal damage during an earthquake. In some cases, prior to the Loma Prieta earthquake, the vulnerabilities of this infrastructure were known, but either there were other more pressing priorities to address or no budget to do anything about it.

When shaking started, for some, it was unfortunately fatal. However, for those that survived, the earthquake was a good thing.

Seeing the Oakland Athletics win the World Series when playing resumed, after a roughly two-week delay, might have somewhat lifted the spirits of the residents on the east side of the San Francisco Bay. However, losing the World Series would not have been as bad as losing access to water because of an earthquake. Following the earthquake, the East Bay Municipal Utility District (EBMUD) that supplies water to more than one million people living along the east side of San Francisco Bay decided to assess the vulnerability of its entire distribution system. Results did not look good. In part, because most of the utility's water at the time came from further east and reached the thirsty East Bay by crossing the Oakland hills through a tunnel built in 1929. It so happened that the east-west tunnel intersected the Hayward fault, one of the most dangerous faults in California. Although forgotten by most—and overshadowed by the 1906 San Andreas Fault earthquake—the 1858 earthquake that occurred along the Hayward fault was one of the most destructive in California's history.[35] The past six earthquakes on that fault have happened on average 150 years from each other (ranging from 95 to 183 years). Seismologists expect a magnitude 7 earthquake to occur on that fault anytime[36]—in fact, looking at the average, simple math shows that it is overdue.

The EBMUD study showed that the tunnel was likely to be ruptured by slippage along the Hayward fault, cutting the water supply to eight hundred thousand people for a period of up to six months.[37] By 2007, nearly twenty years after Loma Prieta, construction of the complex engineering solution to overcome this deficiency was completed. Looking at it with today's eyes, it may be seen as outstanding foresight and evidence of thinking ahead of time. However, in retrospect, this seismic improvement program largely happened in response to the Loma Prieta earthquake.

In response—again.

Likewise, shortly after the Loma Prieta earthquake, California Governor Deukmejian appointed an Independent Board of Inquiry to focus on bridges. The Board was tasked to report on why so many bridges collapsed or suffered significant damage during the earthquake, and to recommend what to do with the more than twenty-two thousand bridges in the state (half of those being maintained by the California Department of Transportation, also known as Caltrans) to prevent future similar destruction.[38] The Board's final report stated: "The fiscal environment at Caltrans in the last two decades seems to have inhibited giving the necessary attention to seismic problems. Many items ranging from research on earthquake engineering to seismic retrofitting were placed in low priority because of the limited possibility of funding due to budget constraints."[39] Reading between the lines, this was a polite way to tell the governor that Caltrans needed a round "tuit" when it came to fixing the state's bridges, but that the government—and thus the governor—always had more important things to do. Decades before, after the 1971 San Fernando earthquake during which many bridge spans simply fell off their supports, Caltrans had embarked on a project to add restrainers to tie spans to their supports; yet, it took seventeen years to execute this $54 million program—effectively $3.2 million per year, out of a state budget of approximately $50 billion.[40] When it came to the Department of Transportation's budget, the priority in California was not to fix bridges but to invest in roadwork to relieve traffic congestion—obviously, still a work in progress fifty years later.

Why is so much being said about what California does, and so little about what other states do, when it comes to earthquakes? Simply because most other states have not had damaging earthquakes in recent history— the "not so recent" ones being long forgotten. Therefore, while some of these other states had the prescience to retrofit key bridges, others still have not found a round "tuit" in their pockets—but they sure will not fail to do so after the next damaging earthquake.

In response . . .

Examples also abound for all other kinds of "earthquakes" considered here. For instance, there has been a back-and-forth dance in coastal regions to bury or not bury power lines to minimize power outage due to line damage during windstorms. In 2019, Florida passed legislation requiring power companies to move outage-prone portions of their distribution network underground.[41] Some utilities, like Florida Power and Light already have 40 percent of their lines buried, as this is an effective solution to eliminate damage to the power lines from flying debris and falling tree limbs during hurricanes. In contrast, North Carolina decided against doing so after assessing that it would be too expensive, translating into possible electricity rate increases of 125 percent.[42] Note that buried pipelines are in principle watertight, unless damage occurs to the PVC pipes in which they run, which could happen in the case of uneven soil settlement. Repairing buried power lines is substantially more time consuming and expensive, but it will always be possible to "cross that bridge when getting there."

It may or may not be wise to wait for a disaster before acting. Arguably, to make an enlightened decision, one would need to know "what are the odds" that the Big Bad Wolf will come? Answering this question requires diving into the world of statistics and probabilities, for a quick dip, hopefully without drowning.

ON THE DISASTER TRAIL

Kiwi Airport Security
At one point during a three-month stay in Christchurch, New Zealand, in 2010, I had to take a domestic flight from Christchurch to Wellington on Air New Zealand. After grabbing my boarding pass at the check-in pod in the old domestic terminal (replaced since by a more modern facility), I proceeded to the departure doors. I spotted the X-ray machine and metal detector ahead—in my mind, standard airport equipment in the post 9/11 era—and lined up waiting for my turn. When the officer checked my boarding pass before pushing through the X-ray machine the plastic bin in which I had deposited my laptop, wallet, and keys, he looked at me and said, "You are at the wrong place. Your flight leaves from there." He pointed at Gate 6, behind me. Sure enough, there was my gate, outside of the security-controlled area. It had never occurred to me that there could be a departure gate where nobody needed to go through a metal detector and an X-ray machine before boarding the plane. Anybody could have bought a ticket online, printed their boarding pass, and walked inside the Wellington-bound plane without even as much as an inquisitive stare. The entire Osama Ben Laden fan club of New Zealand could have boarded the twin-engine turboprop and it would have been greeted by the same smiling Air New Zealand flight attendants as everyone else. No need to hide anything in shoes, water bottles, underwear, or body cavities.

After landing, I mentioned to my host in Wellington my surprise at, not the lax security, but rather the total absence of any security check. He smiled and said, "Why would any terrorist want to blow up kiwis? That would not make much of an impact now, would it?" All a matter of perspective, evidently.

Statistical Hocus-Pocus

LYING WITH STATISTICS

When it comes to understanding hazards and how to cope with them, having a good understanding of statistics is valuable. Unfortunately, understanding statistics is not always as simple and intuitive as calculating batting averages. In fact, human intuition and experience too often lead to false expectations when it comes to statistical outcomes.

Imagine that a coin has been flipped five times and that on each of the five times, the outcome was tails. If one had to predict the outcome of flipping a sixth time, what is the best bet? Most people would argue that the next toss is more likely to be heads.[1] They reason that hitting tails six times in a row is quite a rare event, so after five "tails" in a row, "heads is about due." Yet, this is not only false, but a misconception so common that it has been given its own name: the "gambler's fallacy."[2] In reality, past results do not in any way affect the outcome of the next toss: when flipping a coin, there is always a fifty-fifty chance of each outcome. Of course, the coin could also land on its edge and not fall on either side, but that is an amazing feat of equilibrium with a very low probability of occurrence when tossing the coin on a hard surface, so that possible outcome is discounted here as it is beyond the point.[3]

The gambler's fallacy exists because human intuition has some interesting biases when it comes to games of chance. One of these biases is to treat random events as if they were not random. The human experience in many ways suggests that long sequences of repeated identical outcomes are never sustained. If it is sunny for ten days in a row, then a rainy day ahead is expected; make that twenty sunny days in a row and the rainy day ahead is believed to be even more of a sure bet. Therefore, in the above example, since tails comes up five times in a row, then, in fairness, it should now be heads' turn. Unfortunately, a coin being tossed has nothing to do with fluctuations in weather, and a coin could not care less about human experiences and biases; it is only a stupid piece of metal that always has an equal chance of landing on heads or tails—or on its edge once in an eon to stun everybody.

Likewise, imagine a lottery game that requires picking six numbers between 1 and 70. The gambler's fallacy in this case is to think that picking the set of numbers 1, 2, 3, 4, 5, 6 is less likely to win than 4, 11, 27, 33, 59, 65. Ask someone next to the lottery ticket vending machine, "I have never played this game before. Should I pick 1, 2, 3, 4, 5, 6 for my lottery ticket?" and watch reactions. Most people would confidently argue that it is better to pick a nonsequential set of numbers, or state outright that one would have to be nuts to pick a series of consecutive numbers. However, in reality, the probability of winning is the same as selecting nonconsecutive numbers. In fact, for any combination of six numbers, there is the same 1 in 130,000,000 chance of winning, no matter how laughable the sequence of numbers selected. The only sure way to win is to buy 130,000,000 tickets to cover all possible combinations, which, incidentally, might be a good idea if the jackpot on a given week exceeds 130,000,000 times the price of an individual ticket, as long as the grand prize that week does not have to be split with another winner. Interestingly—as there are always exceptions to every rule—some contrarians (math or psychology students?) purposely pick combinations that most people avoid. Notably, 2,014 people won the grand prize of 5,000 times their bet when North Carolina's Pick 4 lottery winning combination on June 22, 2019, turned out to be 0, 0, 0, 0, with a record payout of $7.8 million.[4] Given that the odds of winning are 1 in 9,999, and that the lottery is held twice daily, the odds for a winning combination with four identical numbers are 1 in 999, or once every 2.73 years; this relatively short return period creates "visibility" and might explain why such "quads" are among some of the most popular bets placed in this specific lottery. Indeed, the 1, 1, 1, 1 winning combination in 2012 was North Carolina's previous largest payout, at $7.5 million.[5]

Psychologists have had a ball trying to explain why the gambler's fallacy exists in the first place and why it is so hard to overcome—from psychological to neurological studies, and everything in between. To put it bluntly, statistical literacy is not ingrained in human genes because it apparently has not been essential for survival—like ignorance in itself has never been a barrier against evolution of the species. As a viral online joke says: "We hate math, say 4 in 10—a majority of Americans." Given that none of the World Statistics Congresses since 1887 have been held in Las Vegas, it might be fair to assume that it is likely easier for a trade association who wants to hold a convention there to get comp rooms and massive hotel discounts if its members do not fully grasp why slot machines are also called one-armed bandits.

It did not take long for the advertising industry to realize that hazy statistics can be used advantageously to influence a mathematically challenged public. How to lie with statistics is an art and a game that can be played in many ways. Some of the most basic tricks consist of:

- Conducting surveys on samples having built-up biases, which then casts the results in misleading ways: For example, when the 1950s Kinsey report on human sexuality came out, it presented surprising statistics on the number of American males who reported having had sex with prostitutes (69 percent) and homosexual experiences (37 percent). Closer scrutiny of the data revealed that the sample of those interviewed, possibly due to the difficulty in gathering sexuality data in the 1950s, included a substantial number of prison inmates, including incarcerated sexual offenders.[6]
- Playing on the fact that most people confuse median and mean: A mean is simply the sum of a series of numbers divided by how many numbers there are in that series. Take $250,000 + $500,000 + $10,000,000. If these were the prices for three homes sold in one week in a certain city, then the mean price of homes that week would be reported to be $3,583,333—a value severely affected by the one mansion with a $10 million view, but a value nonetheless that real estate agents would not hesitate to quote to qualify the other two homes as massive bargains. The median is the middle point for which half of the numbers in the series are greater, and half are lower. In the above example, it would be $500,000—the number the Chamber of Commerce is more likely to use when it wishes to report affordable housing to attract businesses. To confuse matters, since most people do not know the difference, both medians and means are casually called "averages" depending on what point one wishes to make.
- Presenting incomplete comparisons: For instance, self-driving cars are being promoted as being safer. This begs the question: "Safer than what?" or "Safer under what circumstances?" The National Highway Traffic Safety Administration reported in 2015 that 94 percent of accidents are due to driver errors,[7] so many have concluded, logically, that if the human factor is eliminated, cars will be safer. This is a big claim. For sure, nearly 100 percent of accidents are due to driver errors because it is quite difficult nowadays to have a car crash unless there is a driver in the car. Yet some have

reported that self-driving car accidents have occurred at twice the rate for regular cars.[8] Still, is driver error a comprehensive basis of comparison? How about all the other factors? Drilling down into the data shows that roughly 5 percent of accidents nowadays are due to mechanical failures (namely, failures related to tires, brakes, steering, suspension, transmission, and engine) or environmental conditions (obstructed view, glare, ice, snow, fog, etc.).[9] However, if self-driving cars turn out to be safer when cruising well-marked highways on sunny days, will they also be safer in heavy rain, on ice or snow-covered roads, on gravel or muddy roads, in construction detours, at intersections operated by a police officer, or when hackers remotely take control of the car?[10] Will there be other new problems? Recent surveys indicate that the majority of people would be afraid to ride in a self-driving car.[11] Indeed, for anybody who has suffered the frustrations of working with personal computers or who has owned "smart" but unreliable electronic products, it might be a significant leap of faith to trust driverless cars operating with hardware and software created by the same computer programmers. As structural engineers like to say, if buildings were designed the way computers are designed, the world would be in serious trouble: after a building has crashed, it is too late to close all of its windows, reopen them, and expect everything to return to normal as if it was a laptop.

- Headlining selected facts and neglecting the impact of various variables and parameters: A study was reported to have found that death rates due to accidents double for each ten miles per hour increase in speed limit. That would be an attention-grabbing headline for sure, but useless in itself. It means absolutely nothing because the conclusions have been separated from their context. Such a headline may suggest that reducing speed on highways by ten miles per hour would have a dramatic impact on safety, but that would be perverting the fact that the numbers cited actually refer to fatalities in terms of pedestrians struck by cars—definitely not statistics related to freeway speeds. In this case, the doubling in fatalities jumps from 31 percent to 60 percent for victims hit by cars at speeds of 35 and 45 mph, respectively.[12] The fatalities also triple from 10 to 31 percent if comparing 25 mph and 35 mph, but only increase by 25 percent if comparing 45 mph and 55 mph. In other words, the fact that the data follows a highly nonlinear curve is conveniently omitted by the headline. Note that results also vary most significantly if looking at fatalities for thirty- and seventy-year-old pedestrians, with the older victims being more frequently fatally wounded, irrespective of speed. It also is interesting that according to different data sets, the fatality rate (all ages) at 45 mph is either 40, 50, or 80 percent.[13] The results also vary from country to country, as roads and sidewalks in the United States, England, and Germany are definitely not the same. Hence, everybody knows that speed kills, but heralding simplifications is also harmful.

- Presenting data using clever graphs that mislead by design: An example is using axes that do not go to zero, or graphic elements whose sizes betray the nature of the data. This is well illustrated by the beautiful maps of the world that come with all countries coated in the same dull-gray latex material that is used to hide numbers on lottery tickets. By using a penny to scratch-off the latex that covers specific countries, world travelers reveal the map colors of all the countries that they have visited. This is equivalent to putting pins on a world map, but presumably more fun and visually appealing. However, it is also misleading. Take two budding globetrotters who have happened to visit four cities. The first one visited Shanghai, Moscow, Montréal, and San Francisco, and the second one Tokyo, Stockholm, Bogota, and Cairo. The first one will get to scratch-off almost half the map, with China, Russia, Canada, and the United States revealed in full colors on the gray map, giving the impression of having traveled half the world already. By comparison, the second one will scratch-off countries of smaller size in square miles, giving the exact opposite impression, even though it involved traveling more miles and visiting cities that are more populous. The same kind of visual trickery is used all the time when presenting statistical results.

- Using correlation between events to establish relationships that do not exist: There is obviously a one-to-one correlation between days when sidewalks are hot and days when people suffer from sunburn, but it would be crazy to claim that hot sidewalks produce sunburn. Such strong correlations are used all the time to convincingly link unrelated phenomena. For example, it is often said that graduates of Ivy League schools are more likely to become the wealthiest members of the country, compared to graduates from other universities. This does wonders for these universities' reputation and recruiting, but it fails to consider that each of the Ivy League schools entering class has a large number of students already coming from the country's wealthiest families.[14] They could fail tons of classes, collect less than stellar grades, graduate by the skin of the teeth, and still have a wealthy future ahead—some may even become president of the United States, on slim academic accomplishments.

- Hiding the variability that exists within samples or because of sample size: In particular, for marketing purposes, instead of asking one thousand people to participate in a blind taste test to determine if they prefer the product from company A or B, and reporting the results for the total number of participants, it is better to organize one hundred such blind tests on groups of ten people each. As if flipping a coin ten times, it will turn out that, in many groups, half the people will indicate that they prefer A and the rest B. In some other groups, six will prefer A, four B (or prefer B over A). In a smaller number of groups, maybe seven will prefer one over the other. Inevitably, when repeating the experiment often enough, with many more groups of ten, it is bound to happen that in one of those groups, participants will prefer either A or B in an even bigger percentage. It is just "luck of the draw," but this will make it possible to report that "in a blind test" (meaning, in one of these many groups, but not saying it like that), more people preferred A over B, with precise percentages cited to back this up.[15]

- Using large variability to mask reality: For example, millions of people write with the hope of becoming successful novelists. It is difficult to know what the average income of fiction writers is, but the extremes are well known—and top earnings well publicized. In 2018, James Patterson was the highest-paid author, with income from book sales of approximately $86 million.[16] Yet, a survey in the UK revealed that, on average, authors earned less than the minimum wage.[17] This is not surprising. There are more than three hundred thousand new books published in the United States alone each year—or more than eight hundred per day.[18] About fifty thousand of those are fiction—thus, almost one hundred fifty per day.[19] Hence, for every Harry Potter success story, there are tens of thousands of titles that go nowhere. Likewise, for nonfiction, in spite of numerous best sellers, the sheer number of unsuccessful books drags the average sales number to less than two hundred fifty copies per title.[20] To make things worse, a quarter of all Americans do not read any book (and that ratio is increasing fast), and most people read less than five books per year.[21] The odds of becoming rich by pouring blood, sweat, and tears into writing a novel are similar to those of hitting the jackpot by buying a lottery ticket for a few dollars and no work at all. Yet thousands of people write novels (instead of buying lottery tickets). Obviously, some do it because they are compulsively driven and because they have something to say, but those who do it driven by the appeal of fame and fortune are seriously misguided by the pernicious attraction of extreme values in statistical sets having enormous variability.

In fact, statisticians have grown so masterful at transforming the truth by playing games with the numbers that this is often derided by jokes. A case in point is the story of the statistician who was arrested by TSA agents at the airport because they found a bomb in his carry-on luggage. When asked why a serious scientist like him would do something so stupid as trying to bring a bomb on board an airplane, he replied that he had read that the probability of a bomb being on an airplane was 1 in 10,000, which made him uneasy about taking an airplane. However, for that given probability, the odds that there would be two bombs on an airplane can be calculated to be 1 in 100,000,000 (equal to 1/10,000 times 1/10,000), which is a much lower risk and made him feel much safer, which proved the benefit of carrying his own bomb.

The examples in the bulleted list above are only some of the tricks of the trade used to make the numbers say whatever is desired simply by casting statistical results in different lights. So, why does all that matter when it comes to earthquakes and other hazards? This deserves some scrutiny.

THE CURSE OF INCOMPLETE INFORMATION

Given the regular abuse of statistics to manipulate the public (to make people think favorably of anything, be it a specific brand of toothpaste or a politician), it should not be surprising that most people have become skeptical—if not cynical—when facing claims that are presented as facts supported by statistics. Even when those statistics have been developed by reliable sources based on complete data sets, the news media reporting on the findings often distort reality when presenting the information—accidentally or intentionally twisting reality by some of the methods outlined above—which does not help. If that happens when all the facts and all the data are known, and statistics only consist of tallying the information, imagine what happens when it comes to predictions that must rely on statistics to forecast the future by extrapolating from existing data. It should not be surprising that even fewer people pay more than passing attention in that case.

Then, one step further, who cares when these predictions are based on incomplete or sparse data? That, unfortunately, is what scientists predicting extreme events and hazards must deal with.

At one end of the spectrum, those predicting weather-related events benefit from continuous streams of data fired from arrays of instruments scattered across the world. There are weather stations collecting air temperature, barometric pressure, wind speed, humidity, precipitations, cloud cover at multiple elevations, jet stream activity, ocean temperature, wave directions, swell patterns, ocean currents, and the number of butterflies flapping their wings—and all in real time. Data is collected by ground-level stations, antennas, airplanes, geo-stationary satellites, and grandpa's arthritic joints. Yet, the fact remains that nobody can exactly predict where and when hurricanes or tornados will strike. Consider the following:

- On Tuesday August 27, 2019, the National Hurricane Center of NOAA predicted that Tropical Storm Dorian would hit Florida a bit south of Cape Canaveral, five days later, at 2 a.m. on Sunday, September 1.

- By the end of the day, it revised the severity of the predicted wind speeds, upgrading the tropical storm to a hurricane status, with the eye of the hurricane arriving instead at 8 p.m. Sunday.

- The next morning, on August 28, it further increased the severity of the upcoming impact, upgrading the storm to a major hurricane to hit the coast farther north by one hundred miles, on the upcoming Monday (that is, still five days ahead). The Florida governor declared a state of emergency for the counties in the path of the hurricane.[22]

- By midafternoon on August 29, NOAA predicted that Dorian would be a Category 3 hurricane, with the eye of the storm hitting the coast near West Palm Beach, two hundred miles south from the previous prediction.

- By early morning Friday August 30, the hurricane was still forecast as a direct hit on West Palm Beach, with Category 4 wind forces, but with the eye of the hurricane arriving only by Tuesday morning, and subsequently weakening along a northern path mid-width of the state.

- By Friday evening, the forecasted northward path was revised to "hug" instead the entire eastern coastline of Florida, with sustained disastrous hurricane-force winds all the way.

- Then, the Saturday morning NOAA forecast predicted that the eye of the hurricane would not hit Florida after all, but rather curve up north and stay nearly one hundred miles away from the coast, reaching the latitude of Cape Canaveral sometime during the night from Tuesday to Wednesday. This provided a temporary sigh of relief to all Florida residents, as the tropical-storm winds of roughly 70 mph forecasted to be felt on the coast, while still strong, were far less devastating than the near record-level winds of 180 mph at the eye wall, since Dorian had grown by then to be a Category 5 hurricane.[23]

- Throughout the Labor Day weekend, the predicted path of Hurricane Dorian remained sensibly the same, forecasted to be at least a Category 4 when it would be skirting the coast, but the predicted distance of its path away from the coast as it moved North progressively shrank, bringing it closer to Florida than forecasted a couple of days earlier. By Monday morning, the predicted northward path of the eye of the hurricane had become so close to the coast of Florida that a mandatory evacuation order was issued for the hundreds of thousands of people living east of the intracoastal waterway, over a distance of roughly three hundred miles, from Palm Beach County[24] (which includes Boca Raton and Palm Beach) up to the northern state line. This called for evacuating the barrier island parts of nine counties along the coast, including the cities of Palm Beach, Melbourne, Daytona Beach, Saint Augustine, Jacksonville, and everything in between. Similar evacuation orders were issued for the coast of Georgia and South Carolina.

- While everybody in Florida waited for Dorian, the Category 5 hurricane decided to park itself over the Bahamas, battering the islands for a solid forty hours, dumping more than thirty inches of rain there while wind gusts of up to 220 mph made water levels surge by twenty-three feet.[25] While the eye of the hurricane was nearly stationary, moving no faster than one mile per hour over the Bahamas,[26] the various computer models crunched new predictions nonstop and reached somewhat of a consensus that the storm would eventually get in gear as a Category 4 hurricane along a northwestern path, hugging the Florida coast so closely that any small deviation from the predicted path could spell disaster on land.

- By noon Tuesday September 3, when the storm did get in gear, it was declared to be a Category 2 hurricane, with maximum winds of 110 mph. While a 110 mph wind is not a small breeze, the downgrading from 220 mph brought a breath of fresh air to those that were bracing for the worst along the Florida coast. Considering that 50 percent smaller wind speeds result in four-times-lower wind forces acting on buildings and other infrastructures, that was indeed a significant relief. Yet, somehow, the hurricane lost more strength than predicted, as if none of the computer models had previously taken into account the fact that, in all of its rampage, a stationary hurricane also eventually cools the ocean, thereby depriving the storm of the warm water that serves as its fuel. The southernmost counties along the coast that had previously declared mandatory evacuation lifted these orders by midday that Tuesday.

- In the twenty-four hours from Tuesday evening to Wednesday evening, the storm moved north, staying offshore far enough to spare Florida from destructive winds or storm surge. The whole coast of Florida was left unscathed, leaving the news media scrambling for something newsworthy to show; the wind and rain had been so banal that the few fallen trees and instances of localized flooding (on a few streets and parks here and there) were given extensive and repeated live coverage throughout the day. Across the state, all curfew orders were lifted and displaced people returned to their homes.

- All in all, as far as Florida was concerned, Dorian amounted to a lot of anxiety and a near-miss. It did mess up some parts of the Outer Banks in North Carolina in the end,[27] and later, as a tropical storm, part of Nova Scotia in Canada,[28] and—of course—it was disastrous for the Bahamas. However, for Florida, the bullseye of the storm for more than a week: nothing.

Note that the NOAA forecasts focus on where the center position of the hurricane will most likely be, which is what everybody is interested to know, but they also provide a "cone of uncertainty" that is intended to illustrate uncertainty in the prediction. This cone is created by a set of circles drawn along the forecast track at the positions where the center of the hurricane is expected to be over each of the next five days ahead. It is significant that the radius of these circles is 198 miles for the forecast five days ahead and 68 miles two days ahead.[29] These radii have been determined by comparing past predictions and actual path of past hurricanes and calculating how large the circles should be if the actual path of past hurricanes was to fall within these circles 60–70 percent of the time. This implies that when forecasting any hurricane path, even the prediction that the eye of the hurricane will remain within the cone of uncertainty is in itself wrong at least 30 percent of the time.[30] Not surprisingly then, on August 27, the 400-mile diameter of Hurricane Dorian's five-day circle of uncertainty encompassed the entire Florida east coast. A day later, it engulfed all of Florida and Georgia's Atlantic coast. From Thursday to Saturday morning, except for the panhandle part of the state, all of Florida was within the cone. By Sunday evening, the entire Atlantic coast of New Jersey, Delaware, Maryland, Virginia, North Carolina, South Carolina, Georgia, and all of Florida (except for the bit of the state south of West Palm Beach) were encompassed within the cone of uncertainty.[31] The path of a drunken sailor might have been more predictable than that.

The meanderings of Hurricane Dorian provide a good example of uncertainties in landfall predictions that ping-pong from one impact point to another—driving everyone crazy in the process because a difference of fifty miles matters tremendously considering that the most destructive hurricane winds develop roughly over a radius of forty miles from the center of the hurricane. Wild fluctuations in predictions of where a hurricane will intersect the coast makes the planning of emergency preparedness and response—to say the least—challenging.

No doubt, Category 5 hurricanes will hit Florida head-on in the future and each strike will produce extensive and heartbreaking damage—it is not a question of "if" but "when." However, as far as human nature is concerned, ten days of flip-flopping predictions on landfall, wind speeds, and storm surge levels often hardens cynical responses instead of raising awareness to the risk—a behavior well captured by the fable of the boy who cried wolf. In fact, this tendency to ignore valid warnings is so ingrained in our DNA that psychologists, philosophers, scientists, and experts from many other disciplines refer to it as the "Cassandra complex,"[32] in reference to the princess from Greek mythology who was cursed with the ability of making 100 percent accurate prophecies that nobody would ever believe. Indeed, the last day of Dorian's visit to Florida was not even over when some of those who had defied the evacuation orders and had stayed home through the storm started posting videos on YouTube, bragging that hurricanes are nothing to get excited about, peppering their "I told you so" commentaries with a pride in not buying into the end-of-the-world herd mentality—like proud conspiracy theorists. These are the folks who do not believe that it is possible to die during a hurricane, because nothing in their life experience has approached this reality—and it is exactly because they have not yet personally experienced the devastating punch of 150 mph winds coupled with fifteen feet of water in the street that they could be the ones most likely to die in a future hurricane.

Obviously, cynics will not miss the opportunity to underscore that, given the inability of meteorologists to reliably predict if tomorrow will see rain or sunshine, only fools would bet the bank on a five-day weather forecast of any kind—even more so when dealing with forces of nature unleashed, like hurricanes. But the point remains that given the debatable accuracy in predicting the path of an extreme event like a hurricane that lasts days and for which the data that goes into the models are measured both directly and remotely on a continuing basis with hundreds of thousands of sensors, what confidence can possibly be developed for hazards at the other end of the spectrum, where the data needed to make predictions cannot be directly accessed and is grossly incomplete?

This is the case for those predicting extreme geological events like earthquakes and volcano eruptions, as they must contend with the fact that the totality of human history spans a mere few thousand years, which is effectively nothing on a geological timescale. For example, the Rocky Mountains have been growing upward for eighty million years.[33] Within that mountain range, the much younger ("only" ten million years old) Grand Teton's summit, at 13,775 feet, is growing on average one millimeter per year,[34] in fits and starts, as earthquakes happen and push it up a notch at a time.[35] By comparison, the Richter scale that quantifies the size of earthquakes (as the length of the wiggle recorded by a Wood-Anderson seismograph) was invented in 1935—less than a century ago.[36] That is about eight hundred thousand centuries too late to get a full set of data and reliable statistics on the size of all the earthquakes that created the Rocky Mountains.

Relatively little is known about these earthquakes. So how much actual data is there to predict when future earthquakes will actually strike and make these mountains rise a bit more in the future? Focusing on the part of the Rocky Mountains located in Colorado, the largest "recorded" earthquake in that state occurred in 1882[37] at the north boundary of the Rocky Mountain National Park, although it was qualitatively "recorded" by written descriptions of the event rather than quantitatively measured by an instrument. Based on interpretation of the information provided in this written record, geologists have estimated this earthquake to have been of magnitude 6.6, with an error of plus or minus 0.6—or, in other words, of a magnitude somewhere between 5.8 and 7.2 when considering the uncertainties. Therefore, since a magnitude 7.2 earthquake is twenty-five times bigger on the logarithmic magnitude scale than a 5.8 one, getting highly reliable predictions based on statistics calculated using data points with that kind of uncertainty is like trying to predict when the drunken sailor mentioned earlier will fall, and the severity of the injury.

SWEEPING HAZARDS UNDER STATISTICAL RUGS

Those who get heartburn when merely thinking about mathematics will likely need some Ultra Strength Tums or Rolaids through the rest of this chapter, but the numbers are not necessarily that overwhelming when put in context.

A major challenge lies with conducting statistics using incomplete data. Imagine a student who got a grade of 72 percent in Professor Bruneau's Steel Design class. If the student is fully satisfied by the knowledge of having passed the course, the story stops there. However, if the student is curious to know if 72 percent is a grade close to the class average, there is no way to know unless a comparison is made. Given that Professor Bruneau has not posted the class average for the exam, the student may ask a couple of friends what grade they received. If the two friends respectively got 88 percent and 66 percent, this would correspond to an average of 75.3 percent. If the student had asked two other friends that instead got 95 percent and 79 percent, the average would have been 82 percent. Big difference. It is not possible to know with certainty what the real class average is from a sample of only three grades. However, if the distribution of all the grades in the course follows a known distribution, such as a normal distribution (commonly called a "bell curve"), it is possible to infer the probability that the real average is close to either 75.3 percent or 82 percent, or to the average of any other three numbers for that matter. Calculating the certainty in these estimated averages—that do not correspond necessarily to the truth—is exactly what Bayesian statistics is all about. As one would expect, the greater the number of data points used in that estimate, the closer the estimate is to the truth. Unfortunately, in this case, Professor Bruneau can attest that the distribution of grades in Steel Design has never followed a normal curve—the actual curve looks more like a three-hump camel than a bell. Therefore, the students would be kidding themselves; all their estimates of the average would be more in error than expected, since they were calculated assuming an incorrect distribution of the real data (yes, it can be tough taking Professor Bruneau's class).

Likewise, imagine that an earthquake fault has produced one major earthquake every one hundred years, like clockwork, with the only thing changing from event to event being the magnitude of the earthquake—say ranging from magnitude 6 to magnitude 8. If it were possible to record every one of those earthquakes over millennia, then it would be possible to perform statistics on that data and determine the average size of the earthquakes produced by that fault, and the probabilities of getting the larger earthquakes. However, if the civilization near that earthquake fault has existed for only two hundred years, then only two earthquakes have been experienced. Can the size of earthquakes expected in the future be deduced from these two earthquakes? Can they be reliable as representative of the average for all events that have happened there over time immemorial? How much more reliable would the predictions be if civilizations had existed near that fault for three, four, or five hundred years? How many events does it take for the average of this small sample to approach the real average?

This is the statistical game played when forecasting extreme events.

Thankfully, some information on earthquakes that have occurred prior to civilization can be inferred from the geological traces they have left. This detective job is what paleoseismology is all about.[38] Clues left by large earthquakes can be found in the landscape (using aerial photographs or laser sensing), or in trenches dug across faults where evidence has been buried by thousands of years of sediment deposits. These findings can provide some useful "guestimates" on past earthquakes, but with nowhere near the resolution provided by modern instruments. Like asking more friends about their grades and getting "somewhere between 65 percent and 85 percent" as an answer—it can help to some degree, with a hefty dose of additional "statistical massaging."

Not surprisingly, that ends up providing predictions that are, at best, somewhat informative. For instance, the Hayward fault is considered to be one of the scariest in the world. According to the USGS, it is a seventy-four-mile-long "tectonic time bomb"[39] due to produce a magnitude 7.0 earthquake any time soon. Paleoseismology studies have identified twelve large earthquakes that have occurred on that fault over the past 2,000 years, from 100 to 220 years apart. The last one of those—and the only one since the establishment of Western civilization—is the 1868 earthquake, more than 150 years ago. Would anybody wish to live right next to that fault? Yes, actually, three million people do (some even literally live on top of it). When the next magnitude 7 earthquake happens there, and the fault ruptures from Oakland to San Jose in California, it will damage one million homes, produce $100 billion in financial losses, ignite firestorms that will burn down eighty thousand single-family homes, leave part of the San Francisco Bay Area without water for up to three months, and leave ten feet of broken glass piled-up in the streets of downtown San Francisco (and presumably people buried below all that glass).[40] The number of deaths and injuries is anybody's guess.

More statistics: The odds of a severely damaging earthquake in the Bay Area over the next thirty years are currently pinned at 75 percent.[41] With those odds, taking earthquake insurance coverage would seem to make sense. Not so easy. After the 1994 Northridge earthquake, even though only half of the $20 billion in residential damage was insured, the massive number of claims delivered quite a blow to the insurance industry. Within a year, 93 percent of the insurance companies in California had stopped issuing homeowners insurance—or severely restricted them—to minimize *their* future losses rather than those of the policyholders. The State had to create a not-for-profit, publicly managed entity, called the California Earthquake Authority, to fill the void.[42] According to the free earthquake insurance premium calculator on their website, the cost to insure an average $1,000,000 single-story home in Berkeley, California, would be $4,000 a year—with a $150,000 deductible! Roughly double that premium to bring the deductible to $50,000—there is no option to reduce the deductible further.[43] With such great deals on insurance premiums, how many Californians have earthquake insurance? Barely 10 percent.[44] Many believe that FEMA will come to the rescue. It actually will, but only to some degree, as FEMA might limit the size of its checks to roughly $35,000 per claim,[45] and may try to keep actual payouts at less than that. [46] With real estate going at more than $700 per square foot in Berkeley,[47] that $35,000 check from FEMA would "cover" fifty square feet of the $1,000,000 house—the size of a cute walk-in closet.

That is for a 75 percent chance of earthquake damage over the next thirty years. If statistics showed that there was a 99 percent chance instead, would it make any difference? Some may argue that statistics—and predictions based on statistics—are not credible. Yet, even if predictions could be 100 percent accurate, would it really matter? It depends on the odds.

UNDERSTANDING AND MISUNDERSTANDING THE ODDS

Would there be volunteers for the following experiment? Imagine a six-lane boulevard in a North American suburb, the type of road where a squirrel dashing across at rush hour is certain to be mashed into a paste between the threads of a tire, yet a road on which hardly anybody drives in the middle of the night. This experiment seeks volunteers to cross the road blindfolded, at 3 a.m., when the chance of being hit by a car is calculated at one in a million. The reward if making it safely to the other side of the six-lane road is a small sum of money, the exact amount determined by increasing the value of the reward until a volunteer comes forward. If $50 is not sufficient to find a volunteer, the reward is increased to $75. Still no volunteers, then $100. At some point, someone will volunteer, because given these odds, there is a 99.9999 percent chance of collecting the money. Definitely better odds of winning than in all lottery games, except that in this case there is also a 0.0001 percent chance of ending up like the squirrel. Now how big would the reward have to be if the volunteer was asked to repeat the crossing every night for a year? How about for fifty years? If it paid $100 per crossing, this would amount to $36,500 per year. Some jobs that pay less than that have a bigger risk of casualties: according to the US Bureau of Labor Statistics, one out of every one thousand fishermen lost their lives in 2017, for an average yearly salary of

$28,310,[48] so being paid $36,500 per year is relatively good for a task less risky than chasing a tuna.

Would the ability to recruit a volunteer change if the same probabilities were explained differently? In a first approach at describing the odds, if there is one-in-a-million chance of being hit during a single crossing, this means that, on average, someone is going to get hit every 1,000,000 crossings. At 365 crossings per year, this implies that this experiment will produce one victim roughly every 2,736 years. Presented this way, it is likely that volunteers could still be found to collect a living wage for one minute of work per day.

Now, there is another way to recast the same information differently. If the volunteer is selected to be eighteen years old at the start of the experiment—always good to keep minors off the street—and asked to retire at sixty-eight years of age, that will amount to a fifty-year career that officially consisted of "crossing the road to get to the other side." What are this volunteer's chances of enjoying retirement? Doing the math, it turns out there is a 2 percent chance this volunteer will be hit by a car once during this period of employment. Is the volunteer still up for it? How about if the number was 10 percent? There might no takers, unless the salary was significantly increased, because the odds are far less attractive presented this way. Yet, this 10 percent is the same as a 1-in-200,000 probability of a fatal crossing, or roughly one victim every five hundred years.

By the way, for those who have an interest in understanding how this number is achieved, the math works out this way:

- If the return period for a hazard is once every 500 years, this is the same as saying there is a 0.2 percent chance per year of it happening on any given year (1 divided by 500 equals 0.002).
- This also means that there is a 99.8 percent chance it will not happen (1 minus 0.002).
- The probability that it will not happen on two consecutive year is 99.6 percent (obtained from 0.998 times 0.998).
- The probability that it will not happen at all in fifty consecutive years will be 0.998 multiplied by itself 50 times, which is equal to 0.904.
- This leaves a chance that it will happen once over a fifty-year period as 1 minus 0.904, equal to a 9.6 percent probability (rounded up to 10 percent in the previous paragraph).

This is how return period and probabilities of exceedance for earthquake levels are determined in buildings codes. It means nothing more than that. There is no implicit warranty that two large earthquakes cannot occur on consecutive years. It is only statistics. A low probability is not an impossibility; there is always somebody winning the lottery, even though the probability of winning is low—incredibly low, for that matter.

When telling someone that there is a 10 percent chance something horrific will happen to a building that will be owned for fifty years, it may spark interest in buying insurance to cover for that risk. Tell that same person that the same building that will be there for fifty years can face something horrific that happens on average once every five hundred years, and watch the yawning. Yet, statistically, it is the exact same risk—only packaged differently. Packaging matters.

Things get further complicated when questioning—statistically again—the confidence level in these predictions. When the beautiful and talented meteorologists that deliver the weather forecast after the news say that there is a 10 percent chance that it will rain tomorrow, that is the calculated odds that it will rain. If that statement is followed by the rare qualification that this prediction is correct 90 percent of the time, that is the confidence level.

When a poll predicts that Savy Pierrot will win 56 percent of the vote during an election, and then states that "the margin of error of that prediction is—say—plus or minus five percentage points, 19 times out of 20," it essentially says different things at the same time. First, it establishes that the poll accuracy is limited because the pollsters only contacted a limited number of people. If 400 people were called Monday evening, then 225 of them indicated they would vote for Savy Pierrot (56 percent). Second, it states that different results could have been obtained if an altogether different set of 400 people had been called that very same evening. There is randomness in that process, so maybe only 195 people (49 percent) in the second group of 400 would have expressed their support for Savy if they had been called. Or 240 (60 percent) in another group. And so on. In other words, the poll result is not an absolute value, because there are fluctuations in the number that would be obtained when calling completely independent groups of 400 people. Statistics can be run to quantify that error. The experiment can be made by calling 100 different groups of 400 people and seeing how much scatter there is in the results. This would provide 100 different results. Using made-up numbers to make a point, say the results from all 100 polls ranged from 150 (37.5 percent) to 325 (81.2 percent) votes for Savy. This huge spread can be quantified by stating that the error is roughly plus or minus 22 percent—a reality, but one that would certainly make the polling company look "unreliable" to its clients. However, out of those 100 poll results, because most of the results are usually bunched up closer together, it may be wise not to count the "outliers." If the five poll results most distant from the mean are discarded, then maybe the spread becomes 204 (51 percent) to 244 (61 percent) votes for Savy, which makes the error become only plus or minus 5 percent. Then, to acknowledge that the outliers have been thrown out, the polling company will state that the results reported are true nineteen times out of twenty (that is, 95 percent of the time because 5 of the 100 polls have been discarded).[49] The polling company can even "tighten" the error if it is willing to say that the results of its polls are true only 90 percent of the time. Does it mean that Savy Pierrot will win? *Probably*, but certainly not a sure thing, as history has often shown—besides, poll results can get significantly less reliable if the group of people sampled is not truly representative of the population, but that is a different story.

The same goes with every prediction in life, from natural hazards to engineered products. If a gizmo purchased from NOTACME is advertised to be effective 99 percent of the time, one would need to be able to collect feedback on personal experiences from a large number of purchasers of the product to be able to perform statistical analysis that would either confirm or discredit NOTACME's claim. Doable, but tough. Demonstrating this with numbers, to be 99 percent sure of the results (the confidence level), with an error no greater than plus or minus 0.25 percent in the results (the confidence interval), one would have to collect data on approximately 1 percent of all customers (the sample size divided by the total number of units). This is equivalent to saying that, by conducting a survey of ten thousand customers out of the one million who purchased the gizmo in question, it can be determined that the gizmo worked as intended for 99 percent of the users, with the caveat that there is a 1 percent chance that this survey gives wrong answers, off by at least 0.25 percent. In short, to have an ironclad certainly that the survey results are 100 percent representative, without any error, with a bet-the-house absolute confidence in the results, one would have no choice but to sample all the clients. And that is when dealing with quantitative data that does not change over time—otherwise, that is another ballgame altogether, like when trying to rely on polls to predict election results.

If all those numbers are nothing but a blur at this point, not to worry. Even the stars of the medical profession get it wrong most of the time,[50] as frequently demonstrated.[51] To illustrate, a group of 160 gynecologists was asked how many women who test positive from the results of a routine mammography actually truly have breast cancer, if: (1) 1 percent of all woman have breast cancer; (2) 90 percent of the women with breast cancer test positive, and (3) 9 percent of the woman who do not have breast cancer, receive a false positive, like a false alarm. In a multiple-choice question with four answers to choose from, 79 percent of the gynecologists (who typically have more than twenty years of schooling) said that women who test positive have a 90 percent chance of having cancer, which is the wrong answer. Based on the above statements, out of 1,000 women who take the mammography, 10 would have cancer, of which 9 would test positive, and 990 would not have cancer, of which 89 would test positive. Therefore, of the 91 that tested positive, only 9 actually had cancer, or 10 percent—maybe a sad reduction in the number of potential clients for all these gynecologists compared to what they thought was the right answer. Fortunately, 21 percent of the gynecologists picked the right answer—although as is the case in multiple-choice exams, each of the four answers had a one-in-four chance (25 percent) of being picked by those who had no clue what the correct answer was. Since 25 percent is more than 21 percent, this means that medical doctors did worse in answering this question than, say, kindergarten kids who would have picked answers randomly—which provides an interesting perspective on the value of getting a second opinion when it comes to medical diagnostics. (If baffled by the math, do not worry: most people need multiple reads to figure this one out).

OBFUSCATING REALITY (AND THIS SUBTITLE)

There are over 450 types of sharks in the world.[52] As incredible as it may seem, at least to those who have never been scuba diving, most types of sharks are harmless. Some are shy, some are curious, some are bottom feeders—like the nurse shark, which some have called the "couch-potato of the shark world."[53] Only about a dozen types of sharks are deemed dangerous, and most human attacks have been by three species only,[54] which includes the great white shark that so often makes the headlines. The same goes with lawyers. There are dozens of different types of lawyers, with practices covering family laws, criminal laws (prosecution and defense), corporate laws, constitutional laws, immigration laws, tax laws, real estate laws, civil laws, employment laws, administrative laws, personal injury laws, appellate laws, bankruptcy laws, malpractice laws, and many more, all with many subdisciplines. Like sharks, only a few are dangerous and known to attack engineers.

The same way many will avoid swimming at dawn or dusk when sharks are known to feed, engineers take precautions to avoid troubled waters. For example, members on a code-writing committee were discussing ways to help engineers understand the intent of seismic design provisions for bridges by explicitly describing in the code itself the expected performance of structures designed by these provisions. One suggestion was to describe the level of damage expected to occur in reinforced concrete columns in terms of crack widths corresponding to various bridge damage levels. If designing a critical bridge for which immediate access by all vehicles following a rare severe earthquake was the objective, then cracks in columns would have to be quite small, contrary to those in less important bridges that could be allowed to suffer more damage. This idea, while valid and effective to help engineers appreciate the expected extent of damage for various design scenarios, was killed by the specter of lawyers walking around bridge columns with measuring tapes following earthquakes, checking crack widths with the resolve of gold-diggers. The approach adopted by the code instead refers to maximum displacements reached during the earthquake—something that occurs only for an instant during the earthquake itself and that only engineers can calculate using computer programs, leaving no evidence after the fact.

Likewise, when dealing with extreme events, statistics is the most effective lawyer repellent. This is absolutely not the reason why statistics are being used, but it is a convenient side effect. Statistics eliminates certainty. No certainty, no guarantees. No guarantees, no broken promises—and thus, no actionable failures. Or maybe this presumption is wrong and, to the contrary, it will lead to more protracted, complex, and convoluted court cases. Time will tell.

THE EARTHQUAKE VANISHING ACT

Magicians love to make things disappear. At the top of his career as an illusionist, David Copperfield even made the Statue of Liberty disappear.[55] Yet, no magician has yet succeeded in making a magnitude 7 earthquake disappear. That exceptional distinction belongs to seismologists, with a helpful dose of statistics. Indeed, thanks to the wonderful world of probability theories, statistics can not only be used to masquerade blatant lies as credible truths, but it can as easily make expected magnitude 7 earthquakes disappear from the engineering criteria used to design buildings. The trick to accomplish this feat is no secret, because it has been published in the scientific literature, but it is admittedly far less exciting than making an elephant disappear.

The trick goes as follows. The "North American Craton," also known as Laurentia, is essentially the massive geological rock shield that covers all of North America east of the Rockies.[56] It is nowhere as prone to generate earthquakes as the more seismically active West Coast of the continent, but large earthquakes happen there too, every now and then, for reasons that are far less understood than along, say, the ring of fire around the Pacific. Within the craton itself, there are some relatively well-defined and active sub-zones where significant earthquakes happen with enough regularity to have an impact on the national seismic maps that are generated for use in the engineering design of infrastructure. However, over the entire craton, it is also recognized that sizeable earthquakes can pretty much occur anywhere, albeit far less predictably and far less frequently. Interesting arguments arise among seismologists on what the largest earthquakes are that could occur there, and how frequently.[57] The Canadian seismologists tasked to develop seismic design maps for the National Building Code of Canada had to make decisions in this regard. To do so, they compiled data on all of the world's cratons having similar geological features, and determined that magnitude 7 earthquakes reasonably can be expected to occur anywhere along the North American Craton,[58] particularly given the evidence that earthquakes of that size in stable cratons have occurred in modern times (for example, the 1988 magnitude 6.7 Tennant Creek earthquake in Australia's Central Craton).[59] They also determined—among many statistics established—that such magnitude 7 earthquakes occurred at a rate of 0.001 per year per 50,700,000 km² (roughly two million square miles), and that earthquakes of magnitude 6 or greater were four times more likely to occur. Given that the North American Craton is roughly 10,700,000 km², this would make it one magnitude 6 or greater every one thousand years. Magnitude 6 earthquakes can produce quite a lot of damage, as demonstrated by the magnitude 6.3 Christchurch earthquake that killed 185 people in 2011 in New Zealand—of all places, one country where earthquakes are expected to occur and that is, generally, better prepared for that eventuality than many others.

Note that other seismologists performing similar studies would typically come up with different numbers—sometimes more critical, sometimes less. When it comes to the design of critical facilities, such as nuclear power plants, offshore platforms, and nuclear waste disposal sites to name a few, where the design earthquake is intended to be the one with a ten-thousand-year return period, these debates are important. This lack of consensus among experts is not uncommon or unsettling when seismological models are created—it is accounted for, and defined as, "epistemic uncertainty."[60] Interestingly, seismologists have been brilliant and deliberate when expressing the differences in predictions obtained from their various models, by plotting their results on log-log graphs, rather than "normal" diagrams. In a log-log graph, the distance between each major tick-mark corresponds to a tenfold increase. In other words, in such a log-log graph, the major tick marks along the vertical axis correspond to 0.1, 1, 10, and 100, instead of 1, 2, 3, and 4. This makes a 100 percent disagreement between, say, the numbers 10 and 20 graphically look like only a 14 percent disagreement[61]—which is quite a clever way to save face when comparing models that predict such dramatically different results.

However, even such variations in return period estimates are not significant when it comes to the earthquake vanishing act. What matters is the size of the area over which the earthquake is "smeared," and the distance from a potential epicenter to any point on the map where engineered infrastructure of interest is located. This is because anybody can survive a magnitude 7—or even a magnitude 8—earthquake, provided it happens far enough from where one stands. In fact, as mentioned earlier, there is on average of one magnitude 8 earthquake occurring somewhere on the globe every year—in most years, nobody hears about it because it happens far away from civilization.

Coming back to the North American Craton, when considering that a magnitude 7 earthquake can occur somewhere in the North American Craton, what matters is not the temporal dimension, but the physical dimension. The chances that any specific location on the massive craton will be close to the epicenter when it happens are so low that it effectively makes no difference whether it is considered or not when developing the probabilistic seismic maps of, say, the National Building Code of Canada. In other words, the magnitude 7 exists, but it is its impact that vanishes/disappears.

As a result, if one thinks of a magnitude 7 earthquake that happens every four thousand, one thousand, or even five hundred years as a dart thrown on a map of the Canadian part of the North American Craton, (which is effectively all of Canada minus the Atlantic provinces and the land from the Rockies to the Pacific), the chances are greater that it will happen in the middle of nowhere, or in a place where there is not much infrastructure to damage. Canadian moose, geese, and beavers will get shaken up, but nobody will care—in a sort of earthquake engineering version of the metaphysical "If a tree falls in a forest and no one is around to hear it, does it make a sound?"

However, if the dart unfortunately bullseyes on Toronto, seismologists will be doing a lot of hand-waving and fancy footwork trying to explain to the Torontonians looking at their damaged buildings, bridges, arenas, and infrastructure, that they were simply "sheer out of luck"—fubar style—because even though the dart had a negligible probability of ever landing there, it was not a zero probability. Such is the perverse effect of messing with statistics. Like other events with an insanely low probability of happening—be it dying from an asteroid impact or having the Detroit Lions win the Super Bowl—an insanely low probability does not mean it cannot ever happen. Do not bet on it, do not wait for it, do not prepare for it, but do not look for culprits when it happens.

To repeat, absolute certainty does not exist. Everything is possible, only with different probabilities of occurrence. Severe earthquakes are rarer than smaller ones; winning $1 million does not happen as frequently as finding a coin in the street.

At the same time, not all low-probability events have the same consequences. Parking the brand-new car in the driveway to see it crushed by a falling tree during a wind storm is not as bad as being hit by lightning, which itself is not as bad as losing everything and all loved ones in a collapsed building.

It is all a question of "Risk."

ON THE DISASTER TRAIL

Sports as a Proxy—Football for All

It was while watching a documentary on World War I that it jumped out at me. There, on the scratched black-and-white film, arms waving out of every door and window, the train was leaving the station with a bunch of exuberant kids on their way to play and win the ultimate game. Same crazy optimism on the cruisers' decks as they left port.[62] As an ice-hockey historian described it in part 3 of *Hockey: A People's History*[63] (a ten-hour Canadian documentary that only Canadians might care to watch),[64] going to World War I was better than any sports match; it was an irresistible drive to be part of the home team and fight side-by-side on a great adventure.

In any competition between two teams of even strength, the odds of victory are roughly 50 percent. Yet, all that youthful exuberance was about the thrill of victory. In all that excitement, few seemed to acknowledge that in the game of war, there is also the agony of defeat—as can be seen strikingly emphasized in the 1970s *ABC Wide World of Sports* intro.[65] The recruits on their way to fight in Europe acted like a mob of sports fans on their way to a football game, cheering for the home team and expecting no less than a full and dominating victory.

In that moment, watching the old war footage, it became clear that a characteristic of youth is an uncanny ability to see the world only from its own point of view. Undoubtedly, German trains were likely leaving home with the same testosterone-laden drive to head-butt other rams, expecting no less than a full and dominating victory.

Fortunately, nowadays, all that head-butting energy can be released through organized sports, where the misguided belief that a win is guaranteed can be crushed without life loss or—in most cases—debilitating injuries. Even better, professional sports can provide that same outlet to passionate sports fans while keeping them safer than players.

From that deep observation, it is tempting to formulate the hypothesis that the desire of a nation's youth to go to war is inversely correlated to that country's professional sport infrastructure. In other words, lots of professional football, baseball, basketball, soccer, and hockey teams, equals lots of opportunities for frenzied youth to spend their rambunctious energy and learn that the home team can lose big time on any given day against any given team—maybe suffering some emotional pain, but without losing limbs or life to learn that lesson. Conversely, the countries whose citizens are most eager to become soldiers—or even terrorists in the absence of a national army—sorely lack professional sports.

Therefore—clearly outdoing all the wishes of past Miss Americas in the quest for world peace—I hereby postulate that to demilitarize the world and reduce the risk of wars, it is essential to establish multiple professional sports leagues in every country on earth, and use conscription to draft all kids into amateur sports teams. Dress everybody in well-padded uniforms and let them bang each other to exhaustion on a football field or an ice rink or equivalent—a well-controlled warring environment that is safe for the planet and its citizens.

Problem solved! And maybe the first Nobel Peace Prize awarded for simultaneously enhancing both the safety of the world's citizens, and their fitness.

Life's Casino

RISK TOLERANCE AND RISK AVERSION

When a contractor was hired to fix a small leak in the roof of the NOTACME Corporation headquarters building, it did so expeditiously. However, as work started, it was found out that the roof underlayment contained asbestos. As far as the contractor was concerned, this was not a problem but a blessing, because a site condition not mentioned in the original contract can be handled by a change order—in essence, extra work at a higher profit margin, adding the special protection necessary to deal with asbestos. When the employees learned that asbestos was present in the roof, they requested that the entire roof be replaced. Taking this request into consideration and investigating, NOTACME learned that removing asbestos from a roof is one of the most expensive asbestos abatement operations, due to access and containment requirements, and the cost estimate for that operation came up at $120 per square foot.[1] For NOTACME's modest 100' × 100' building, that added up to $1,200,000, plus disposal fees, permit fees, and other ancillary expenses. In the process, NOTACME also learned from government agencies that a roof containing asbestos is not hazardous, that the best course of action is to leave it in place because it does not affect conditions inside the building,[2] and that asbestos is not a banned substance in the United States.[3] Besides, NOTACME learned that, incredibly enough, asbestos is a product found in rocks and that natural erosion of those rocks releases a small amount of asbestos fibers in the air nearly worldwide.[4] NOTACME gladly informed its employees that the roofing material presented no risk. Yet, no matter how many facts NOTACME provided, many employees found plenty of websites stating that asbestos can cause asbestosis and different types of cancers (mesothelioma being one),[5] and that one fiber of asbestos in the work place is one fiber too many. The fact that the legal system has typically found non-persuasive the claim made in many lawsuits that a single inhaled fiber of

184

asbestos is sufficient to cause cancer,[6] the NOTACME employees took this rejection of the "one-fiber theory" as clear evidence that the justice system is in bed with big business. That asbestos roof had to go.

Three months after the discovery of asbestos in the roof and no action from the NOTACME management, the employees met with the management. They threatened to sue the company and go on strike until the issue was resolved. The crisis had reached its climax, and NOTACME had to budge: given the alternative of millions in lost revenues due to work stoppage, it agreed to hire the contractor to replace the roof at once.

The employees were ecstatic. The risk of death by cancer caused by asbestos in an unsafe workplace was at once removed. All went to celebrate this victory at the nearby bistro. Congratulations, backslapping, cheering, fist pumping, and singing were sustained for hours by profuse amounts of wine and beer. By midnight, the party was over. Some staggered to their cars, struggled to put the key in the ignition, lit up a cigarette, and drove home completely plastered. Others did the same on their motorcycles— wearing a T-shirt, shorts, sandals, and no helmet.

What is wrong with that picture?

HAZARD VERSUS RISK

Before going any further, we must clarify the confusion that often arises between hazard and risk, which are often used interchangeably or listed as synonyms in dictionaries. In essence, to borrow a definition from experts in occupational health and safety,[7] a "hazard is any source of potential damage, harm or adverse health effects on something or someone." As such, a knife, electricity, a wet floor, and a coconut in a palm tree are all potential hazards. What ensues from this is that a "risk is the chance or probability that a person will be harmed or experience an adverse health effect if exposed to a hazard." Avoiding the risk requires not sleeping under a palm tree (although the claim that falling coconuts kill 150 people every year is apparently an urban legend and only a few actual documented cases of such an exotic way to die apparently exist).[8] If the relationships with the hazards are handled properly, the risks can be managed.

Some risks are, in theory, easy to manage: Who would use a hair-dryer in a shower, dive in shallow waters, light a cigarette next to a gas pump, or text while driving? For most people, the risk is obvious and therefore avoided—others will end up upside down in a ditch, crushed by an airbag, cell phone firmly in hand. However, there are many instances where risk is not an easy thing to assess and to manage. In fact, this is so true that "risk manager" is a professional career in itself and its practitioners include (and engage) experts from various disciplines. There exists a Society for Risk

Analysis, a Risk Management Society, a Risk Management Association, a Professional Risk Manager International Association, a Global Association of Risk Professionals, a Federation of European Risk Management Associations, and many more international, national, local, and discipline-focused groups that tackle the broadest imaginable range of risks. Every possible risk has been scrutinized, researched, dissected, and quantified. Financial risks, health risks, societal risks, nuclear war risk, and so on. Multiple studies have continuously attempted to explain how biases, beliefs, social and cultural factors, trust and credibility, and many other factors affect the perception of risk by individuals, and why that perception often (if not always) differs from that of experts. All done in a rigorous, scholarly way, for the benefit of academic and professional scrutiny—which is infinitely more credible, reliable, and actionable than the (hopefully more enjoyable) approach taken here.

Given that risks are part of nearly all human endeavors, there exist many context-dependent definitions of risk. It has been said that past efforts by the "definition committee" of the Society of Risk Analysis to come up with a single, consensual, exact, and all-encompassing definition of risk have failed.[9] This should not be surprising—as the saying goes, if you need something done now, do it yourself; if you can wait a few weeks, delegate the task to a trusted employee; if you are willing to wait forever, ask a committee to do it. So, for now, to keep things simple (manageable?) and yet reasonably accurate and general, suffice to say that, "risk is the possibility of an unfortunate occurrence"—to borrow one of the seven definitions of risk provided in the glossary of the Society for Risk Analysis.[10] In the present context, the "unfortunate occurrences" and interest are at a massive scale—not of the "toe stubbed on the bed leg" type, but rather of the "massive casualties and losses from a disastrous event" type. More specifically, the focus here is on risks that have a low probability of occurring, but that can have massive consequence when they do—typically called low-probability, high-consequence events in the academic literature—because academics are notoriously famous for never coming up with catchy names for whatever they study.

To appreciate how risk is handled for low-probability, high-consequence events, it pays to look at how human nature deals with risk in a general sense, as it reflects trends ingrained in human nature. In all human activities, some risks are avoidable, some not, and it is sometimes amazing to see the risks that are taken for the pleasure of it.

RISKS FOR REWARDS

Everybody takes risks for rewards. Why? Because of the rewards, obviously.

Risk is always a matter of perspective. Most people would not think of jumping from an airplane with a parachute for fun—but some do. Likewise, most of the roughly forty thousand members of the US Parachute Association (USPA) would not think of BASE jumping for fun—but some do.

BASE jumping is the extreme sport that consists of parachuting off a building, a bridge, a tower, a cliff, or any point that seems too close to the ground ("BASE" stands for "Building, Antenna, Span, and Earth," although some have claimed it should stand for "Ballsy Awesome Suicidal Exhibitionists").[11] The USPA reported nineteen deaths from over three million "normal" parachute jumps in 2012. By comparison, statistics show that one in sixty BASE jumpers typically die from that sport[12]—which makes it unlikely to become an Olympic discipline anytime soon. Many of those still alive are typically not deterred by the stories they can tell of hours spent in intensive care, neurosurgeries, or at funerals.[13] If to the normal parachutist, the combination of adrenaline, goose bumps, butterflies in the stomach, increased blood flow, perspiration, and constricted blood vessels produced by freefall[14] is a rush worth the jump, then BASE jumpers must be adding a spiking fever to the cocktail. Statistics on accident rates simply do not matter.

Although the comparison with extreme sports is—well—extreme, the same could be said for most human activities. The fifty-five million people who enjoy the breath of fresh air and awesome views that come with alpine skiing[15] are not deterred by the numerous skiing-related injuries that occur every year. Interestingly, those who enjoy a breath of fresh air and the more mundane lower altitude views that come with bicycling face the same risk of severe injuries as those practicing alpine skiing.[16]

Or, for an even more mundane example, the thirty unprovoked shark attacks that have occurred in Florida every year[17] have not prevented some of the more than 125 million tourists visiting Florida[18] from making 810 million day visits[19] to its beaches. Likewise, one would be hard pressed to attempt convincing an avid snowboarder to abandon the slopes to play monopoly instead based on the fact that only 1 person in 100 million per year dies while playing board games, compared to one death per 2.2 million snowboarders.[20] Relative safety is not what attracts people to philately, coin collecting, or gardening—although the latter can be deadly to those allergic to bee stings and contact with certain plants.

All activities have risks, and free individuals undertake activities in exchange for rewards. Whether the rewards consist of money, medals, grades, food, fun, excitement, self-esteem, love, acceptance, titillation, power, casino chips, or Pokémon eggs, does not matter. If the rewards are deemed to be worth it, nobody calculates risks—or cares about risks. It is exactly because people are willing to take risks for rewards that syphilis, gonorrhea, and AIDS continue to exist. That is, people are willing to tolerate high risks as long as the risks are self-imposed.

When risks are imposed by others, strangely, a completely different zero-tolerance policy seems to apply. Nobody cares about risk calculations and statistics: zero risk becomes the only acceptable norm—which is never a problem when the dollars to reduce that risk to zero will be paid from someone else's pocket. This is the mindset that has, in part, given rise to the asbestos abatement industry, with $3 billion in revenues per year in the United States alone.[21] The same mindset that has planted a poisonous kiss on the cheeks of the Keystone XL pipeline[22] and handicapped the nuclear industry in the United States. Yet there is a cost. A study reviewing the US federal government expenditures on various regulatory programs enacted in the mid-1980s calculated that the Occupational Safety and Health Administration regulations of asbestos cost $89.3 million per life saved.[23] This was cheaper than its other regulations on formaldehyde, which ran at $72 billion per life saved, but significantly more expensive than its regulations on concrete and masonry construction, which cost a measly $1.3 million per life saved.[24]

When medical doctors routinely start their voice recorders and recite the boring list of possible complications, infections, debilitating conditions, and death risks—and their probabilities—when consulting with a patient before performing any surgery, they effectively make it part of the contract that it is the client that wishes to assume all these risks in hope of the rewards. Although it should be clear from the outset that a breast enhancement surgery or vasectomy usually does not happen without consent, the medical professionals apparently wish to make it absolutely clear—and on record—that they are only willing to perform the surgical procedure (for a small fee) because the patient begged for it, irrespective of all the stated risks and dangers.

So, to recap, there is a definite double standard when it comes to risk. Individuals have a definite willingness to take risks—even high risks— when these are self-imposed. Some will even resist—in the name of life, liberty, and the pursuit of happiness—well-intentioned attempts by others to impose measures intended to reduce the high risks of some activities. They will actively fight against laws that would make motorcycle helmets mandatory or that ban smoking in public places. Yet the same individuals have a zero tolerance for risks imposed on them by others and will not balk at the prospect of massive costs to abate minuscule risks when those costs are borne by others. The golden rule does not apply here. The new twist says, "It is OK to do unto others as you would NOT have them do unto you."

One should not confuse things, as generalizations can sometimes be misunderstood. Without doubt, activism is not necessarily bad. A lot of it serves worthy causes. If the 10,000 lives lost per year in drunk-driving crashes[25] were only those of the intoxicated drivers, few people would care. Except for the poor parents of teenagers (because teenagers are apparently victims of an evolutionary flaw that has made the development of sound judgment seriously lapse far behind that of their physical ability to take self-destructive risks), most sober people would not shed a tear for dead drunk drivers if they were not exposed to any risk by them. However, since thousands of innocent people are killed every year by alcoholics, this high-probability risk justifiably warrants being eliminated. In the foreseeable future, emerging technologies will likely eliminate that risk,[26] and devices intended to prevent powering of an engine by anyone who had too many mojitos will become standard equipment—possibly as cheap to implement as alternating wipers, power windows, and pine air freshener trees hanging from the mirror. Even then, though it will be a positive outcome that thousands of senseless deaths will have been eliminated, the risk of death resulting from drunk driving accidents will never be zero.

Can anything be?

ABSOLUTE SAFETY IS A MYTH

As has often been said, life is the only disease with a 100 percent casualty rate. Everything else is less certain. Although one may find solace in the belief that everything is ordered for the greater good, all the evidence is to the contrary. Life is a chaotic process. No computer can predict what will happen next. Things get complicated by the fact that the behavior of some types of chaotic systems can be affected by the prediction itself. If a renowned climatologist looks at a blue sky and predicts that it will rain in ten minutes, this forecast will not suddenly make gray clouds appear—literally—out of the blue. On the other hand, if a renowned and highly respected economist predicts that the stock market will drop by 10 percent within a few hours, it might stir a panic that will affect the market instantly and precipitate its dive. The weather and stock market are both chaotic systems, but the first one is insensitive to predictions while the second is not—although the volatility in response to the prediction depends, of course, on the credibility of the person making the prediction and whether that person was dead serious or dead drunk when making the prediction.

In light of the challenges in predicting what a chaotic system will actually do, anyone promising absolute safety is either sadly deluded or an exceptional liar. Absolute safety is a pie in the sky.

The fact that an offshore platform is designed to resist earthquakes and iceberg collisions that are deemed to happen on average once every 10,000 years[27]—and 100-foot-tall waves too—does reduce the risk significantly and makes it possible to sleep at night. Yet that is not absolute safety. Which is in part why oil producers evacuate staff from their offshore platforms when it is predicted that they will be in the path of an incoming hurricane.[28]

And, of course, even if none of the natural hazards could ever destroy a building, or a bridge, or an offshore platform, or a nuclear power plant, there will always remain the ultimate risk: an operator error, thanks to an overworked, distracted, or distraught employee—or one of the Homer Simpsons of this world.

WHAT IS A PREDICTION WORTH, ANYHOW?

A lot of research dollars and efforts are invested in developing the ability to predict where and when earthquakes, hurricanes, tornados, and other extreme events will strike. These are valid endeavors and valuable tasks to better understand the world and what is at stake. However, what if, thanks to massive scientific breakthroughs and the astounding insights of geniuses, it became instantly possible to predict with 100 percent accuracy when and where earthquakes will strike. Would the world be better off? How would humans react—or how would everything that has been said so far change? The answer, as always, is "it depends."

What would most people do if an official warning came that "A magnitude 8 earthquake having its epicenter right under your feet will strike in exactly one minute"? What? Say that again. "A magnitude 8 earthquake having its epicenter right under your feet will strike in exactly fifty-five seconds." Damn it! The clock is ticking.

As a reflex, those living in a bungalow might throw themselves (and the kids) out the door at once. Once out, some will wonder if they have enough time left to run inside and grab their laptop or some other prized possession. Not having prepared anything ahead of time for such a rare occurrence, or given the topic any thought until now, the daredevils storming back inside are more likely to come out with the goldfish bowl and wedding photos than precious paperwork or survival kits. That is all nice and dandy, but what about those living on the tenth floor of a building, on the fifteenth floor, on the twentieth floor, and so on? Panic? Jump out the window? Brace in a doorway? Duck under a desk? Freeze in place and hope for the best? These are all things that can be done within the first few seconds of shaking, when one realizes that an earthquake is happening, without any warning. So, for many people, a one-minute warning is not a particularly useful prediction. It may allow time for the dentist to stop drilling, for trains to slow down, for people to exit elevators, for some operations and facilities to initiate safe shutdown,[29] but it will not change the outcome for things at risk of failure or collapse.

How about: "A magnitude 8 earthquake having its epicenter right under your feet will strike in exactly one hour"? What? Say that again. "A magnitude 8 earthquake having its epicenter right under your feet will strike in exactly fifty-nine minutes and fifty-five seconds." Damn it! The clock is ticking. Some will have time and the presence of mind to walk into an open field and ride out the waves safe from any falling hazard or collapsing buildings or power lines—that is, if there is an open field nearby. What about those downtown? Panic? Not having prepared anything ahead of time for such a rare occurrence, nor given the topic any thought until now, who knows if their building can survive a magnitude 8? Should they "ride it" in place or should they rush to the streets or to the subway? How many people can be packed onto the baseball stadium field, and will the gates be locked? A slightly more useful prediction but, again, maybe not for everybody.

How about: "A magnitude 8 earthquake having its epicenter right under your feet will strike in exactly one day"? What? Say that again. "A magnitude 8 earthquake having its epicenter right under your feet will strike in exactly twenty-three hours, fifty-nine minutes, and fifty-five seconds." Damn it! The clock is ticking. Pack the car? Not having prepared anything ahead of time for such a rare occurrence, nor given the topic any thought until now, many will rush to the highways, get stuck in massive traffic jams, at a standstill for hours, possibly running out of gas, not sure where hotel rooms will be available. Is this a more useful prediction?

How about: "A magnitude 8 earthquake having its epicenter right under your feet will strike in exactly one hundred years, to the second"? What? Say that again. "A magnitude 8 earthquake having its epicenter right under your feet will strike in exactly 36,524.24 days, five hours, fifty-nine minutes, and fifty-five seconds." Oh. No need to worry then.

So where is the "sweet spot" between one hour and one hundred years where the prediction would be optimum in terms of making a difference?

Two days? Mayhem. Massive evacuation traffic jams again—except for those cowboys determined to ride the storm and guard the castle.

One week. Ghost town again, minus the cowboys.

One month. More time for packing, more orderly evacuation, better braced cowboys. Yet, nobody has time in a month to do much, if anything, to change the earthquake resistance of the buildings and infrastructure. Some of these structures will survive the shaking and some will collapse, depending on their year of construction, type of structural system, and care taken during their design and construction to address the issue of seismic survivability. Lives will be saved—except for cowboys crushed by debris—but the extent of damage will remain the same as if there had been no prediction at all. A small victory, but no less a disaster.

How about thirty years? That should be enough time to strengthen/ retrofit the infrastructure, shouldn't it? Would that make any difference? Maybe. Financial buzzards could start factoring future damage into present dollar values and build that into the cost of mortgages and loans, but that would not prevent the disaster. Hard cash would need to be spent on upgrading the infrastructure to ensure its satisfactory performance during that predicted earthquake, so many of those considering making these investments will have tons of questions that would first need to be answered. Is this really a 100 percent sure prediction? Then, predicting magnitude is fine, but what about the amplitude of the design parameters that engineers use in their calculations; these can vary by a factor of ten among a bunch of earthquakes having the same magnitude. How long will it take to do any work to strengthen the infrastructure and will there be a shortage of labor if everybody waits to the last minute before doing something about it? What is the point of investing money in something like that when one could die from so many other causes over a thirty-year period?

Besides, why worry about damage from one earthquake when, even in the absence of any threatening hazard, the American Society of Civil Engineers' Report Card for America's Infrastructure[30] gave the nation's existing infrastructure a grade no better than a solid D+, thus deserving of a serious spanking for lack of effort. This report called for trillions of dollars in investment to bring the existing deficient infrastructure to a safe and acceptable level, and it has been making this call for decades with only timid action by all successive governments. Twenty percent of all highway pavement is in poor condition and $160 billion is wasted every year in time and fuel on badly congested roads;[31] $45 billion is needed to repair more than two thousand aging dams that risk failure;[32] $123 billion is needed to repair the more than fifty-six thousand deficient bridges across the country;[33] $1 trillion is needed to address the water needs of the nation while fixing the two hundred forty thousand water main breaks that waste over two trillion gallons of treated drinking water each year;[34] and many more problems exist with levees, energy production and transmission, solid waste, hazardous waste, schools, and so on. For decades, all of this has been known, and the funding has not been sufficient to address these pressing issues to the extent the problems warrant. So why would there be sudden action taken to do something about one earthquake predicted to happen at some point (any point) in the future? When people are perfectly happy to drive cars with four patch-covered tires that have no threads left, they probably could not care less about a flashing warning light on the dashboard indicating an airbag failure; the risk of an accident is not at the forefront of their mind.

From that perspective, the concept of "let the earthquake clear up the place and restart brand-new" does not seem too bad an idea. Aren't earthquakes Black Swans anyway?

ON THE DISASTER TRAIL

To Be or Not to Be (in a Crane)

In construction, a large number of temporary structures are used. In particular, some relatively "flimsy" bridges can be used for traffic detour while an existing bridge is being replaced, or a tall tower crane may be used on a high-rise construction site. Typically, these temporary structures are supposed to be there anywhere from a few months to a couple of years—although I am aware of some temporary structures that have been "temporary" for more than seven years.

In the early 1990s, I had casually mentioned as part of a technical presentation that temporary structures should also probably be designed to resists earthquakes—at the time, most were not. Some people were rather upset by that comment. They argued that because temporary structures only exist for a short period of time—say a year—there is only a remote probability that they will be struck by an earthquake. That certainly is one way to look at it. From that perspective, nowadays, most of the time (but not always), in many countries (but not all), temporary structures are simply not designed to resist any earthquake forces,[35] or are designed to resist an earthquake that has a 10 percent chance of being exceeded during the existence of the temporary structure—say a year, instead of the fifty-year span considered for regular structures. There is an engineering logic in computing the size of the design forces based on the "lifespan" of the temporary structure—if looking at it from the perspective of each temporary structure on its own. This is like recognizing that smoking is dangerous but saying that the risk of getting cancer by smoking an entire pack of cigarettes once, on a single day, is infinitesimally low.

However, looking at it from another perspective, if a crane operator is hopping from one tower crane to another, from project to project, one year at a time, this adds up to an entire career spent in temporary structures. Who would like to spend a lifetime working in the elevated cabin of a tower crane knowing they are not going to be safe during an earthquake? This is where the packs of cigarettes, day after day, add up over a lifetime.

When Taipei 101 opened in 2004, at 1,671 feet to the top of its spire, it was the tallest building in the world—and remained so until 2010.[36] One of Taipei 101's lesser-known features is that the restrooms in one of its eighty-fifth-floor restaurants have floor-to-ceiling windows that force users to admire the view while standing at the urinals—thankfully, the surrounding buildings in sight are ten- to thirty-story dwarfs. Another lesser-known fact not mentioned in the Taipei 101 tourist brochure is that on March 31, 2002, an earthquake struck Taipei. At that time, construction was in the process of erecting the steel frames between the fifty-third and fifty-sixth floors. One of the four tower cranes in use at that level failed and dropped 750 feet to the ground. Five people were killed, including the crane operator.[37] That is the curse of risk calculations for temporary structures.

(Note: Prompted by Japan following the 1995 Kobe earthquake during which many tower cranes suffered damage,[38] the International Organization for Standardization published a new standard in 2016 (ISO 11031) for the design of tower cranes to ensure that they will not collapse and endanger the public during severe earthquakes.[39] Progress is a process.)

Black Swans

PREDICTIONS THAT FAIL

As the saying—often attributed to various authors, humorists, quantum physics experts (Neils Bohr), baseball gurus (Yogi Bera), and many others—goes, "It's hard to make predictions, especially about the future."

All the might of statistics and all the wizardry of risk managers is useless when it comes to predicting something that has never occurred before. For instance, it would not have been possible to calculate the odds of having a bunch of terrorists simultaneously hijacking commercial airliners to crash them into national landmarks. As early as 1998, the CIA had warned the president that al-Qaeda terrorists were planning to hijack commercial airplanes.[1] Hijacking was not a novel idea and, as mentioned earlier, the fact that some groups had planned to crash airplanes on landmarks was already known, but nobody foresaw the exact set of events that unfolded on September 11, 2001. Yet, even if someone had imagined such a scenario—as a real possibility rather than as a bad screenplay for an even worse movie—any person taking a passing interest in the idea could not have calculated the probability of it happening. Nor could anyone have calculated the odds that this would have led to the invasion of Iraq.

This kind of unpredictable event that is extremely rare and of catastrophic consequences is typically called a "Black Swan Event."[2] It can also be recognized by the fact that following the black swan event itself, armchair quarterbacks of every ilk offer their expertise to explain that it was all foreseeable and predictable—as there is never a shortage of geniuses willing to share predictions that are solidly anchored in hindsight and nothing else.

Throughout Western civilization, nobody could conceive that swans could be any color other than white. It was simply impossible. This remained an unalterable truth until 1697 when black swans were discovered in Western Australia. There it was. Mathematicians (e.g., Karl

Popper)[3] and philosophers (Hume)[4] have used this historical event to emphasize that a single observation unpredicted by the prevailing theory (or belief) is sufficient to invalidate that theory (or belief). The global financial crisis of 2007–08 is possibly the most famous contemporary black swan event, both because it occurred at a time when the term was being popularized in a bestselling book about the fragility of financial markets ("The Black Swan: The Impact of the Highly Improbable," by Nassim Nicholas Taleb),[5] and because nobody from the financial elite saw it coming. All banks, investment firms, financial institutions, and even—as Taleb himself likes to remind everybody[6]—Nobel Prize winners in economics were caught with their pants down. As unexpectedly as for the terrorists that had used airplanes as missiles, nobody anticipated that top Wall Street investment firms would end-up being accused of having deliberately defrauded their own clients[7]—and nobody predicted that all, except one,[8] of the top executives that were part of the worldwide financial meltdown would evade jail and get rewarded with millions of dollars in bonuses.[9]

As bad as these events were, there can be worse. The problem arises when experts in many domains develop forecasting models entirely based on past data, and rely exclusively on these models to make decisions. Some have called this folly—or smug arrogance—because, by definition, past data does not include things that have not happened yet. As such, this reliance on highly sophisticated forecasting tools is misleading their users into a false sense of control and security, thereby increasing their exposure and vulnerability to the black swan events.

The plain truth is that some events are unpredictable, mostly because people are unpredictable. Even the most cautious parents who had the foresight to put plastic protectors over all electrical outlets to prevent their kids from filling them up with playdough, will be surprised when their little devils wade in the toilet bowl, put the cat in the washing machine, or try to catch a skunk—creativity that they had not foreseen in their wildest dreams.

Given that black swan events of catastrophic consequences cannot be predicted, the only sensible way to prepare for their occurrence is to build robustness into the systems they can affect. Trusting forecasting tools based on past data will provide no protection—and, in fact, provides a false sense of security that can lead to reckless actions, that can in turn increase exposure and vulnerability. The only way out is to assume that something terrible and unforeseen can happen and plan accordingly.[10] As the saying goes, hope for the best but plan for the worst.

Unfortunately, almost the opposite often happens, where people plan for the best and ignore that the worst could even happen. For example, in an attempt to obtain authorization to operate the Shoreham nuclear power plant it had constructed, the Long Island Lighting Company decided to conduct a drill to demonstrate that evacuation of the population around a set radius from the plant was possible following a nuclear incident. The Three Mile Island nuclear accident had occurred somewhat halfway during construction of the Shoreham project, so it had become a political hot potato and the target of much grassroots opposition. With the local emergency response agencies refusing to participate in this evacuation simulation, the Long Island Lighting Company paid 1,800 of its employees to drive buses and tow trucks, and play the role of traffic officers and emergency responders.[11] The Long Island Lighting Company plan thoroughly documented that all the drills it conducted worked smoothly and showed the effectiveness of their evacuation plan. However, serious doubts were expressed throughout the hearings for the nuclear power plant as to whether the bus drivers paid to participate in the drills would as eagerly dash into the radioactively contaminated zone during a real emergency, or would even bother to show up at all. What bus drivers would rationally rush to pick up children from schools within a certain radius of the power plant, and drive them to predesignated safe havens, rather than rush to take care of their own kids and leave the world to fend for itself—all for nothing more than minimum wage? Somehow, the state of panic that will exist when trying to evacuate the population where the air has been fouled by a radiological incident cannot be quite replicated by an evacuation drill executed with pure air and no stress. A beautiful sunny day is never quite as worrisome as a day when rain is washing down radioactive particles. The county and the state rejected the evacuation plan,[12] and as no evacuation plan was ever approved, the $6 billion plant had to be decommissioned and dismantled.[13]

One unfortunate problem with the black swan theory is that it has been a victim of its success. The fate that awaits every highly popular concept is that it eventually becomes a buzzword—meaning that it becomes profusely misused to the point of becoming meaningless. Even more so, nowadays, when every person that contributed to creating a crisis of inordinate proportion will rush to call the event a black swan, as if to disengage from any responsibility. This fundamentally misses the key point in the definition: a black swan event is something that no sane person could reasonably have expected to happen in real life. Negligence, incompetence, ignorance, and lack of foresight cannot hide under a coat of black swan paint.

As such, the COVID-19 pandemic of 2020 is definitely not a black swan event, contrary to what some have claimed.[14] Governments and corporations worldwide had been warned repeatedly of the risk of a pandemic and of how rapidly it could spread. Not warned by unknown scientists toiling in obscure labs, but warned by some of the most reputable national and international agencies and organizations, as well as by high-profile speakers with international name recognition.

None other than Bill Gates, founder of Microsoft and one of the richest men on earth, in a 2015 TED talk (that can be seen on YouTube[15]) warned that the world was not ready to face a pandemic. This was in the aftermath of the Ebola outbreak that killed more than ten thousand Africans. The world had been lucky to escape a global Ebola pandemic because a person infected cannot spread the virus before developing symptoms, and when the symptoms appear, they are brutal. Ebola triggers a cascading meltdown of the immune system, blood vessels, and vital organs. The virus kills on average 50 percent of those infected,[16] which is, ironically, what makes the heroic task of confining an Ebola outbreak possible. During a security conference in Munich in February 2017, Gates further warned that within ten to fifteen years, a rapidly propagating pathogen would kill thirty million people in less than a year.[17]

Beyond philanthropic billionaires, many other organizations forecasted a world pandemic and its consequences, and stressed the need for preparedness. For decades.[18] In all serious analyses of the situation, there was never any doubt;[19] when it came to the topic of a worldwide pandemic, it was never a matter of "if" it will happen, only a matter of "when" it will happen.

In fact, in October 2019, barely a few months before the COVID-19 pandemic started, a Washington think tank on national security ran a simulation to determine what could happen during a "highly transmissible coronavirus" outbreak.[20] A coronavirus scenario was used because both the SARS and MERS viruses at the root of recent epidemics were also of the coronavirus family. The scenario assumed that governments would be too slow to react and would impose travel bans and border closures only after the virus had spread worldwide through international air corridors. It assumed that government would have to pump massive amounts of money into the economy to prevent its total collapse, and that development of a vaccine that could effectively end the crisis would take up to a year. One of the main conclusions of the pandemic simulation conducted by these invited experts in global health, biosciences, national security, emergency response, and economics was that early and preventative actions are absolutely critical if there is any hope of minimizing the mess created by a pandemic. Not a reassuring thought in an era where trust and cooperation between countries, levels of government, companies, and citizens has been eroded to an all-time low.[21] The key conclusion from the scenario pandemic was that "leaders simply don't take health seriously enough as a U.S. national security issue."[22]

In fact, such simulations have been played multiple times in the past by numerous organizations. In 2001, the "Dark Winter" exercise was conducted to simulate an epidemic caused by a malicious smallpox infection. In this "senior-level war game" focusing on a biological attack, the roles of National Security Council members were played by former senior government officials.[23] In this case, a smallpox vaccine exists, but the US supply at the time was estimated to be only seven to twelve million doses and production of new vaccines to replenish the stocks would have taken two to three years because the facilities to produce smallpox vaccines had been dismantled after 1980.[24] The simulation was stopped after it had stretched over thirteen virtual days, during which the disease spread to twenty-five states and fifteen other countries, leaving the world with massive casualties and under the threat of a "breakdown in essential institutions, violation of democratic processes, civil disorder," and compromised governments that had lost the confidence of their citizens.

In 2018, the "Clade X" exercise similarly considered "an outbreak of a novel parainfluenza virus that is moderately contagious and moderately lethal and for which there are no effective medical countermeasures."[25] Again, national security and epidemic response experts[26] played the role of National Security Council members. On the first day of the tabletop exercise, they were briefed about the recently discovered spread of a new virus, called "Parainfluenza Clade X" that killed 10 percent of those infected and for which a vaccine would take twelve months to develop. The scenario considered that each infected individual transmitted the illness to two to three other people, and that the virus incubation period was approximately five to seven days. Impressively, that is similar to the playbook for the COVID-19 virus that appeared two years after that simulation. Throughout the Clade X pandemic scenario, air traffic shut down, a quarantine was imposed by the federal government, "national state emergency" and "public health emergency" were declared, the capacity of intensive care units was exceeded, public demand for surgical masks and respirators surged. Four months after the initial outbreak, almost nine million were infected and three million dead, with 10 percent of those in the Americas. Projecting ten months ahead, twenty million people had died in the United States, the gross national product had dropped by 50 percent, the stock market collapsed to 10 percent of its previous peak value, and the health care and health insurance system had gone bankrupt. Civil unrest led to widespread looting and violent clashes, with the police and the military securing borders. Some governments collapsed. A vaccine finally became available, but only five million doses per month were to be produced. Who gets it first?

Approximately 150 people were invited to attend the 2018 table-top exercise, and videos of the more than five hours of National Security Council discussions that took place through the entire simulation are freely available online for anyone interested in watching.[27]

With all the knowledge in hand, and the "close calls" of the 2003 SARS and 2014–16 Ebola epidemics,[28] it is simply not possible, in all seriousness, to call a pandemic—including the COVID-19 one—a black swan event.

So, to be 100 percent clear:

- A future world pandemic is not a black swan event (be it Ebola or any other deadly virus not yet known to exist);
- A future destructive magnitude 6 earthquake near New York City or Boston is not a black swan event;[29]
- A future destructive magnitude 7 earthquake near Charleston, North Carolina,[30] or near Salt Lake City, Utah,[31] is not a black swan event;
- A future destructive magnitude 8 earthquake near Memphis, Tennessee,[32] or anywhere along the US West Coast is not a black swan event;
- A future destructive magnitude 9 earthquake near Seattle[33] is not a black swan event;
- A future destructive Category 5 hurricane anywhere along the Atlantic coast or the Gulf of Mexico is not a black swan event;
- Any of the destructive hazards described earlier in this book are not black swan events—except for the extraterrestrial invasion.

Any of these events have the potential to wreck the economy to various degrees and to inflict massive deaths and injuries. Those who will be hit by any one of these events will suffer in predictable ways. If they survive, hopefully, they will have become more knowledgeable and be able to build on their firsthand experience of the disaster to make judicious decisions to prepare a better future. However, none of them will be legitimately able to claim that they have survived a black swan event, simply because none of these events are black swans. The threats posed by all of these events are well known; the cold facts have been repeated multiple times by experts from many disciplines; the risks have been surgically assessed by academics and other credible sources using specialized tools and models. In some cases, the majority of those who had to make decisions—from individuals to government officials—decided that the best course of action was to "cross that bridge when we get there." That is perfectly fine—but, after the disaster, leave the poor black swans alone.

INCONSEQUENTIAL SWANS

Not all unforeseen events are black swans either. Things that nobody could have expected happen all the time, but not always with significant consequences.

For example, a magnitude 6.9 earthquake occurred on December 23, 1985, in the Nahanni National Park Reserve in the Northwest Territories[34]—nothing dramatic given that nobody lived within a hundred miles of the epicenter. The shaking, which started with the first in a string of earthquakes in October that year and peaked with the December event, startled the roughly three thousand people that lived scattered over the seventy-five thousand square miles in and around the Dehcho First Nations administrative region.[35] The seismic waves found nothing of significance in their path to damage, but they rattled the confidence of seismologists who, prior to the 1985 earthquakes, considered this region to be a "relatively quiet earthquake zone."[36] In developing the seismic maps of the National Building Code of Canada, to be conservative, seismologists had predicted that earthquakes as large as magnitude 6 could happen there.[37] It turned out that not only is a 6.9 roughly ten times larger than a 6.0, but the ground motions recorded on December 23 showed peak horizontal ground accelerations greater than 2g (that is, twice the gravity force), setting a world record at the time.

Indeed, every now and then, earthquakes happen where none have been observed before. This was the case for the 1988 magnitude 6 Saguenay earthquake that struck in the middle of a national park, mostly shaking up all the resident moose, Canada geese, and other wildlife (it is unknown whether any were traumatized in the process). It cannot really be called a black swan event either because, while it produced some minor damage in cities up to two hundred miles away, none of it was significant.

Typically, where no previously known significant earthquake activity has occurred in the past, none is predicted to occur in the future, but once an earthquake happens there, a new seismic region is born. Evidently, earthquake engineering is an expanding business.

SIMCITY ON STEROIDS

When driving on a highway, if a little ten-year-old kid riding in the back-seat sees warehouses lined up along the road and asks, "Is this industrial zoning?" one can safely bet that this kid has been playing SimCity. SimCity is a role-playing game in which one takes a virgin piece of land, divides it into residential, commercial, and industrial zones, and starts building the infrastructure of a virtual city[38]—as in some sort of über urban planner power trip. The player then sets up the city budget and tax rates to attract people and grow the city. This is an open-ended game with no goal other than trying to build a satisfactory outcome, either by maximizing population, profitability, aesthetics, quality of life, or any other self-imposed objective. Although there is no end to the game, the key is to find an equilibrium to avoid chaos and urban decay—without using any of the cheat codes that can be found online and that can provide infinite water and electricity or infinite money to spend.[39] The rules that govern the behavior of the SimCity residents and infrastructure interdependencies are hidden, but can be inferred to some degree by trial and error and some common sense. Practically speaking, this video game is an agent-based model. Agent-based models are computer programs that simulate the actions and interactions of individual agents (such as, people) that follow simple decision-making rules that are deemed to be rational to achieve and maximize, one small incremental step at a time, a specific benefit such as reproduction, wealth, social status, and so on, depending on the model.[40] Agent-based modeling has been used to study how specific situations can evolve over time, for a multitude of applications, ranging from the spread of epidemics, to traffic problems, to warfare (obviously), with variable levels of success.[41]

Tens of thousands of players have exchanged tricks on social networks to figure out how to achieve a successful SimCity. Some have reported that once the basics of water and power are provided, neighborhoods of concentrated dirty manufacturing industries only need lots of policing and shorter distances for freight trips to be "happy" so it is pointless to invest further resources there, whereas achieving a happy residential neighborhood requires loading up on schools, hospitals, parks, non-polluting industries, low traffic noise, low crime, and no garbage. Some successful players have argued that to beat the system, the strategy to have a successful city is to surround it by over-polluted failed industrial cities to which all pollution and garbage is exported.[42] Others have figured out that each individual virtual citizen in the simulation responds to the same simple rules, with the population distributing itself as a homogeneous mass moving around no differently than water, sewage, or traffic, taking jobs in the first random building they encounter that is hiring, irrespective of skills or prior employment, easily moving from a commercial to an industrial job.[43] Astute folks have even unmasked faulty rules that can lead to illogically dense traffic in a long dead-end street with only one house at its end,[44] traffic jams on small roads running parallel to empty mega-highways,[45] or intersections forever blocked by masses of pedestrians going in circles and nowhere else.[46]

Providing a step-by-step guide on how to "win" in SimCity by building prosperous and stable cities would be a book by itself. The point is that the outcome of any agent-based modeling construct depends on the assumptions built into the model. The expectation when running an agent-based simulation is that if the rational response of each person or agent to a given input is accounted for by the model—such as having a percentage of the population move away when taxes are increased—then running the model over time will result in an accurate simulation of the outcome. In other words, the belief is that running the program will effectively aggregate all of the cause-effect and action-reaction rules that have been embedded into the model, taking into account average responses and variations in responses from agent to agent, and provide a clear and reliable prediction of the future. As such, the program will consider at any point in time the respective response of each individual to taxes, public services, pollution, infrastructure quality and effectiveness, and their tolerance and reactions to shortages or abundance of each, and act accordingly—moving to or away from a city. The beauty of an agent-based model is that it computes evolution in accelerated time—something that is hard to do by any other means. The drawback is that the resulting "path of evolution" that the program will simulate depends entirely on the specific rules embedded in the model. Changing one rule changes the outcome. To make things worse, if the evolution that is simulated is a societal model that relies on a number of rules that define how an individual behaves, it does not take long to realize that missing an important action-reaction rule, or a sudden unpredicted change in how people behave over a short or long time, will throw a wrench into the entire prediction process. Like having the in-laws unexpectedly move in can change the dynamic of a marriage.

Experts on agent-based modeling and decision-making models will argue that everybody makes rational decisions and reacts to actions in ways that are in their best interest. Ask them why it is, then, that many people are smoking in spite of overwhelming evidence that it is a major health hazard, and they will respond that people who smoke have made the rational decision that the enjoyment they gain from smoking is worth the resulting reduction in life expectancy. This is tantamount to saying, with an ironclad confidence in the ability to model human behavior, that those who jump off a bridge (without a parachute) have reached the rational decision that it is a logical action in response to a broken heart or other setback—or maybe that the jumpers (still without a parachute) have made the rational decision that the enjoyment they gain from the sensation of flying as they fall is worth the resulting reduction in life expectancy. Political scientists will be quick to counter that people make decisions against their best interest all the time—some will emphasize that it is actually the entire purpose of political science to figure out how to manipulate people to vote for things that are clearly against their best interest.

Ask a prominent economist why SimCity-like models have not been developed to predict economic cycles and recessions, and the likely response will be that most factors that have a large impact at that scale are unpredictable.[47] Nobody can predict the impact of changes in national and international political regimes on the economy and stock market, or rumors from out of nowhere that can suddenly lead to a bank rush or some illogical financial panic, or when a real estate market bubble will pop. How to model the fact that an international crisis, a pandemic, or the emergence of a green movement might suddenly change the rules of the game or maybe even shift the short-term fixation on immediate profits of some companies into a long-term view focused on survival?

Sociologists and psychologists who have studied the mass hysteria that can be created by unfounded fears and rumors would concur. Who could have predicted that the tense political and religious quarrels of the inhabitants of Salem would escalate into a full-blown, hysterical witch-hunt that culminated in two hundred accusations, thirty guilty verdicts, and the hanging of nineteen citizens[48]—likely all innocent victims, as any self-respecting witch would have used witchcraft to escape death. Yet are the Salem trials and executions worse than the modern anti-vaccination movement that has led to the resurgence of previously eliminated diseases?[49] If someone wished to construct an agent-based model to predict the impact of the next mass hysterias, delusions, and groupthink disasters, what fortunetellers should be consulted to figure out the rules that "rational agents" would follow in such cases?

All of that is to say that unpredictable events are—well—unpredictable.

ON THE DISASTER TRAIL

Let It Be

I was stunned. A local radio station that had simply played, beginning to end, the "red album" and "blue album" had the gall to call it the greatest Beatles special ever aired. It was 1979 and the two greatest-hits double-LPs had been out since 1973. Playing them back-to-back without a single commentary—but plenty of commercials between songs—was no feat. Anybody with a little bit of "Beatles knowledge" could have done better, and for some unexplained reason, I sat at the typewriter and gave it a shot. By the time I was done, I had created an eighteen-hour series embracing the totality of their recorded opus, from the Star-Club in Hamburg up to the breakup, with commentaries or anecdotes for each song. It even included seamless mixes of the Beatles' covers of songs by other artists with the original versions of these songs, to highlight the sharpness added by the newer versions. That was the "ultimate Beatles special" I would have liked to have heard in the first place.

It was a fun challenge for a mini-Beatlemaniac like me to create that fictitious radio show, but what was I supposed to do with it? Was it good enough to put on the airwaves? There was only one way to know, and it was to try to sell it. For sure, the local community radio would have taken it for free, but if I started at the bottom, I would never know if anyone higher in the radiophonic hierarchy would have been interested. I therefore had no choice but to start at the top and work my way down until I found the perfect match.

With no prior experience in radio, my prediction was that I would be laughed at and kicked out of every station—possibly until I landed back to the community radio. This low expectation was a direct consequence of my one prior experience as a door-to-door salesman. It did not help that the product then was a glossy fishing/hunting magazine—something nobody needs. My coworker who covered the houses on his side of the street sold eight subscriptions on his first day. He was a "foot-in-the-door" salesperson that literally put his foot between the doorframe and the door—old-style—such that the potential customer could not escape by slamming the door. My single sale was to a couple that happened to be coming back from a fishing trip, still unpacking and more than a bit giddy from the experience—and possibly from alcohol.

With nothing to lose, ready to face as many rejections as there were radio stations, with the latest Bureau of Broadcast Measurement (BBM) ratings in hand, I showed up unannounced at the no. 1 station in town. The program director seemed amused and talkative. He wanted to know more about the breadth of my musical tastes and interests, which thankfully was much broader than the Beatles. When he asked what was the latest LP that I had purchased, it was a bit awkward. Facing the program director of a station focused on pop and rock, I stuck to the truth and admitted that it was a country music album—actually, the only country music I ever purchased in my entire life up to this day for that matter—because it was a concept album on the secession war that was so brilliantly done that I could not resist.

"White Mansion?" he asked.

"Exactly!" I replied surprised.

He could not believe it. He thought nobody else in town knew about this album—so did I, for that matter. He not only bought my Beatles special, he also hired me as a part-time content producer, developing shows on progressive rock and retro music. Absolutely no agent-based simulation could have predicted that outcome.

The eighteen-hour series *L'Épopée des Beatles* (The Beatles' Saga) ran during the fall BBM that year and the ratings it received crushed the competition—which is evidently easier to do when headlining with the world's best rock band. My prediction that I would tumble down the stairs of the radio world all the way to near bottom had failed badly, but it was an unpredicted event with a positive outcome—hence, not a black swan event.

The Brain

WHOSE BRAIN?

Serious research in neurology has been conducted in the past decades to determine why teenagers typically engage in more risky activities than any other age group. From these expensive studies, neuroscientists discovered what every parent of teenagers could have told them for free: overnight, the brain of a healthy kid transforms into a mass of Jell-O in which synapses and neurons are drowning and thus unable to perform normally. Every part of teenagers' bodies grows at accelerated speed into adulthood, except the brain, which lags behind. As a result, teenagers can physically do anything adults do (and often better), but have no pragmatic perception of the future and therefore have not fully grasped that death can be as close as, for example, the distance between the beer in their right hand and the steering wheel in their left one. They may have an acute perception of risk when filling out questionnaires designed to determine if they understand the consequences of their actions, but that is theory. When it comes to practice, they still engage in incoherent and risky behavior,[1] be it binge-drinking, unprotected sex, smoking anything that burns, sexting, truancy, speeding, snorting smarties,[2] performing unrehearsed daredevil stunts and tricks, and listening to Justin Bieber's music, to name only a few. Statistics confirm this all the time: this segment of the population loves to partake in high-risk behavior and does not bother with the nuance of "calculated risks."

Circumventing the mysteries of why teenagers act brainless,[3] the focus here is on so-called "normal" adult brains that are considered fully developed and on their occasional irrationality.

OUR SEGMENTED BRAIN

To say the least, knowledge on how the human brain functions is far from complete—and what is being discussed here is not the proverbial joke that

207

members of one sex are clueless in understanding how the other sex's brain works. Yet, in spite of all the advances of the past decades in mapping how thoughts travel and jolt neurons here and there in all the nooks and crannies of the brain, no reliable roadmap exists and many of these roads are at best foggy pathways. Yet one characteristic of the human brain has been long proclaimed and believed by all: it has a split personality.

Going back millennia, traditional Chinese medicine considered the mind to consist of: (i) the "Yi," which provided wisdom, correct judgment, and clear thinking, and qualified as being active, calm, and peaceful; and (ii) the "Xin," connecting the mind to the heart, feelings, emotions, and desires, and qualified as passive, excited, energized, and confused. The sum of the Yi and Xin creates humanity and personality.

The contemporary version states that there is a right and a left side of the brain, where the right hemisphere harbors the intuitive artistic side of this schizophrenic relationship and the left hemisphere is the domain of the logical thoughts and mathematical mind. Much of that perception arose from the study of patients having suffered severe brain damage—victims who were willing to let other people tinker and probe inside their heads in hope of relief and remission. Neuropsychologist/neurobiologist Roger Wolcott Sperry famously did so (the tinkering bit, that is), receiving a Nobel Prize for his work on "split-brain" patients.[4] The idea, in its most simplistic "distinct cleavage" form, quickly found its way into popular culture—resulting in the tagging of individuals as being "left-brained" or "right-brained," which is more polite than saying nerdy or artsy.

The two-minds idea is now deeply entrenched in contemporary views, including statements such as this one: "The intuitive mind is a sacred gift and the rational mind is a faithful servant. We have created a society that honors the servant and has forgotten the gift."—a quotation widely but incorrectly claimed to be from Einstein.[5]

The truth is evidently somewhere in between. It is generally recognized that both sides of the brain are typically orchestrated to work together if successful results are to be achieved in any activity.[6] Yet, irrespective of facts and legends, all of this underscores that fact that being human requires juggling the demands and expectations coming from two very different sides of our personality (or brain).

One side, after stepping on the scale, asserts that it is time to start going to the gym, and to replace French fries and ice cream with Brussels sprouts, broccoli, and other weird veggies. The other side fully recognizes that it is the right thing to do, but after calculating the effort and discipline required to do it, concludes that the priority is to first finish off the box of Twinkies—otherwise they will go to waste, which would be a shame—and then see if there is a gadget on the shopping channel that could "bring the gym to the living room couch" instead.

Therein lies the eternal tug-o-war between wisdom and willpower: The knowledge of what should be done versus the fact it does not get done for a number of reasons—valid or not.

The business model of many profitable gyms is built on that dichotomy: Wisdom will bring many to pay upfront for a year membership, but willpower (or lack thereof) will make them progressively reduce the number of visits per month, eventually down to zero. This is in part why gyms that can host three hundred people can get away with selling six thousand memberships.[7] Likewise, an entire industry of self-help books thrives because it offers the low-hanging fruit of easy initial commitment for a few bucks to appease the demands of wisdom, knowing full well that the willpower to implement the advice will be lacking. At worst, this will make it possible to sell another book down the line, fueling a sort of perpetual motion.

PERCEPTION, MISPERCEPTION, AND OUTRIGHT IRRATIONALITY

Assuming someone can muster the willpower to act, other brain-created obstacles stand in the way. Taking action is good, but what action exactly?

Everybody is aware that there are many optical illusions that can fool the brain. For example, there is the "scintillating grid illusion" (a version of the Hermann grid illusion)[8] that is created by a grid of gray lines drawn on a black background, with white dots placed at all the intersections. Staring at a white dot makes the dots farther away look black. When eyes wander across the grid—as fixing a single point does not come naturally—the dark dots appear and disappear at various locations across the board, producing a scintillating effect and driving the mind nuts without the help of 1960s music or hallucinatory drugs. Straight lines can look curved, static wheels seem to turn, shapes of equal size appear to differ, and a myriad of other tricks[9] are all illusions that fool the brain.[10] Among hypotheses proposed to explain why this happens, a simple one is that survival requires the rapid recognition of familiar patterns, and that through the long process of human learning, simplified models are created by the brain as "shortcuts" to accelerate things.[11] The illusions are created because these shortcuts lead to expected answers, which are wrong in the case of the illusion. The point being, here, that the brain can be fooled all the time—as any respectable illusionist would be able to prove in the blink of an eye.

Likewise, intentionally or not, humans can follow other paths that lead to erroneous judgments. Indeed, humans can be convinced of pretty much anything. Joseph Goebbels (minister of propaganda in Nazi Germany, as mentioned earlier) built his career on the fact that a lie repeated often enough can become a truth—a principle that nowadays produces a lot of confused people when reporters and politicians across the entire ideological spectrum accuse each other of spreading "fake news."

The list of irrational things that humans have held to be true at various times throughout history is endless. If anything, this suggests that the human brain is wired with an ability to believe just about any nonsense, as long as it is presented with authority or in ways that please expectations. While it is possible to forgive those who lived in the Middle Ages for believing that tomatoes were a creation of the devil,[12] that black cats were witches in disguise,[13] and that Catholic priests were trustworthy, it is much harder to explain in this era of infinite access to knowledge what motivates the Monty Pythonesque logic of flat-earthers and the proponents of loony conspiracy theories.

Talking to someone who is wed to the moon landing conspiracy theory[14] can be a surreal experience; it feels like talking to someone who lives in a parallel universe. They typically are lost in an alternate reality that has been constructed in a way that blinds them to facts. Their brain has been conditioned to certain "shortcuts." As a result, they are navigating a mental Hermann grid where all the white dots are seen as black. Attempts to hold a rational debate are futile, although some make it a point to do so.[15]

While conspiracy theorists may seem off the chart in their beliefs, they are not that different from everybody. It turns out that there is some seriously odd "wiring" in the brain—everybody's brain—that makes people interpret all information in ways that confirm their beliefs. The phenomenon is called confirmation bias.[16] Not only do humans prefer and seek information that reinforces their existing mindset, but they also avoid, discredit, distort, or conveniently forget information that conflicts with their views—a process often called selective exposure.[17] When the beliefs are political, it leads to divided factions that become further divided and an unwillingness to compromise, which can lead to war and death. When the beliefs are those of religions or cults, it leads to entrenchment in self-righteous proclamations of unalterable truths, which again can lead to war and death.

Although it does not always have to escalate to extremes, this "odd brain wiring" is pervasive in all activities. At insignificantly low levels of consequences, confirmation bias and selective exposure can simply lead to ridicule and unintended consequences. In 1979, Monty Python produced the silly movie *Life of Brian* which tells the hilarious misadventures of a poor sap mistaken for the "real deal" by a mob searching for the Messiah. Despite his best efforts to chase them away, it eventually leads him to crucifixion—a story unfolding in parallel to the life of the "real deal" shown far in the background in some scenes to confirm that Brian is not Him. It is nowadays considered by many to be one of the best comedy films of all time.[18] When the soundtrack of the film offended one South Carolina Presbyterian minister (who did not see the actual movie), he shared his concern with the wife of Senator Strom Thurmond (who did not see the movie or hear the soundtrack), who in turn shared her outrage with her husband,[19] who did not need to see or hear anything to take the necessary measures to have all showings of the film suspended in South Carolina. Many other religious groups also publicly condemned the film.[20] Of course, as censorship always does, this turned out to be better publicity than what any money could have bought.

While censorship is fundamentally an attempt to control the thinking of others, expurgating nonconforming ideas is at the same time a form of selective exposure by removing from one's realm any material that is not compliant with ideas already anchored in the brain. This helps perpetuate perceptions even if counter to evidence from reality. It should stand to reason that when tangible proof is provided about certain facts, people should immediately renounce previously erroneous ways, but no. Unless the evidence is brutally physical—like putting the hand of a guru on a hot stove to see if his claim that the mind can control pain holds true—in most cases, the first reaction of someone confronted with any proof is to question the validity of the proof. In fact, the more some beliefs are questioned, the more entrenched those beliefs become. People defend their beliefs against all odds, like kids weaving convoluted explanations to patch up glaring holes in the charming narrative supporting the existence of their beloved Santa Claus—except that kids eventually figure it out, some never forgiving their parents for having fooled them with this enchanting fairy tale.

All cults thrive on this wonderful characteristic of the human brain. In the 1950s, psychologists infiltrated a doomsday cult that awaited to be saved by aliens on a specific date when the end of the world was predicted to happen. Of course, the date came and passed, without a single flying saucer in sight, but instead of realizing that they had been duped, members of the sect concluded that the world had been saved by their piety and continued efforts to recruit new members.[21] Interestingly, the promise of accessing heaven via an alien spacecraft rescue is a recurring theme of cults, with multiple hilarious examples throughout history,[22] but some tragic ones too—such as when the thirty-nine disciples of the Heaven's Gate cult committed mass suicide, convinced by their "spiritual leaders" that leaving their bodies was the way to access a spacecraft hidden behind the Hale-Bopp comet when it reached its closest distance to earth.[23]

However, spacecrafts are not required for wacko beliefs. Nor are wackos needed. Some psychiatric studies report strong evidence that cult victims include "normal people" without any previous long-standing conflict that would have made them more vulnerable. While psychiatric and addictive disorders can be a factor leading some folks to join cults, many simply were searching for spirituality or personal development.[24] As one former cult member testified, she came from a "loving, middle-class, Midwestern family,"[25] but her brain was "rewired" by the cult leaders.[26] Possibly the same kind of rewiring that made sixteen executives from Hydro Québec—a government-owned power utility company with more than twenty thousand employees—join the Order of the Solar Temple, after one of the company's vice presidents invited the cult's leader to give motivational "management" lectures on topics such as "Business and Chaos," and "The Real Meaning of Work."[27] Built from an eclectic mix of New Age philosophy, extraterrestrial salvation, and archaic Christian rituals, the legacy of the Order can be summarized as arson, mass suicides, and murders,[28] including the killing of a baby identified to be the Antichrist[29]—supposedly saving the planet in the process.[30] Having ordinary people fall for that stuff is a testimony to the manipulation skills of cult leaders.[31]

All the above illustrates the challenges in convincing someone of hazards and associated risks when the mind is not predisposed to receiving this message. This mental block can be attributed to many causes, ranging from ignorance of reality to wishful interpretation of data. The "Nah! There's never any earthquakes here," or "there's never any hurricanes here" is at one end of the misperception spectrum.

At the other end of the spectrum is the individual who has studied all the information available and has reached a conclusion that supports a desired outcome. After the storm surge from Hurricane Matthew flooded a Palm Coast neighborhood, the owners of a house where water had luckily stopped right at its doormat put it for sale, hoping to move to a less flood-prone area. Two years after Hurricane Matthew, an interested buyer parked in front of the house and bragged to the neighbors that he had studied all of Florida's historical data and determined that hurricanes had never happened in Palm Coast before, which is why he was house-hunting in that wonderful neighborhood. One must be uniquely cocky—or downright blinded by his own beliefs—to preach to those who had survived Hurricane Matthew two years before that hurricanes do not occur in that part of Florida. This kind of naive confidence is akin to putting stamps on a baby's forehead and mailing it to grandma on the other side of town, on account of having read that the mail service is highly reliable—incidentally, mailing babies is a practice that was banned by the US postmaster general in 1913.[32]

Fortunately, over time, biases and prejudices can be overcome. First, as the saying goes, seeing is believing. That inescapably works for those surviving a disaster—albeit, too late to have a positive impact for that event. Second, the brain's resistance to changes in beliefs, which is presumably the result of an evolutionary mechanism to maintain group cohesion and preserve self-identify,[33] can be abated over time—sometimes a long time, though. This may require instilling a culture of scientific thinking, where rigor and curiosity matters more than beliefs, and where discovery is recognized to be exhilarating rather than threatening—always an immensely uphill battle in an environment where beliefs are admired as virtues.

Eventually, at some point in time, which varies depending on the individual, all the facts are gathered and the risks are correctly identified and acknowledged. When that finally happens, if one is willing to assume the risk after having duly considered all that information, either because the foreseen benefits are deemed worth the gamble or because one is willing to transfer the risk to insurance coverage, at least, the brain has not been fooled. It can be deemed a rational and respectable decision—as long as it does not harm others.

The challenge comes in the significant and deliberate effort it takes to collect evidence and make such a rational decision with all the inalterable facts in hand—even more effort when the evidence is counter to expectations or wishes.

Effort?

Hmm . . .

This brings it all back to the tug-o-war between the "Yi" and the "Xin."

THE MAÑANA SYNDROME

Compounding the problem, ingrained in human nature, to various degrees, is a natural tendency to procrastinate, to unnecessarily postpone decisions or actions. This is sometimes called the "mañana syndrome," where mañana, the Spanish word for "tomorrow," has become a common expression meaning not only "later," but "unpredictably later," to the point where it may actually never get done.

Decades of psychological studies on the topic have allowed identifying many possible causes for procrastination,[34] including some that are particularly relevant in the current context. At a fundamental level, tasks are usually performed in expectation of a reward, be it money or pleasure. When the reward is perceived to be too far in the future, it is hard to contextualize. This is called temporal discounting or delay discounting.[35] Even when it comes to the "satisfaction of a job well done" or "for the greater good," it turns out that "a bird in the hand is worth two in the bush." The further in the future the reward, the less appealing it is. The fact that most people will prefer a smaller reward today than a larger one in the future is a phenomenon called hyperbolic discounting.[36]

A lot can be said about the many different reasons why people procrastinate; that people are more likely to procrastinate when dealing with more abstract goals (for example, lose weight) than concrete ones (spend thirty minutes on the treadmill); that rewards in the future to individuals who perceive themselves as living in the present are strangely perceived as happening to someone else (as if the person spending the retirement fund in the future cannot possibly be the same one investing the dollars today); that some people have an optimistic view of their ability to do it all later when the deadline approaches (such as, not studying at a steady pace but rather cramming the evening before the exam); that some procrastination is attributable to wishful thinking that the problems will automatically take care of themselves over time without the need for intervention; that some people suffer "paralysis by analysis" because they are unable to make decisions for fear of making the wrong choice; that people sometimes prefer to avoid decisions or actions when the task to be done appears overwhelming, creates anxiety about negative outcomes, or requires talking to unpleasant people; that perfectionists will delay doing something unless they are absolutely certain that it can be done flawlessly by unattainable standards; that people with low self-esteem and low self-confidence harbor such a fear of failure that they prefer not acting to receiving negative feedback and criticisms on their accomplishments; that people who are driven to self-handicap or self-sabotage themselves can use inaction to save face (for example, failing an exam on account of not studying does not carry the same judgment on intellectual fitness as failing because of inability to understand the material); that some people have no interest seeing the task done (particularly when feeling they have been set up for failure by being assigned a task above their ability or feel they will be criticized no matter the outcome and irrespectively of their effort); that some people are simply lazy, plain and simple, and prefer watching TV, playing video games, texting, or whatever else provides instant gratification, distraction, or teenage-like rebellion; and much more.[37]

What leads to procrastination is a most interesting and wide-ranging topic worthy of more exploration. Unfortunately, doing so is apparently an overwhelming task, likely to be criticized, with the risk of focusing too much or not enough on it, and might be best to do in a different context. At the same time, not expanding on this subtopic here may not be so critical, as it will probably take care of itself in the subsequent chapters—and there are many other important and more enjoyable things to do, so . . . mañana.

As a result of all the above wonderful dilemmas and conflicts at play in the cerebral cortex, things do not often get done to mitigate ahead of time the conditions that can lead to a disaster. When the inevitable happens, then the brain shifts into a different gear.

BUYING GOOD INTENTIONS

One of the cornerstones of Christianity is the forgiveness of sins. Although the original "your sins are forgiven" statement in the gospel did not split hairs, the church spent much of its existence trying to determine which sins can easily be forgiven and which cannot[38]—including in that latter category the life-threatening offenses of blasphemy, worshiping false gods, and homosexuality. How the clergy kept busy sorting all of the things that humans do into venial sins, mortal sins, and all manner of categories is a whole story in itself.[39] The important point is that those at the top of the hierarchy realized that some people, when becoming aware of their own wrongdoing—no matter how one defines *wrong*—feel remorse and need closure. Be it the soul or the brain, something in human nature calls for redemption of some sort to restore peace and return to equilibrium. Depending on whether the feeling of guilt is a burning pain or a mere annoyance, some people seeking redemption will "go all in" and seek a new life to have a positive impact—no matter how *positive* is defined (as sometimes it is purely imaginary or even harmful)—while others will prefer a quick fix solution.

Given that the consequences of wrongdoing most often cannot be undone, the expedient solution is to forgive in exchange for something else, making for a convenient transaction. It can also become a lucrative one. Any system of confession and penance between a person and his/ her conscience that involves an intermediary empowered to dictate the terms of the penance is fraught with risk. Penances that consist of repeated prayers are gentle—even when it involves hundreds of repetitions—but money talks. Monks tallied the penitential tariff to be paid for each sin in books to be used by confessors. By the eleventh century, it was already possible for the rich to reduce their required number of days fasting during Lent by offering donations to the church.

Throw a bit a corruption into the mix and it becomes a racket. For example, Pope Leo X demanded that all bishops and cardinals actively sell "indulgences"—like bonds that the nobility could buy at various prices in remission of a smorgasbord of sins—as a way of financing Rome's extravagant expenses under his regime. In spite of this source of extra revenue, Leo's lavish expenditures increased faster than his income, and he left the church deeply in debt when he died.[40]

Likewise, a modern way to soothe the brain following disasters is by contributing dollars to relief efforts. Cynics might see this as purchasing peace of mind, unconsciously done as it may be. Nonetheless, on the positive side, these dollars are directly spent on relieving the plight of the disaster victims. On the negative side, they are rarely directed to activities that would prevent repetitions of similar disasters in the future.

Benefit concerts are one popular expression of how individuals can help with money as a proxy for helping by actions. These are typically put together by megastars who can pull in big dollars from an adult crowd that has plenty of disposable income, or from a less wealthy but larger crowd of star-struck fans. The first of these in modern times was the 1971 Concert for Bangladesh,[41] put together by George Harrison and Ravi Shankar, with a last-minute surprise participation by Bob Dylan. Music icons Harrison and Dylan took over the Madison Square Garden stage after a fifteen-minute repertoire of Indian compositions led by the sitar virtuoso Shankar that would have rapidly emptied the Garden had it been played at the end of the show instead.

The objective of the Concert for Bangladesh was to raise money for what was turning into a humanitarian crisis, while world governments were not doing much other than watching and commenting from the perspective of their geopolitical interests and ideological biases.[42] In November 1970, Cyclone Bhola hit East Pakistan with massive rains and a storm surge that arrived at high tide, flooding many villages and destroying crops.[43] Mismanagement of the relief effort by the central government gave munitions to a secessionist movement that escalated into a civil war that displaced more than seven million people.[44] The genocide of millions was followed by a rainy season that again flooded the region, and the creation of Bangladesh a few months later.[45] All that triggered by a cyclone—which goes to show that natural disasters should scare politicians.

Examples of other major benefit concerts following disasters include: "The Concert for New York City" in 2001, organized by Paul McCartney with a lineup of eighteen megastars, following the September 11, 2001, attack that destroyed the World Trade Center towers;[46] the "Tsunami Relief Cardiff" charity concert to raise money for the victims of the tsunami that hit countries around the Indian Ocean in 2004;[47] "Shelter from the Storm: A Concert for the Gulf Coast" following the 2005 Hurricane Katrina;[48] "12-12-12: The Concert for Sandy Relief" following the 2012 Hurricane Sandy;[49] "Hope for Haiti Now: A Global Benefit for Earthquake Relief" following the 2010 Haiti earthquake;[50] "Hand in Hand: A Benefit for Hurricane Relief" in 2017 to provide relief to victims of Hurricanes Harvey and Irma;[51] and many more—above and beyond the ones unrelated to specific

Sixteen feet offset in fence near Point Reyes created by slip of the San Andreas Fault in 1906. *U.S. Geological Survey*

Formerly straight road now permanently offset due to fault slippage during 2010 Darfield earthquake near Christchurch (five months before the more famous 2011 Christchurch earthquake)

Parkfield, the earthquake capitol of the world. *Photo by Linda Tanner*

This steel frame is all that is remaining of the Emergency Operations Center of the town of Minamisanriku destroyed by the 2011 Tohoku tsunami. The building was fully underwater; twenty-four of the thirty occupants made it to the roof; only a few survived by climbing the antenna or holding on to the top of stair guardrail. *Richard Eisner and EERI*

As the tsunami wave pushed debris inland (from right to left in this photo), a blue truck wrapped itself around this column of the Shizugawa Hospital in Minamisanriku. *Richard Eisner and EERI*

Damage to Biloxi-Ocean Springs Bridge, one of several destroyed by Hurricane Katrina in 2005. *Photo courtesy of Jerome S. O'Connor, P.E.*

Example of beach erosion due to storm surge during a small hurricane that did not even make landfall

This formerly ocean front home is now ocean front, ocean back, and ocean sides

Hard hat as part of school uniform for kids living on Sakurajima Volcano. *Osumi River and National Highway Office, Ministry of Land, Infrastructure, Transport and Tourism, Japan*

Volcano tourism inside Whakaari/White Island, New Zealand's most active cone volcano (on a dormant day), rising 5,250 feet from seafloor, 1,050 feet above sea level

Collapsed double decker Cypress expressway in Oakland, California, during the 1989 Loma Prieta earthquake. Vehicles on the lower deck were crushed by the collapsing upper deck.

Second floor is now first floor for this apartment building in the Marina District of San Francisco, after the 1989 Loma Prieta earthquake

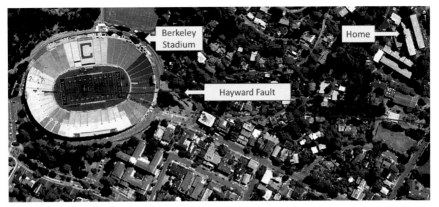

Home sweet home, 550 feet from the Hayward fault on which also sits the U.C. Berkeley football stadium *US Geological Survey—annotations by Michel Bruneau*

Unique view from eighty-fifth story of Taipei 101, dwarfing all other buildings in sight. *Photos by Dan Dowden*

Oroville Dam spillway damage, first shown with outflow of 100,000 cubic feet per second on February 15, 2017, then after cutting outflow to zero on February 27, 2017. *California Department of Water Resources*

At 7 a.m. in Christchurch, a drunk decided that this heavily damaged and closed pedestrian bridge was good enough. He made it to the other side, proving that there is a god for drunks.

Thank you for visiting this
Remembrance Space.

You may sit in any chair -
choose one that speaks to you
of those who died as a result of
the 22nd February earthquake.

You are welcome to stay as long
as you wish.

You may leave a comment
in the book.

Artist's statement:
185 square metres of grass depicting new growth,
regeneration.
185 white chairs, all painted twice by hand as an act
of remembrance.
This installation is temporary - as is life.
Pete Majendie

Temporary memorial artwork in
remembrance of the victims of the 2011
Christchurch earthquake

Growing damage in fenced-off Christchurch Business District in the years following the 2011 Earthquake

Deadly shower of bricks from an unreinforced masonry building. This example speaks for itself.

This building was resilient and survived the earthquake well, but it was crushed by the unreinforced masonry wall that fell from a far less resilient neighbor, which shows that resilience is a team sport.

Clown Doctors of New Zealand: Not exactly what I expected to find in a disaster area, but this clown was dead serious when it came to using the medical art of joy and laughter to fight traumatic stresses.

Where road A1A used to be in that part of Florida before the ocean progressively reclaimed the land after successive storms—now a sand trail only accessible by four-wheel drive vehicles

Surprise, the sand is gone! Example of beach erosion following a storm

Ruins of the only structure left standing near ground zero after an atomic bomb exploded 2,000 feet above Hiroshima. This "Genbaku Dome" is preserved as part of Hiroshima's Peace Memorial Park and is also designated a UNESCO World Heritage Site

disasters by name but raising money for charitable organizations that generally get involved in recovery activities following disasters.

By the way, short of a full concert, some projects focused on the release of special music albums and videos, such as the "Rock Aid Armenia" project to raise money for the victims of the 1988 Armenian earthquake. The highlight of this project was a rerecording of "Smoke on the Water" with members of Deep Purple, Pink Floyd, Led Zeppelin, Queen, Yes, Black Sabbath, ELP, Iron Maiden, and Asia playing together this four-chord classic—at the time, a CD for the true collector to find, but conveniently within reach on YouTube nowadays.[52]

Arguably, some have criticized all the above activities on account that having a big party with lots of sing-along does little but stroke the egos of the participating artists and might be a bit too giddy when the circumstances rather call for mourning, but they generally do bring in the cash. As in all such things, the money raised did not always reach the victims it was intended to help. Benefits from the July 1990 show/CD/DVD of "The Wall—Live in Berlin" by Roger Waters and more than a dozen guests, intended to go to the Memorial Fund for Disaster Relief, actually turned out to be losses.[53] Even George Harrison met hurdles when trying to forward the concert profits to Bangladesh, in part because his manager forgot to file for tax-exempt status with the US Internal Revenue Service, but eventually, a decade later, most of the funds found their way there through the "George Harrison Fund for UNICEF"[54] still active forty years later.

Nonetheless, on balance, charity is always positive and all the above certainly achieved much good and—addressing the topic at hand—provided disaster relief. Yet there apparently has been no benefit concert to provide third world citizens with money specifically targeted to the demolition of brittle buildings that are sure to collapse and kill all their occupants in future earthquakes and to fund the reconstruction in their place of safer buildings. Even less so for fixing buildings and infrastructure in wealthier countries. Therefore, after an earthquake crushes bodies in the debris of collapsed buildings, the relief comes in international aid, people recover in some fashion and bury their dead, and then, in many cases, rebuild the exact same way as before, setting up the stage for a repeat disaster in the future.

To be fair, focusing on Bangladesh again, UNICEF and other charity organizations invest heavily in programs toward eliminating child marriage, child labor, malnutrition, and violence, and provide social protection and education to children and women to reduce their exploitation. These are important priorities and essential steps, because only a safe and educated population can start dealing with other risks.[55] With respect to disasters specifically, UNICEF works to find measures to prevent the outbreak of waterborne diseases due to unsanitary conditions in refugee camps located in flood-prone lands when the cyclone season is looming.[56] As always, the problems of the present must be urgently addressed if there is to be a future to be planned.

IT IS DEFINITIVELY SOMEONE ELSE'S FAULT

What do the cities and regions of Salò, Calabria, L'Aquila, Offida, Irpinia, Sicily, Perugia, Lazio, Emilia-Romagna, Ancona, Friuli, Umbria, Campania, Basilicata, Abruzzo, Porto San Giorgio, Asti, and Molise have in common? Sure, they are all located in Italy, and visiting each one of these would make for a grand tour of the country. More relevant to the topic at hand though, is that these are all places in Italy where strong earthquakes have struck during the twentieth century alone. One and a half dozen memorable ones occurred in only one hundred years. To make matters worse, these quaint cities and regions are packed solid with stone and masonry buildings—the type of unreinforced construction known to be most vulnerable to earthquakes. Hence, when each of these earthquakes struck, buildings were destroyed, people died, and many more were injured. Each time, pictures of narrow streets filled with masonry rubble filled the front page of the Italian (and international) newspapers and magazines—frequently enough that it should be impossible find an Italian unaware of the country's severe seismic risk (short of running into a disciple of the 1970s Pinball Wizard). And, each time, the Italians picked up the rubble and rebuilt their homes nearly the exact same way, and most of the time at the exact same location, or close. Why? Maybe because there is no place to hide from earthquakes in Italy, or maybe because masonry buildings have been the way of life in Italy. Tradition?

What is even more amazing in Italy's history is that some cities are true "repeat customers" when it comes to taking a seismic beating. Like the city of L'Aquila, where it is estimated that two thousand died during the 1349 earthquake, possibly another five thousand during the 1703 one, six thousand again in 1786,[57] and roughly thirty thousand in 1915.[58] This is why the following story is somewhat surreal.

From November 2008 to April 2009, a swarm of small tremors centered around L'Aquila made the population edgy—particularly one of magnitude 4 on March 30. So, on the following day, the Italian Civil Protection Agency convened a meeting with its Major Risks Committee in an attempt to assess the situation. However, since nobody can predict earthquakes—as should be obvious by now—the committee concluded that there was no way to know if this swarm of small earthquakes could be the prelude to a major one. Members of the committee left town, leaving it to the deputy head of the agency to call a press conference, during which he somehow translated the committee's opinion into words that were construed by some to be reassuring, to the point where citizens who were planning to temporarily leave their home elected instead to stay. Unfortunately for them, a magnitude 6.3 hit a week later. Thousands of stone masonry buildings suffered damage, many collapsed, 309 people died, more than 1,500 were injured, and more than 65,000 were left homeless.[59]

In the aftermath of this catastrophe, something unprecedented happened: Six of Italy's top seismologists and earthquake engineers, and one civil defense official were indicted for manslaughter for not having warned all residents of the impending disaster, and for not having told people to leave town prior to the earthquake.[60] A petition signed by more than four thousand of the world's scientists was addressed to the Italian president, to tell him in no uncertain terms that nobody can predict earthquakes, and that therefore nobody (and especially not scientists) can be held responsible for their occurrence. That petition was ignored. By October 2012, the Italian court found all the accused to be guilty of manslaughter in connection with their prediction about the earthquake.[61] The verdict was not an indictment of the inability of science to predict earthquakes, but rather of the responsibility of scientists to share their expertise. It was felt that the tragedy could have been averted if not for the lack of communication by the experts—who had left town without talking to the public—and the poor communication of the risk by the civil defense official who was reported to have said: "The scientific community tells us there is no danger, because there is an ongoing discharge of energy. The situation looks favorable." He also allegedly told people that they should go home and have a glass of wine. Two years later, the Italian appeals court reversed the verdict for the six scientists, but not for the civil defense official, who was condemned to spend two years in jail, pending any further appeals—a far cry from the stiffer penalties sought by the angry mob gathered outside the court.[62]

Now, irrespective of the outcome of the court case and its multiple appeals,[63] it is worth pausing for a moment and debating why Italian citizens, many of them living in stone masonry buildings all over the country, would for a moment fault someone else for the losses incurred during an earthquake—as heartbreaking as these losses may be.

In all complex human activities, a large number of intertwined factors typically contribute to any specific outcome. It is often impossible to untangle all the cause-and-effect chains that are semi-randomly weaved together, to truly understand how all events unfolded and all players collided on the road to that specific outcome, but simple explanations—right or wrong—are always easier to understand. Human nature loves it that way, and when the simplification consists of labeling one person as being responsible for it all, that pretty much settles it—at least, for a while. When the outcome is a positive one, weasels of all kinds have mastered the art of having others "recognize their genius" and make them the reason for the successful outcome. Obviously, all these skillful manipulators make sure to vanish when the outcome is negative and the finger-pointing experts start to exercise their talent to find the perfect scapegoat.

Multiple simple theories to explain a complex chain of events may compete for a while, or not, depending on circumstances. Sometimes, people can find their "Yoko Ono" right away, which greatly simplifies everything and allows for the perfect explanation—circumventing all needs to dissect the problem further. As such, The Beatles did not break up because they had been playing music together for over twelve years, in grueling conditions and penniless for the first five, in grueling conditions and wealthy for the next seven. Not because they met as teenagers fixated on girls and rock'n'roll and grew up to become adults with diverging interests in other important endeavors (in addition, of course, to sex, drugs, and rock'n'roll). Not because they became entangled in creating and running the money-bleeding Apple Records enterprise. Not because they wanted full license to exercise their creativity in their own terms, even if one or all of the other band members disapproved. No. Too complex. The Beatles broke up simply because John Lennon fell in love with a Japanese avant-garde artist named Yoko Ono, presumed to be a nutcase, who enticed him away from the Beatles with more sex than drugs and rock'n'roll—as if John Lennon could not be a nutcase in his own right. The Yoko scapegoat explanation is much simpler: Sold! The "Someone Else's Fault" syndrome at its best. It always solves the problem.

Some prominent social and philosophical anthropologists, who study and theorize on the evolution of human behaviors to explain how civilization has occurred, believe that the use of the scapegoating mechanism goes at least as far back as the arrival of *Homo sapiens*. Ever since, when tensions have arisen within or between groups, the murdering of a convenient scapegoat has allowed a release of the tensions, a temporary resolution of the crisis, and thus a return to social peace—an efficient mechanism that eventually became a major cultural characteristic of all societies, and notably their religions.[64] Typically, the scapegoat is someone at the wrong place at the wrong time, who is certainly not the source of whatever caused the conflict. It helps to lay accusations if that person conveniently looks or acts differently from the rest of the group in crisis or enjoys envied powers or privileges. Through some subconscious mechanism, the community deceives itself into believing that this targeted victim is a horrible person who has transgressed a prohibition, violated a sacred rule, or failed in a duty, and therefore deserves to be killed or severely punished—depending on what the term *civilization* means in the time and place. With the eradication of the victim comes the illusion that the problem is solved, and the community returns to flourishing—for a while, before the cycle repeats itself.

In other words, scapegoating is a psychological defense mechanism.[65] It makes it possible for individuals or groups to cope with anger or frustrations created by problems of their own making by projecting their culpability onto other individuals, unwarrantedly blaming them instead for being the root cause of the problem.[66] No matter the problem—unemployment, low income, poor sex life—when the emotional energy expended becomes too painful to bear, it is unconsciously transferred to someone else.

Coming back to earthquakes, now that the root cause of the problem is known—namely someone else's failure—it makes perfect sense to return to living in unreinforced stone masonry buildings, in Italy or elsewhere. After all, many unreinforced masonry buildings have survived earthquakes in the past—smaller earthquakes maybe, but survived nonetheless—so there is no proof that the one that is home sweet home will collapse during a future earthquake. Furthermore, some unreinforced masonry buildings are more likely to survive earthquakes than others, so one's cozy dwelling could survive just as well—not that anyone necessarily bothers to investigate what might be the assessed seismic resistance of a building before signing on the lease or on the mortgage papers. That is what insurance is for, isn't it? Finally, earthquakes never happen here, but even if they did, there is no point in worrying—there is a far greater chance of dying in a car accident, truly.

Besides, when things do turn for the worse, it will be "Someone Else's Fault" and the best scapegoats will not be those ants who trained on their own nickel to become scientists and engineers and toiled days and nights to predict extreme events and to build a more resilient infrastructure. Rather, they will be those grasshoppers who are addicted to living at our expense, who play and sing all day: those we elect.

ON THE DISASTER TRAIL

Who Dominates the Yellow Pages: Where the Buck Stops
A Canadian friend who visited the University at Buffalo one day elected to stay in a hotel close to campus. When I saw him again the next morning, he asked me, "Do you happen to know which profession fills the largest

number of pages in Buffalo's Yellow Pages?" The University at Buffalo north campus is suburban, unlike many other universities in denser urban surroundings where tons of businesses and restaurants hug the campus boundaries; when someone has spent time counting pages in the Yellow Pages, you know there is no life around the University at Buffalo campus. For the benefit of the younger generation that has never owned or seen a landline phone in private homes, the Yellow Pages was a massive book that "Ma Bell"[67] gave every customer and that contained the phone numbers of all businesses in the area, grouped by types of business. It was so thick that ripping the phone book in half was a stunt performed by some as a show of strength. The book went the way of the dinosaurs, but some can still be seen in the Smithsonian's museum or in black-and-white gangster movies as a useful tool to beat suspects during police interrogations.

It turns out that my friend had counted the number of pages filled by medical doctors, engineers, car repair shops, insurance brokers, pharmacies, and every other group that had a sizeable number of pages, but the one group that had more pages than everybody else, and by a factor of two over the nearest contender, was lawyers. To top it all, there were also ads for law firms on the back-cover page, on the inside of the cover page, the inside of the back-cover page, and—as no space was wasted—on the thick spine of the book itself.

When the "Someone Else's Fault" syndrome rules the day, there is apparently money to be made.

The Mirrors We Elect

WINDY PANHANDLE

Even though there are plenty of rocket scientists in Florida, it did not take one to figure out that something was wrong with the state's building code in 1992. With sustained winds of 140 mph and gusts up to 165 mph, Hurricane Andrew damaged more than one hundred twenty-five thousand buildings when it blew across Miami-Dade County (not counting the mobile homes that were utterly destroyed in its path).[1] The devastation it left behind was the signal many needed to finally recognize that the building code in effect at the time needed a serious overhaul of its design wind speed, which is the value used to determine the wind pressure that buildings must be designed to resist, by law. So legislators worked steadfastly—albeit slowly, as always—to update the Florida building code, which, like all building codes, sets the *minimum* standards for construction—nothing less, nothing more. When it was finally adopted by the Legislature in 2000, and came into effect in 2002, sustained wind speed to be considered in the design of buildings on the shore and some distance inland ranged from 120 to 150 mph, with the higher values used in South Florida (which was still recovering from Hurricane Andrew at the time).[2] The new design wind speeds became the benchmark across Florida.

Except in parts of Florida's Panhandle.[3]

Following 2018's Hurricane Michael, which wrecked multiple buildings across the Panhandle and obliterated the beachfront community of Mexico Beach, journalists—as always—started looking for someone to blame for this disaster. That is when they remembered Charles W. Clary III, the first architect ever elected as a Florida state senator.[4]

Senator Clary was the founding principal of a respectable architectural firm[5] whose portfolio includes multiple educational facilities (kindergarten to college), civil and community projects, recreational projects, office buildings, and various resorts such as the Waterside Village at Mexico

Beach, Florida.[6] He had received multiple awards for his architectural works and civil service. That was apparently of little interest to reporters; it was Clary the politician that was in their sights. On October 16, 2018, the *Miami Herald* presented him as the architect and convincing advocate of an amendment to the 2000 legislature bill that exempted all new construction in Florida's Panhandle and located a mile or more from the beach from having to comply with the requirement that storm shutters or impact windows be used[7] (unless windows were considered broken by debris and treated as openings, producing an impractical wind design challenge).[8] The argument supporting the exemption was that the Panhandle region is densely covered by trees that serve as a buffer that attenuates the higher velocities that would otherwise be felt.[9] Given that the community of Mexico Beach is not inland but rather—not a surprise—right on the beach, it is not clear why the *Herald* singled out this issue, except maybe to support the statement it made earlier in the article that, for years, local politicians wielded the probabilistic assessments of engineers and scientists (based on the limited data set of past hurricane occurrences) "to successfully water down or stave off the local application of stricter windstorm codes" compared to the rest of the state.

Whether it was accurate reporting or not is irrelevant here, because it misses the point of what the true role of politicians in disaster mitigation is. Appreciating this role first requires briefly reviewing what makes a politician tick.

NEEDED, CREATED, AND CONVENIENT DELEGATION OF RESPONSIBILITY

Much could be said as to whether there is a better model of government when it comes to preventing disasters. The list of existing political systems is long and full of nuances. For those who have never seen it before, the "Parable of the Isms," dating back to at least 1935 and loosely attributed to various uncertain origins, modified and enhanced by many others since, summarizes the philosophy at the core of most forms of government. It goes somewhat like this (alleged origin in parentheses below):[10]

- Feudalism: You have two cows. Your lord takes some of the milk (Anonymous).
- Socialism: You have two cows. The government takes one and gives it to your neighbor (Silas Strawn, 1935).
- Communism: You have two cows. You give them to the government, and the government then gives you some milk (Silas Strawn, 1935).

- Fascism: You have two cows. You give them to the government, and the government then sells you some milk (Silas Strawn, 1935).
- Militarism: You have two cows. The government takes both and drafts you (Anonymous).
- Capitalism: You have two cows. You sell one and buy a bull (Silas Strawn, 1935). Then put both of them in your wife's name and declare bankruptcy (Pat Paulsen, 1968).
- Nazism (dictatorship): You have two cows. The government takes both and shoots you (Silas Strawn, 1935).
- New Dealism: You have two cows. The Government takes both, shoots one, milks the other, then pours the milk down the drain (Silas Strawn, 1935).
- Totalitarianism: You have two cows. The government takes them and denies they ever existed. Milk is banned (Anonymous).
- American democracy: The government promises to give you two cows if you vote for it. After the election, the president is impeached for speculating in cow futures. The press dubs the affair "Cowgate" (Anonymous).
- Russian capitalism: You have two cows. You drink some vodka and count them again. You have five cows. The Russian Mafia shows up and takes however many cows you have (Anonymous).
- Greek capitalism: You have two cows. You borrow lots of euros to build barns, milking sheds, hay stores, feed sheds, dairies, cold stores, abattoir, cheese unit and packing sheds. You still only have two cows (Anonymous).
- Japanese capitalism: You have two cows. You redesign them so they are one-tenth the size of an ordinary cow and produce twenty times the milk. You then create a clever cow cartoon image called a Cowkimona and market it worldwide (Anonymous).
- Chinese capitalism: You have two cows. You have 300 people milking them. You claim that you have full employment, and high bovine productivity. You arrest and make disappear the news reporter who suggested this might not be true (Anonymous).
- Marketing: Congratulations! You are now the proud owner of two thousand millicows (Anonymous).

And the list goes on and on (and it gets crazier[11] and crazier[12]).

In truth, no political system has proved to be better at preventing and managing disasters, and most of them have a horrible record in this regard (or a worse than horrible record). For expediency, the focus here will be on democracies because, per the quote often attributed to Winston Churchill, "Democracy is the worst form of government, except for all the others."

By definition, a democracy is a societal framework in which the population is sovereign in its ability to decide what is best for the collective benefit of the group, and does so by using elections to appoint members of a government to implement this supreme power that is vested in the people. In other words, democracy is a process by which citizens who are too busy trying to make a living and enjoy life choose to delegate the boring day-to-day operations of government to individuals whom they believe share their values, to ensure that these values will be steering all governmental decisions in the absolute best direction. Evidently, beyond "life, liberty, and the pursuit of happiness," nobody can agree on anything. There are almost as many different top priorities and values as there are citizens, and no absolute best direction. Therefore, those seeking an elected political office must become masters of deception to convince everybody that they support all of these values at the same time, even when they totally contradict each other.

It is therefore not surprising that some of those most successful at being elected include individuals willing to do just about anything for money. The Italians, who brilliantly recognized that this is possibly the ultimate defining quality of a politician—and must have a great sense of humor—elected a porn star (Ilona Staller) to the Italian parliament in 1987.[13] When Saddam Hussein, the Iraqi dictator known as "The Butcher of Baghdad,"[14] invaded his peaceful but oil-rich Kuwait neighbor in 1990, the Italian Parliament had a hard time deciding whether it should join the coalition of nations that was gearing up for what became known as the Gulf War to free Kuwait. However, pragmatic Parliament deputy Ilona Staller built on her practical knowledge of what drives frustrated, powerful men to war, and offered to have sex with Saddam in return for world peace. She repeated her offer at the onset of the Second Gulf War in 2002, extending the offer to both Saddam Hussein and Osama bin Laden. None of these offers were accepted and both wars unfortunately happened, but the history books will be able to report that it was not for lack of trying to achieve world peace.[15]

Democracies always get the politicians they deserve.

To put it mildly, there is a reason why politicians have been the butt of jokes since time immemorial. President Ronald Reagan once said that politics is the second-oldest profession, but that it "bears a striking resemblance to the first." Or, as speech writer Paul Begala (who also served as a strategist for the Clinton-Gore ticket) said, and was later made popular by Jay Leno in the 1990s,[16] "Politics is show business for ugly people"—an analogy with acting that could not be more on the nose for those who recite scripted speeches anytime a spotlight hits them, regardless of whether or not they believe the message. Furthermore, among all politicians, those who are the puppets of large corporations, who are willing to kiss as many asses as needed to move up, who have no scruples about spending like sailors on shore leave when it can be charged to an expense account, who have no principles while pretending to be flag bearers of the highest moral values, whose every calculated action is self-benefiting and self-promoting—when they are not outright corrupt, pathological liars, or Machiavellian sociopaths—by no means help the image of the profession. After all, Frank Underwood (in the *House of Cards* series) was not created in a vacuum.

To top it off, in this new millennium's climate of aggressive political divisiveness, where civility has gone out the windows, things are not getting better; putting in print the latest quotes from some politicians is definitely not good etiquette—and might even be insulting to the English language.

Studies that cast a positive light on the character of politicians are few and far between, given that the first words that apparently come to people's mind when thinking of politicians are *scum bags* and *crooks*.[17] On the positive side, personality tests[18] and other analyses have shown that politicians are more agreeable, charming, fearless, confident, and extraverted, but less open to new experiences, than the average person.[19] Whether any of these traits can be considered good qualities depends very much on context. Less flattering are psychological profiles demonstrating that politicians are narcissists[20] driven by power and money and the prestige, status, and authority that comes with it.[21] Not to forget that all the above is only talking about "regular" politicians, not the Machiavellian, authoritarian, paranoid, or totalitarian ones.[22]

What is important here, though, is that politicians in democracies must be elected. As a result, they must be masterly skillful at saying what needs to be said to reach this goal. They may resort to doublespeak to escape difficult situations, but it nonetheless remains that they must advocate some political platform to get votes—even if they must create ambiguity to convince groups entrenched in two opposite positions that they are the best candidate to get their votes and defend both of these incompatible beliefs.

Therefore, when the population is in unanimous agreement on a given topic, a good politician will seize the occasion. Likewise, when the population could not care less about a given issue, no politician will waste time talking about it. Hence, political platforms are tailored to resonate with the population's list of priorities and emotions. It is that simple—and disappointing too.

Yet when things go wrong and roofs start flying off during a hurricane, or buildings collapse during an earthquake, or water reaches roof level in entire neighborhoods, pointing fingers at politicians happens. It is scapegoating at its best. "Why was nothing done by the legislature to prevent this disaster?" "What have X and Y waited for before acting?" "Who is responsible for this debacle?" Always looking for a story, the press will broadcast it all, playing it up with bold headers, often confusing opinions and facts—or expediently focusing on opinions without bothering with facts if under a tight deadline or lax editorial policies.

However, the truth of the matter is that politicians are nothing but mirrors. If nobody cares about hurricanes, or if all have convinced themselves that there is no problem, why would any politician waste time making an issue out of it? A successful politician will not try to sell something nobody wants (contrary to the unsuccessful encyclopedia salesman, thrown out the window in Monty Python's skit). Compounding the problem is the fact that politicians are driven by the necessity to spend money on projects that will show tangible benefits (or soon to be achieved benefits) by the time they are running for reelection, be it in two, four, or six years, depending on where they sit in the political train. "Shovel-ready" projects to build roads, to ease traffic congestions, to expand the subway, to build new infrastructure, are all winners. Changes to the income tax code or to other laws that will end up putting more money in everybody's pockets are winners too. Changing the building code to prevent disasters when a severe earthquake or hurricane happens years—if not decades—after the next election is generally a "ho-hum" accomplishment. Possibly even a losing proposition.

It is worth repeating: If the population does not care, why should a politician care? That is the mirror effect: Politicians only reflect the priorities of the electorate. In that sense, scapegoating them is unfair—although, by definition, scapegoating is always unfair—but it sure feels good to beat on politicians. They are experts at deflecting blame, anyhow, and if not, a few heads falling is good group therapy.

WHO GIVES A DAM

There are usually two acceptable ways for water to exit the reservoir behind a dam, namely, going through the intake pipes to feed turbines that generate power, or through spillways (or by-passes) specifically designed to drain excess water before it overtops the dam itself. Water flowing through any other way is either a looming problem, a big mess, or a catastrophe.

On February 7, 2017, torrential rains in Northern California[23] led to a substantial increase in the level of Oroville Lake behind the Oroville Dam.

The flood gates of the main spillway were therefore opened to let water through. The main spillway is effectively a concrete channel designed to direct water downhill in a straight line, with some devices at the end of the channel to dissipate much of the flow's energy before it is returned to the natural river canyon. Sending water down a spillway is a normal operation to control water levels, and it should have been a routine operation on that day too. However, after a little while, something strange was observed in the flow of water about halfway down the three-thousand-foot-long spillway, and the gates were closed to figure out what was going on. Inspection revealed a crater in the surface of the spillway. The stability of the crater was tested by releasing controlled amounts of water, but the crater enlarged to a length of three hundred feet. Not good—particularly as water kept filling the reservoir in the meantime. Water in the lake was reaching critical levels, so on February 9, the spillway was reopened but only at half capacity in an attempt to prevent the crater from growing too much. Still, by February 11, the reservoir peaked to its all-time height and water started flowing over a concrete structure that served as the crest of a more rudimentary emergency spillway.[24] More specifically, by design, in the emergency spillway, after a smooth flow over a fifty-foot-wide concrete cap that protects the edges of the reservoir along the length of the spillway, water was sent cascading down the mountain, through nature, for the rest of the ride, all the way to the river canyon. Not a pretty "roller-coaster" ride, but it does the job as a temporary measure in an emergency. Unfortunately, this event was "less temporary" than intended by design: the reservoir level kept rising and erosion of the mountain was moving up close to the lip of the concrete cap of the emergency spillway, and doing so faster than expected. Failure was becoming a possibility—the type of failure that would send a deluge downriver. So by February 12, evacuation orders were issued to 188,000 people at risk. The main spillway was then open to full capacity; water jumped out of the spillway channel halfway through (at the crater location), carving a new river into the remaining mountain. Thankfully, sunshine returned in time after the rain, the inflow of water into the reservoir stopped, and the worst was avoided.

While the above story of this "close call" is interesting in itself—or not, depending on which outcome scenario one prefers—it is actually some of the non-technical findings from the independent forensic team report that give the most food for thought.[25] Careful to highlight the fact that "the incident cannot reasonably be 'blamed' mainly on any one individual, group, or organization"[26] at the very start of the report, the forensic team makes the most damning statements: "Although the practice of dam safety has certainly improved since the 1970s, the fact that this incident happened to the owner of the tallest dam in the U.S., under regulation of a federal agency, with repeated evaluation by reputable outside consultants, in a state with a leading dam safety regulatory program, is a wake-up call for everyone involved in dam safety. Challenging current assumptions on what constitutes 'best practice' in our industry is overdue."[27]

In other words, a panel of experts cumulating more than 230 years of experience in dam safety have called a nearly averted disaster at the tallest earth-dam in the United States (at 770 feet), holding up the second-largest lake in the state of California[28] (1,140,000,000,000 gallons),[29] a "wake-up call."

A wake-up call. Really?

As mentioned earlier, approximately every four years, the American Society of Civil Engineers (ASCE) releases report cards on the state of infrastructure in America. In 2017, for its entire infrastructure, the country got a D+.[30] That is not exactly brilliant (in most universities, a student with a D+ average would be expelled from engineering school). In 2021's ASCE Report Card, America's infrastructure improved to a C- grade— getting out of the D range for the first time in twenty years, pulled up by modest improvements in the infrastructures for aviation, drinking water, energy, inland waterways, and ports.[31] On their own, dams—as one part of the country's infrastructure—received a D grade on ASCE's Infrastructure Report Card, both in 2017 and 2021. Dams received this D grade on account of the fact that there were more than 2,300 deficient "high-hazard-potential" dams in need of repairs, rehabilitation, or replacement across the country. Note that 15,498 dams (17 percent of all dams in the United States) are considered to be "high-hazard potential"—a term defined to mean that people will die when they fail.[32] An additional 11,882 dams are considered to be "significant-hazard potential," defined as meaning that people are not likely to die if they fail (although unforeseen casualties are always statistically possible), and that only "significant economic losses" would ensue. Dams across the nation are on average more than fifty years old and the cost to repair the high-hazard potential dams alone is estimated to be $45 billion.[33]

The 2013 ASCE Report Card[34] was a bit more blunt, calling four thousand dams "deficient," half of those being among the fourteen thousand "high-hazard dams." The 2005 Report Card[35] was even blunter, labeling thirty-five hundred dams as "unsafe." Since 1998, dams have held steady at a D grade—possibly because ASCE is more lenient than most university professors and does not give F grades.

So, if the events at the Oroville Dam were a wake-up call, there must have been a lot of snoring through the release of all the ASCE Report

Cards. Parts of the six-hundred-page forensic report of the Oroville Dam failures provide some nuggets of information to explain why that may have been the case, particularly on how politics affects outcomes, by outlining a number of problems that were found to exist within the regulatory context.

Dam construction and operations in most countries—including in the United States—are only possible in compliance with a complex framework of regulations. This is because dam failures can be quite consequential. Yet catastrophic dam failures have occurred worldwide and more regularly than one would think (or wish) over the past century.[36] Given that water weighs a ton per cubic yard, it does not necessarily take the failure of a gigantic dam to produce a large number of casualties.

Take the South Fork Dam in Johnstown, Pennsylvania, built in the mid-1800s, which was 72 feet high and 918 feet long—not minuscule but not a behemoth either. When the earth dam was overtopped after days of severe rainfalls, it rapidly eroded and sent two million gallons of water downhill, wiping out the nearest town, and causing more than 2,200 deaths in the process.[37]

In California, the St. Francis Dam, a 185-foot-tall, 700-foot-long concrete arch-dam forty miles from Los Angeles, failed in 1928, less than two years after its completion.[38] It was a failure so sudden that a 140-foot wave rushed down the canyon. It was still 55 feet tall ten miles farther down when it dissipated into the broader Santa Clara River valley, killing over four hundred people in the process—making it the second-largest California disaster after the 1906 San Francisco earthquake.[39] This prompted the creation of the Division of Safety of Dams (DSOD) in 1929,[40] with the mandate to review and approve plans and specifications for new dams as well as alterations and repairs on existing dams, and to oversee their construction. It operates as an agency under the California Department of Water Resources (DWR).[41] The Oroville Dam itself was built by California's DWR[42] and is owned and operated by the California State Water Project (SWP),[43] which is also a public entity under DWR. To summarize and simplify, the operator (SWP) and the regulator (DSOD) are both divisions of the DWR—in short, all public employees working for the State of California, as one big family. Since the Oroville Dam generates electricity, it is also regulated by the Federal Energy Regulatory Commission (FERC).

As part of its forensic study, the independent team interviewed many key players in all these agencies, to gain an appreciation of the decision-making processes and safety procedures—or challenges related to those—that have led to the near-catastrophe of February 12, 2017. First, they met a DWR imbued with feelings of "overconfidence and complacency" shored by the widespread belief that all the SWP projects had been designed by nothing less than the "best of the best"[44]—like mythological giants. This belief was apparently handed down from generation to generation within the DWR, but was apparently not unanimously upheld, as one of the interviewees alleged that the Oroville spillways had been originally designed by an engineer fresh out of grad school who had never designed a spillway before and who relied on his class notes from hydraulic courses for what turned out to be his first professional design project—in other words, not quite a giant.[45]

The key challenge outlined by the forensic team was:

> DWR generally relied too much on regulators and the regulatory process to identify problems with its aging dam infrastructure, rather than proactively doing in-depth evaluations initiated by DWR and assuming full responsibility for the management and safety of its dam projects, as is required on both an ethical and legal basis. As with other factors, this reliance on regulators and the regulatory process was not atypical for large dam owners.[46]

Unfortunately, they also found that

> in the case of FERC and DSOD, both regulators had heavy workloads relative to their available staff, and due to being government agencies, they faced major bureaucratic constraints related to attracting, retaining, and adding qualified staff, as well as redirecting or terminating low-performing staff. . . . In addition, a further challenge faced by regulators is that, while they may, in theory, be able to reference "tough" regulations when dealing with dam owners for the purpose of managing risks to the public, regulators often have difficulty in gaining compliance from dam owners in practice, because there are limitations to the sanctions they can reasonably impose.[47]

In other words, neither the operator nor the regulators had it within their mandate to check how the spillway designed decades ago would compare to one meeting today's standards, or comprehensively review the dam safety programs in place to possibly identify some of the deficiencies that led to the problems in 2017.[48] Instead, DWR opted to react to problems as they arose or as they were pointed out by regulators rather than to take actions proactively. The forensic team mentioned that this was not atypical for large dam owners because—and here is the real kicker—of "cost control pressures" and other factors that "make it difficult to justify proactively spending money to prevent problems that have not yet occurred."[49] Essentially, "if it is not broken, do not fix it"—but at a grand scale.

Citing the forensic team's incisive analysis:

In addition to cost control pressures exerted ... specifically on DWR, there were also general pressures from the public to control the size and associated cost of California state government, of which DWR is a part. A primary mechanism through which this pressure was exerted on DWR was in controlling the number of staff positions in DWR's organizational chart. Rather than DWR having discretion to hire new staff as it determined necessary, the forensic team was told that new positions could only be added to the organizational chart with approval of auditors in the governor's office, and requests to add such positions were routinely denied. This resulted in a situation in which DWR often had sufficient budgets on paper to meet its objectives, but was chronically unable to fully make use of those budgets due to not having enough staff. This, in turn, also resulted in an increase in the amount of services contracted out to external consultants, which offset some of the cost control intended by limiting the size of DWR, but also entailed DWR needing resources to manage those consultant services.[50]

In simpler terms, citizens elect politicians who promise to cut taxes, and budget cuts are pushed down to government agencies. Critical operations must go on, but with fewer people, less (or deferred) maintenance, and less morale. The attention is focused on short-term needs at the expense of long-term matters (such as proactively preventing problems). Every day, the short-term priority of the SWP is to meet its mandate of providing water to the entire population and the irrigated farmlands of the state, and producing hydroelectric power. And to do it now! The rest? It depends on available resources.

As stated in the forensic report, "DWR needed to balance these goals of 'production' versus safety, and lacked leadership and authority at the executive level which focused on finding the right balance."[51]

It may take shrewd politicians to recognize the desire of the population to cut taxes and to campaign on such a platform, but all they do is mirror the will of the majority of the population—as it so works in a democracy. In other words, some politicians may be spineless puppets and some may even be horrible human beings, but blaming them for disasters—while a convenient stress-relief mechanism, possibly encouraged by mental health professionals to keep heart attacks and depressions at bay—is nothing more than misplaced anger. As stated before, a politician is a mirror.

If the attitude that hurricanes, earthquakes, and other hazards are "not a big deal" is widespread in the population, the disbelief in the true forces of nature becomes institutionalized—or, more accurately, politicized. In such circumstances, it would be illegitimate to claim after a disaster that the government has failed its citizens—no matter how deep the pain cuts—because it would simply amount to a refusal of looking in the mirror.

Incidentally, it is important to state that citizens may be fully justified for wanting to cut down the size of government, be it because of frustrations dealing with some civil servants that are neither civil nor serving, or for having witnessed the careless waste of taxpayers' money that goes on in too many parts of the bureaucracy. However, when budget cuts have been imposed to government agencies in the past, have the uncivil employees lost their job and the wasteful activities ceased, or have the cuts hurt in the most sensitive parts of operations? At the extreme, in times of temporary severe fiscal constraints, some divisions of government agencies could manage to "squeak-by" by spending all of their budget to keep their staff on the payroll, but with nothing to do as there would be no budget left to spend on any of the operations and activities for which these agencies exist in the first place (some may have heard senior managers privately admit having done that at some point). Yes, government agencies are like big families, and, like all families, some are happy and productive, and some are hopelessly dysfunctional.

Not surprisingly, when it comes to shrinking the size of government, deferred maintenance has been a target of choice by managers who seek to meet reduced budgets, and since maintenance of infrastructure can be expensive, it has been a big target. Then again, this should not be totally a surprise given that, according to AAA, 35 percent of Americans "skipped or delayed service or repairs that were recommended by a mechanic or specified by the factory maintenance schedule."[52] In other words, to save $120 per year on oil changes, a third of car owners are willing to bet that they will trade in their car before having to fork the $4,000 it will cost to replace the engine when it is irremediably damaged for lack of maintenance.[53] That amounts to approximately $25 billion in delayed maintenance—mostly overdue oil changes, worn brakes, bald tires, and other things that could cause severe breakdowns or fatal accidents.[54] Maybe this is not surprising considering that 40 percent of the adults under thirty-five apparently do not know how to replace a wiper blade and would struggle to figure out how to add windshield washer fluid.[55] Over the years, auto manufacturers have added red "warning lights" that pop-up when serious maintenance is overdue—called "idiot lights" by those in the industry;[56] however, many drivers cannot be bothered and do not mind driving thousands of miles with an illuminated malfunction light—at least, until something actually breaks down and the car eventually stops by itself.

Likewise, another favorite target when it comes to reducing government spending is to cut down on travel. The ability to make good decisions depends in part on being aware of the state-of-the-art in one's field of expertise, which is usually achieved through continuing education. It is often said that a century ago, half of what an engineer learned became obsolete over a career of thirty-five years. By the 1960s, creation of new knowledge had accelerated, and it was estimated that it took only ten years for half of an engineer's knowledge to become obsolete.[57] In 2002, the president of the National Academy of Engineering ventured that, depending on the discipline, this "half-life" of knowledge had declined to between two and a half and seven years.[58] Not surprisingly, most professional organizations nowadays require proof of continuing education activities as a way to keep current, and therefore as a requirement for renewal of professional practice licenses.

Continuing education is often achieved by attending conferences, participating in technical committee meetings, and completing training courses—which, admittedly, is more costly than staying chained to a desk and learning nothing new. A good way to cut costs is to hold conferences in relatively more affordable locations. Ironically, the cheapest places to hold conferences and meetings in the United States are Las Vegas and Orlando off-season. Hotels and airfares are cheap and almost every city has a direct flight to these destinations, which shortens travel time away from the office and, correspondingly, loss of productivity. Yet, rare are the government agencies that dare to hold their meetings in Sin City or anywhere near Mickey Mouse. The whole idea became irremediably toxic after the US government's General Services Administration—whose very mandate is to "streamline the administrative work of the federal government"—blew $800,000 in an extravaganza training session in Vegas for three hundred employees,[59] complete with room parties, thousands of dollars in sushi, rented tuxedos, a session with a mind reader, and multiple trips to Vegas by management ahead of the conference to "scout" possible locations for the event.[60] Yes, Vegas can be cheap and highly cost-competitive, but there are infinite ways for fools to rack up a bill in no time.

As a result, government meetings are typically held in places unlikely to attract attention. For example, the Committee on Bridges and Structures of the American Association of State Highway Transportation Officials is where the states' bridge engineers meet to discuss changes and updates to a number of the nation's bridge design specifications. It is important business. The committee has typically held its annual meeting in exotic Americana locations, such as Branson, Missouri; Montgomery, Alabama; Burlington, Vermont; Spokane, Washington; and similar places almost forgotten by airlines and not on the radar screen of wild and crazy party animals.

Pushing this logic to the extreme, in tough times, some public agencies have altogether banned travel for their employees. The easiest way for senior managers to avoid the risk that even a meeting in Pohenegamook (one campground, a couple of motels and less than three thousand population)[61] could be perceived as a junket is to deny all travel requests. It also eliminates the need to decide what is justified and what is not—which is ideal for bureaucrats daunted by decisions.

In the meantime, while waiting for federal prisons to open convention centers within their walls, maybe staff can use Google searches to see if free online courses are offered by world-leading experts in dam maintenance, operations, and other valuable topics.

PUBLIC POLICY

It is not clear if anybody ever woke up wondering if the day was going to be a Red, Orange, Yellow, Blue, or Green day and changed their routine in any way as a consequence. Actually, ask if anybody in a crowd can describe a memorable Red or Orange day, and expect to hear nothing but crickets.

The Department of Homeland Security was created a few months after the 9/11 attacks to integrate the activities of twenty-two different federal departments and agencies into a single one, with a staff of a quarter-of-a-million people, a $50 billion discretionary spending budget,[62] and the mission to secure the nation from the many threats it faces.[63] Shortly after it started operation on March 1, 2003,[64] one of the department's first accomplishments was "to provide a comprehensive and effective means to disseminate information regarding the risk of terrorist acts to federal, state, and local authorities and to the American people," in response to a presidential directive.[65] Called the "Homeland Security Advisory System," it took the form of a color-coded scale to warn the public of the terrorism threat level at all times.

Red, orange, yellow, blue, and green warnings corresponded to severe, high, significant, general, and low risk. However, the decision to pick the color of the day was made behind closed doors, following an unpublished criterion, using evidence kept confidential, and therefore providing the public no way to assess the significance of the warning.[66] Furthermore, for the eight years during which the system was used, yellow was the norm. The threat level was raised to orange for a grand total of roughly one hundred days, and to red only once. The blue and green levels were never enacted at any time—which cynics would argue is because terrorist events can be unpredictable, and would make the government look immensely foolish if one happened on a blue or green day.

To presume that Joe Schmo and Jane Doe would have taken seriously such warnings of impending danger (combined with authoritative reassurances that the government was on top of it and taking all necessary actions to ensure security) is tantamount to being unaware of the general level of skepticism of the public toward its leaders. Trust is something that must be earned, and if anything, the public trust in its government and institutions has been severely eroding over time. From a peak of 77 percent of the population trusting the government in 1964, down to 17 percent nowadays,[67] it is quite a dive—but still scoring a tad bit higher than trust in used car salespersons, if that can be of any consolation. Studies show that 90 percent of Americans would ignore evacuation orders if given without explanations.

Yet, in spite of this massive distrust, the government must govern, which means making decisions. Every action of the government that impacts the public—by laws or funding priorities—can essentially be considered to be within the realm of public policy.[68] However, that does not imply that public opinion is surveyed every step of the way. In fact, the government is usually quite busy advancing its own agenda. Senior officials in charge of various dossiers decide what to prioritize—including how to address vulnerabilities—how to spend limited budgets (hopefully wisely), and move forward, hoping for the best, working on assumptions as to what the public is willing or not willing to accept—or pay.[69] Inevitably, as the policy wheel turns, it will eventually roll over nails along the way. As something unexpectedly blows up, a media frenzy may ensue, it may grow to become a political hot potato, an emergency "patch" will be applied to plug the hole, the crisis will abate, the public interest will wane, and things will return to normal. That might be as much public feedback as most processes get along their course.

At the other end of the spectrum, for some hot topics, it is exactly the opposite. The public can be mobilized—actively or passively—in a specific cause in a way that will drive the process to change if a clear majority opinion can emerge out of the process.

Either way, a lot of public policy work has had major positive impacts on the lives of millions, thanks to the blood, sweat, and toil of policy wonks, and the activism of others. Achieving effective public policy is as much art as rigor, patience, and timing. Experts know how this process can be nudged in a desired direction, always one step closer to a target outcome.[70] They are also wise enough to know that a lot of it is smoke and mirrors—such as to defuse or prevent a crisis.

Before 9/11, half of the 126 crossings at the Canada-US border were unguarded at night, with orange cones put in the center of the roadway as an indication that the crossing was closed for the night—which had not stopped many drivers from simply running them over.[71] Nowadays, to make the border more secure, the cones have been replaced by infrared sensors in the road—and, sometimes, for convenience when in the middle of nowhere, a phone booth with dedicated lines for travelers to voluntarily call a US or Canadian customs officer before entering the country.[72] Bilateral discussions on how to continue enhancing border surveillance are ongoing,[73] but nobody has the desire—nor the budget—to build a fence along the 5,525 miles border. Neither side publicizes where the most porous parts of the border are located, but smugglers are known to work some of the weak points—in both directions, depending on the special needs of their clients.[74]

Most things when it comes to security happen under the radar, sometimes in odd ways and definitely invisible to politicians not interested in the details. So where does the politician come into the picture then? It has often been said that the focus on near-term profits and quarterly earnings has hollowed out many companies, cutting the investments needed to innovate and ensure their long-term survival. This has been attributed to CEOs lacking the expertise needed to understand the very product that their firm produced,[75] or other reasons—such as predatory managers parachuted by a board of directors with the mandate to milk the cow dry for greatest immediate earnings. Not surprisingly, politicians can exhibit the exact behavior.

There is always an election around the corner, and image matters. What better way to display empathy than showing up in a disaster zone to act the leadership part in the post-disaster game. Good politicians excel at it because people swallow it all, temporarily forgetting awful character flaws and other faults. Others who have not read the "How to Charm the Media" script typically fail the audition or commit political suicide in the process. Michael Brown, head of FEMA at the time of Hurricane Katrina, is one who definitely did not get a standing ovation. In his own words: "Mishandling of the press during the disaster response was among my greatest mistakes."[76] Within two weeks of landfall, with New Orleans still underwater, he had been so disparaged by the press that he decided to resign.[77] As stated in the bipartisan committee report that investigated why the preparation and response to Hurricane Katrina was so disastrous, "Fair or not ... Brown in particular had become the symbol of all that went wrong with the government's response to Katrina."[78] Transcripts of emails released as part of the congressional investigation did more than raise eyebrows, including one from his press secretary advising him to roll up his sleeves like President Bush, "just below the elbow" to look hard-working, and one from himself stating, "If you look at my lovely FEMA attire you'll really vomit. I am a fashion god."[79] While Brown compared those comments to jokes that surgeons make in the operating room while saving precious lives,[80] it did not endear him to the drowning folks in New Orleans. (Note to future FEMA heads: There is maybe a reason why surgeons never get invited to perform in comedy clubs.)

In summary, disasters can provide opportunities for skillful politicians to steal the spotlight for their own advancement, and sometimes launch investigations that will deflect attention for months—and sometimes sacrifice a few scapegoats—once things return to "normal," the normal after the disaster often being not that different from the normal before it. It was not an electoral issue in 2008 that the government did not have $15,000 in 2004 to fund an emergency hurricane response simulation before Hurricane Katrina, or even that Hurricane Katrina happened. Public policy issues that polarized the 2008 election[81] were mainly universal health care and the fact that the Iraq war was still going on, at the rate of $200 million in spending per day in 2006[82] and more than double that in 2008.

So, if politicians are useless when it comes to public safety, then who is minding the store to prevent disasters? By default, as mentioned at the beginning of this chapter, there are building codes. This makes politicians irrelevant then. Building codes, thanks to their existence, must be taking care of the problem, right? The answer is an unequivocal "not quite."

ON THE DISASTER TRAIL

A Recipe for Disaster

Decades ago, I decided to surprise my wife by baking her a birthday cake from scratch rather than purchasing one. In those days, long before the internet and even personal computers, she kept all of her favorite recipes on notecards stored in a wooden box. So mid-afternoon, when she was out running some errands, I pulled out a notecard that had the words "excellent" and "cake" on the header line, and proceeded to create my culinary masterpiece. The recipe seemed simple enough: Flour, sugar, butter, eggs, milk, baking powder, vanilla extract. That was it. A handwritten bullet list of items. So, I put all the ingredients together in a large bowl, mixed everything until it seemed well blended, and baked the resulting dough. Made glazing similarly following another recipe with much fewer ingredients, and added candles. Voilà!

I still remember her surprise when I brought the "genuinely made from scratch by none other than your husband" birthday cake, and how long she paused before blowing out the candles, incredulous and stunned. I also remember that she was careful not to crush my enthusiasm after her first bite, and that her positive words skillfully balanced encouragement and restraint. Then, my first bite revealed the truth. The cake was full of dried lumps that made it plainly uneatable. Our marriage survived this experience at the antipode of gourmet cooking, but at that very moment, she must have wondered if I was trying to poison her. The "key ingredient" missing from her notecards was the sequence by which the ingredients had to be mixed to achieve a smooth texture—critical steps that the experienced cook that she was had not bothered to write on the notecards, as that concept was already clear in her head. She had never foreseen that a culinary illiterate apprentice would attempt to use any of her recipes.

There were no victims, but I learned the hard way that when things are created with the best of intentions, on paper they can make perfect sense and lead to outstanding results. However, when the underlying assumptions sustaining these constructs are unknown, the concepts, rules, and regulations that seemed perfectly logical when originally formulated can be applied in many unforeseen ways—leading to lumpy cakes and other disasters.

Earthquake Damage Happens Because . . .

BUILDING CODES

Building codes as well as design specifications and standards are developed in many different ways across the world. In some countries, the codes are developed by committees of experts and stakeholders, and government legislation makes compliance with these documents mandatory for all construction—effectively making them national building codes. Governments in other countries, for lack of resources or other reasons, find it more effective and expedient to adopt a code developed abroad, translating it into the local language or, as a minimum, slapping a different cover page on top of it and disguising the content with crafty reformatting to make it look like a national code of their own. In some cases, the borrowing is acknowledged openly, for sake of transparency and convenience.

For instance, in many countries, the "Specifications for Structural Steel Buildings" produced by the American Institute of Steel Construction (AISC) have been adopted—or copied—to become the official standard by which steel buildings must be designed. In fact, given that imitation is the sincerest form of flattery, it is mindboggling that one of the countries that religiously follows the AISC requirements is—of all places—Iran! While the Iranian government has recurrently banned music from America the "Great Satan,"[1] it has been more pragmatic when it comes to designing buildings.

In the United States, the way buildings codes are adopted is somewhat convoluted, not only because each document is often developed from scratch by necessity, but also because the Tenth Amendment to the US Constitution effectively leaves to the people every power not specifically granted to the federal government.[2] As such, regulation of the design and construction of buildings becomes the people's responsibility, which they can elect to delegate to state governments, cities, or towns.[3] In other words, instead of a top-down approach, it is a bottom-up approach, in what some would consider to be the purest form of democracy.

As far as specifying magnitude of loads to consider and quantifying design strengths to provide, there were no building codes for a long time in the United States—as in all countries for that matter. The oft-cited Code of Hammurabi, proclaimed by the king of Babylon circa 1754 BC and engraved in one of the most durable and hardest stones,[4] is more an "eye-for-an-eye" legal statement than a building code per say. It stated: "If a builder builds a house for someone, and does not construct it properly, and the house which he built falls in and kills its owner, then that builder shall be put to death. If it kills the son of the owner, the son of that builder shall be put to death. If it kills a slave of the owner, then he shall pay slave for slave to the owner of the house."[5] And so on. It does not provide much details, but it "tickles pink" those who advocate that all building codes should be so simple, leaving it to the engineers to determine how to design buildings, and to the lawyers and judges to enforce execution—literally.

Beyond the sweeping statements of Hammurabi, prescriptive building codes typically came of age following disasters. Mostly fires. After the AD 64 blaze that devastated 70 percent of Rome over six days—the one Emperor Nero was rumored to be fiddling throughout (a popular myth, with a double meaning, but improbable musically since the fiddle was only invented in medieval times), and even to have started because he wanted to clear space for a new palace[6]—rules were enacted for wider streets and fireproof walls between buildings.[7] Likewise, throughout antiquity, following massive conflagrations, such as the 1666 Great Fire of London that consumed more than sixteen thousand houses,[8] and the Great Chicago Fire of 1871 that turned seventeen thousand buildings into ashes, killing three hundred people and leaving a hundred thousand homeless,[9] some laws were typically enacted to prohibit certain types of construction.

That was all great to prevent the burning of buildings, but beyond that, for the most part, code requirements were enacted to set minimum standards for light, ventilation, plumbing, and other "livability" issues as a way to address sanitation and safety issues created by unscrupulous slumlords that endangered public health.[10] In the best circumstances, the codes were enforced and served their purpose, but not always, as buildings departments in some cities were understaffed—when not outright corrupt.[11]

When it came to mandating minimum structural strength, nothing happened for a long time.

In the era of the master builders, from antiquity to the beginning of modern construction a little bit more than century ago, much was learned by trial and error. The focus of attention was to provide enough strength and stability for the building to resist its own weight and that of its occupants—essentially, a fight against gravity. Blessed were the workers who died in the collapse of medieval and Renaissance cathedrals, as they were rewarded—posthumously—with a free pass to heaven. While it may have been horrible for the five-hundred-foot tall stone tower of the Beauvais Cathedral—claimed to be the tallest building in the world at the time[12]—to collapse during mass in 1573,[13] management recovered brilliantly by calling it a miracle that only two people died in the tragedy. Obviously, there were no billboards along the roads of sixteenth-century northern France advertising personal injury law firms that bragged about their ability to achieve a different kind of miracle, to help accident victims get the justice—and big settlements—they deserved.

As buildings were becoming taller and made of steel and reinforced concrete rather than wood or masonry, many started to design buildings to resist some lateral force considered to represent wind, but the determination of what value to use for that force was based on professional judgment—and certainly not mandated or specified by anybody.

When material was expensive and labor cheap, losing a few lives here and there was "par for the course." Some of the workers building the skyscrapers and bridges of the early 1900s may have been showboating daredevils in front of cameras,[14] but contrary to legend, most were normal people who understood the dangers but needed a job—and were as afraid of height as everyone else.[15] When the Golden Gate Bridge was constructed from 1933 to 1937, it was the first large bridge project to include a safety net running a few feet under the entire bridge roadway during its construction. It saved the lives of nineteen workers who accidentally fell into the net. In an era where one death per million dollars in construction cost was considered normal, those survivors were called members of the "Halfway to Hell Club."[16]

Definitely, leading the profession during the first half of the twentieth century were master builders—structural engineers at the very top of their game—like Gustave Eiffel (Eiffel Tower), Joseph Strauss (Golden Gate Bridge), John A. Roebling (Brooklyn Bridge), and Homer G. Balcom (Empire State Building). Right after that, there were thousands of skilled engineers producing competent structures that have performed satisfactorily up to this day and—barring lack of maintenance—will continue to do so for a long time. Think of all the century-old buildings and bridges in every city's downtown that are still in use—way beyond their original fifty-year design life. Unfortunately, beyond all that, there was the occasional problem, where an undesirable outcome happened not because of a phenomenon yet undiscovered by the profession, but rather due to the inadvertent unawareness by some engineers of problematic factors known to most other practitioners. Everything in life has a statistical distribution, with an average and a standard deviation. Medical doctors may refuse to admit it, but some of their classmates were "not the sharpest knife in the drawer." If all lawsuits were ruled to be frivolous and unfounded, maybe the malpractice insurance premiums would be lower—the American Medical Association reported that obstetricians on average paid $150,000 in annual premiums, and up to $250,000 per year in some parts of the country.[17] It is not a mere coincidence that many health care insurance plans cover the cost of getting a second opinion from a different medical doctor.

Likewise, it can be argued that structural engineering requirements in design standards came about in part to prevent recurring problems that easily could be avoided by listing specific requirements that must be met, stating "thou shall do this" and "thou shall do that." These requirements then could be used not only by engineers, but also by packs of lawyers waiting to feed on errors and omissions—hence the commonly used name "errors and omissions coverage" in professional liability insurance.[18] In fairness, as far as structures are concerned, because engineers are keenly aware of their foremost responsibility to protect the public, the discipline has created a sea of successes, with islands of failures—and not the islands of successes in a sea of failures that have been created by flippant computer jocks.

Note that the above text generically refers to design standards, or codes, or specifications, but nowadays, there are almost as many different design standards as there are types of structures. Evidently, wind is wind, earthquakes are earthquakes, and the laws of gravity are the same for everybody. Likewise, a steel structure is only an inanimate assemblage of beams, columns, and other steel pieces that will deform the exact same way when resisting loads and fail the exact same way when overloaded, whether the structure is supporting people, trucks, machinery, or roller coasters. Unfortunately—except for shareholders of the pulp and paper mills that produced the rolls of paper needed to print all of these documents—there are separate stand-alone documents that specify how steel structures should be designed for bridges, buildings, transmission towers, nuclear power plant, offshore platforms, and many more types of applications. This is in part because these different documents have emerged over time to address the needs of different industries. Also, as odd as it may sound, all that repetition is in the spirit of achieving simplicity, because engineers designing, for example, a steel bridge have to worry about slightly different problems than those designing a steel building. For instance, building designers must worry about properly fireproofing the steel structure to protect it from a fire burning for a few hours—not the case for bridge engineers. Bridge engineers must worry about the fact that the cumulative effect of millions of trucks crossing the bridge could fatigue the steel prematurely if detailing is not done a certain way, because this fatigue could lead to fracture and potentially collapse—not the case for building engineers. Of course, there are exceptions in both cases—landmark bridges should not collapse because of a tanker truck fire, and fatigue should be considered in a multistory bus terminal (which is a building). Still, design standards focus on the specific issues that drive the design of the run-of-the-mill structures that fall within their scope, and do not waste time addressing issues that are usually irrelevant for the type of application at hand. For example, if somebody decides to construct an office building in the middle of a river—no architect's fantasy being denied nowadays—and that building is exposed to the risk of boat collisions, that is not addressed by building codes and the engineer will have to "borrow" from the bridge specifications.

With respect to buildings, in many countries, including the United States, "specifications" (or "standards" or documents by other names) dictate how structural members and their connections (that is, beams, columns, walls, floors, footings, bolts, welds, and many other parts and pieces) made of various materials (be it reinforced concrete, steel, masonry, timber, and many others) must be sized to resist loads. Then "building codes" prescribe the loads that must be considered and the analysis methods that must be used to account for those loads during the design process and cross-reference the specifications that must be complied with in this process.

When the US Constitution's Tenth Amendment was ratified in 1791, lighthouses, windmills, churches, fort towers, and a handful of three-story (and the rare four-story) buildings were the tallest constructions in the nation. There were no building codes then, so there was no compelling reason to delegate to the federal government the power to regulate the design and construction of buildings. Hence, because of a decision made in the eighteenth century, there is no National US Building Code today—and likely will never be. Evidently, empowering the people to regulate construction, as democratic as it may sound, became problematic as construction moved into the modern era. Many states and cities did not have the resources nor the expertise to develop building codes from scratch—and that is still the case today for most of them. As a result, modern building codes and specifications have happened as a voluntary effort with—as voluntary efforts often go—significant variations nationally and over time. To put it mildly, the situation quickly turned into a mess with respect to both specifications and codes, with multiple cities, states, agencies, and other groups all creating their own different documents.

In 1923, the American Institute of Steel Construction published its first edition of the "Standard Specification for the Design Fabrication and Erection of Structural Steel for Buildings," a thin booklet with 9 pages of actual specifications—compared to the 2016 edition, which has 676 pages. This was done for the purpose of establishing "a single code authority that would be recognized by building code authorities and designers to eliminate the confusion that then existed in the construction industry, caused by the numerous and different manuals, each containing proprietary information."[19] In other words, to start cleaning up the mess. That AISC booklet was a specifications document.

As far as building codes were concerned, things were no better. A 1925 report published by the US Department of Commerce, when Herbert Hoover (an engineer) was its secretary, complained that the existing codes suffered from a lack of clarity and consistency and that this confusion was prone to error and the subject of much public criticism[20]—that is, it was another mess. As far as specifying the magnitude of gravity loads to consider in the design of structures, things were relatively straightforward. It is not rocket science to figure out the quantity of human flesh that can be tightly packed in a room in the worst possible conditions and apply a generous safety factor to it. Anybody who has been uncomfortably squeezed into an elevator with seven other people, while noticing the little plaque high above the control panel indicating "maximum capacity 21 persons," will understand what a generous safety factor is—or will conclude that mechanical engineers design elevators for hobbits. Applying that concept to an auditorium is similarly done; imagine the room jampacked with screaming and jumping fans waiting for (say) the Beatles (if in the 1960s) and multiply the number of pounds by 2, or 3. It is the other less understood loads that were a problem, and prone to more variations from one document to another as a result of differing opinions.

Out of necessity, since the federal government could not impose anything on the states, and because most states and cities were thin on resources, engineers took the lead in drafting "model codes" that states and cities could adopt if they so wished. Motivated by natural disasters, and sometimes pressure from the insurance industry, three model codes were created under the aegis of private corporations. In 1915, the Building Officials and Codes Administrators International (BOCA) was founded, headquartered in Illinois. It diligently endeavored to eventually publish, thirty-five years later, in 1950, the first edition of the "BOCA Basic Building Code."[21] This code was generally adopted throughout the northeastern states.[22] The southeastern states came up with their counterpart, as the Southern Building Code Congress International (SBCC), founded in 1940, publishing their first edition of the Standard Building Code in 1945.[23] As for the western states, they came up in 1927 with their own very own "Uniform Building Code" published by the International Conference of Building Officials (ICBO) in California.[24] That was on top of individual codes developed by some cities (for example, New York and Chicago) and various branches of the federal government (the US General Services Administration had its own code, as well as each branch of the military, to name a few).[25] To use the scientific term that aptly describes such a state of organization: it was a jungle.

One may wonder how aggressively these organizations pursued the world market or if they were mostly international in name only—like the World Series in baseball—but they were certainly competing to become the "motherhood and apple pie" of building codes at the national level. While there was a sizeable amount of duplication in these model codes, they focused in more detail on earthquakes (ICBO), hurricanes (SBCC), or snow and wind (BOCA) reflecting the interests of their regional clientele and their geographic origin.[26] That does not necessarily mean that they got everything perfectly right from the get-go.

THE GUESSING GAME

After the 1906 San Francisco earthquake, engineers were divided as to whether it was possible to design buildings to resist earthquake forces. Not only were the geological mechanisms that produced earthquakes poorly understood, but so were the laws of physics and engineering that could explain why some buildings collapsed and others did not. Therefore, amid all that confusion, reconstruction of the devastated city proceeded with an updated building code that—at last, after many years waiting for it—required the consideration of a *wind* force of thirty pounds per square feet (1.44 kPa) for the design of new buildings.[27] No specific earthquake-resistant design clauses were introduced.

Given that many building codes of that time had been designed without any specific requirements for wind resistance (such as the Los Angeles building code, which had no requirement to consider any wind pressure in design until 1924), it was hoped that the new "stringent" wind pressure applied horizontally on the building façade over its entire height would simultaneously address both wind and earthquake effects.[28]

The 1927 Uniform Building Code introduced the first seismic design requirements in North America, partly in response to the Santa Barbara earthquake of 1925. This model code proposed clauses for possible inclusion in the building codes of various cities, at their discretion, and was not binding. In particular, it proposed that a single horizontal point load, F, equal to 7.5 percent or 10 percent (depending on the soil conditions) of the sum of the building's total weight, be considered to account for the effect of earthquakes.[29]

Interestingly, Professor Toshikata Sano of Japan, who visited San Francisco after the 1906 earthquake and probably noticed the very clear horizontal displacements that had taken place across the San Andreas fault, got the idea of using seismic inertia force to design for earthquake action.[30] After all, the very same Isaac Newton that had published the theory of gravitation in 1666 had also demonstrated in 1686, in his second law of motion, that a force is equal to a mass times an acceleration (the "$F = m\,a$" equation that stresses students taking high school physics). Building on that, Professor Sano—most definitely a high school student that got it right—stated that the seismic force is given by the ground acceleration multiplied by the mass of a structure and then recommended that this acceleration be taken as equal to 10 to 30 percent of the gravity acceleration. Given that a weight is, in itself, a mass times this gravitational acceleration, this proposal effectively called for using a horizontal force equal to 10 to 30 percent of the building's total weight—that is 0.1 to 0.3 times its weight. The lower of these two values had been adopted in Japanese building design regulations in 1924. Dr. Kyoji Suyehiro of Japan visited California and reported in a series of lectures that buildings designed using the value of 0.10 in Japan survived the tragic Kanto (Tokyo) earthquake of Richter magnitude 8.2 in which one hundred forty thousand died.[31] That was as good a starting point as any.

Enforceable earthquake-resistant design code provisions in North America were implemented following the 1933 Long Beach earthquake of magnitude 6.4. This earthquake produced damage in Long Beach and surrounding communities in excess of $42 million in 1933 dollars (more than $400 million in 1995 dollars),[32] and the death toll exceeded 120.[33] As briefly mentioned earlier, it was significant that a large number of the buildings that suffered damage were schools and that the total number of casualties and injuries would have undoubtedly been considerably larger had this earthquake not occurred at 5:54 p.m., when the schools were fortunately empty. Nonetheless, this economic and physical loss provided the necessary political incentive to implement the first mandatory earthquake-resistant design regulations—in other words, there is no better way to get parents to do something than to show them how their kids could be killed if nothing is done, and that is what "closed the deal" in 1933. The California State Legislature passed the Riley Act and the Field Act, the former requiring that all buildings in California be designed to resist a lateral force equal to 2 percent of their total vertical weight, the latter mandating that all public schools be designed to resist a similar force equal to between 2 and 10 percent of that weight, with the magnitude of the design lateral force depending on the building type and the soil condition.

From there, things improved every year. Slowly, but still. For example, eventually, in 1994, BOCA, SBCC, and ICBO merged into a single entity, to publish at the dawn of the new millennium a unified document called the "International Building Code (IBC),"[34] which managed to receive significant national endorsement—although still far from the world dominance suggested by its name. This integration broadened the consensus, which is good, but when it comes to model codes and specifications, it remains that cities and states are entirely free to adopt, or not adopt, or adopt in part, or modify the model code as they so wish, by virtue of the Constitution.

In parallel, codes and specifications also improved in terms of the complexity of analysis tools, detailing, and construction quality required to design buildings to resist earthquakes, but also in terms of how large the horizontal earthquake force to consider is. For example, it is now recognized that many buildings in San Francisco would need to be designed for lateral forces as large as 100 percent of their weight if they were to survive undamaged the size of earthquakes considered in modern design. That is not a small thing. A building able to resist a horizontal force equal to 100 percent of its weight could be rotated by 90 degrees and hung sideways from the face of a cliff—something that would be far from functional but that an architect on drugs may end up doing someday, forcing the building occupants to walk on walls. Yet buildings that use certain types of structural systems, when built on rock (rather that soft soils like clay and others), can still be legally designed nowadays considering a horizontal force equal to only a tenth of that, namely, 10 percent of their weight—or less. This is at the cost of many requirements to satisfy and many hoops to jump through, but doable—and done every day.

CHAMELEON CODES

Although Gollum's mother is not featured in the Lord of the Rings trilogy, there is no doubt that she is proud of her child and finds him to be an adorable cutie. That is what mothers do. Likewise, the engineers who have given their blood, sweat, and tears to the task of creating the codes and standards that they cherish may expect the world to embrace them. Yet, the world is a strange and interesting place. As Plato observed a few millennia ago, "Democracy is a charming form of government, full of variety and disorder."[35]

Hence, one form of technological disaster is the resistance to adopt measures that will prevent disasters down the line. Professionals may have had some successes in creating codes and standards ready to be adopted, but that does not mean they are adopted (when it comes to making a jurisdiction adopt codes and standards, as in all things in life, "you can lead a horse to water, but you can't make it drink"). It remains a process, a work in progress.

Some cities and some states have been keen in adopting building codes, and some have ferociously resisted doing so, in the name of freedom—for lack of a better term. By the end of 2010, a FEMA study tallied the number of Midwest jurisdictions exposed to high seismic risk that had adopted a building code—any building code—that included seismic-resistant provisions.[36] Interestingly, it reported, "The State of Missouri relies on the local jurisdictions to adopt and enforce their own building codes. The State only demands that projects for State-owned facilities must be designed in accordance with the latest edition of the International Building Code." In other words, a new two-hundred-seat state courthouse will be designed in compliance with the national model code, but for a two-thousand-seat movie theater, anything is possible; consult the local jurisdiction. What each local jurisdiction will do is up to them, which leaves many at a disadvantage given that small jurisdictions do not always have engineers on staff, let alone a building department.

In the same spirit, in Arkansas in 2016, seismic design requirements were waived for some projects because it made construction cheaper. Note that waivers were also provided to construct in flood zones because it created economic development.[37] It is significant that part of Arkansas is on top of the New Madrid earthquake fault, where three earthquakes close to magnitude 8 have struck in 1811 and 1812.[38] Overall, as reported in the same FEMA document, in the six states that are affected by the New Madrid seismic area, "adoption and enforcement of building codes . . . is spotty at best."[39] FEMA is being polite.

Code adoption tangles are not limited to earthquakes but are symptomatic across the board, irrespective of hazards. Like college students partying on Florida beaches in the middle of a pandemic, convinced of the invincibility of their immune system and not about to let all these fake-news stories disturb their drunken plans, some groups oppose any building code rules and restriction. What earthquakes? What hurricanes? What floods? Why kill the fun with imaginary problems?

As of 2018, Texas, Mississippi, Kansas, Illinois, and Alabama (to name a few) still had no mandatory statewide codes.[40] None.

Earthquake design requirements: Zero.

Hurricanes and flood design requirements: Nada.

Too stressful? OK, how about basic requirements to resist gravity loads: Zilch.

Slightly better, some states are happy to adopt outdated versions of the codes, but are leaving it entirely to local building officials to decide whether or not to enforce them. Georgia adopted an IBC edition six years older than other states;[41] other jurisdictions prefer versions that are even more archaic.[42]

Why not? In fact, why worry so much about building codes if, in the end, the federal government will bail out those who will lose their homes in a future disaster? Logically, the federal government's largesse should be trimmed down when it comes to providing disaster aid to states that fail to adopt mandatory up-to-date buildings codes, but when FEMA suggests that this should be the way of the future, the states have ferociously lobbied against the idea and got their way—irrespective of which political party was in power.[43]

What about progress then? When new knowledge is learned to improve how things should be done, the model codes change, and things are improved with every new edition, published every three years. By inference, do those improvements automatically go into the building codes of the more disciplined states and cities that have adopted the model codes? Sadly, not necessarily. For example, in 2018, the Florida Home Builders Association lobbied for the State to stop automatically adopting these changes, on the basis that "fewer code changes overall (. . .) will keep the cost of a home from increasing superfluously."[44] On the strength of this sweeping belief, the Florida governor signed into law a bill that did just that—against the opposition of the insurance industry, the real estate industry, engineers, architects, firefighters, and building inspectors. That "tickled pink" the National Association of Home Builders, which had the firm conviction that States should not impose mandatory building codes and that the decision on how homes should be built is best left to local officials—who are presumably more knowledgeable and incorruptible. Far less impressed, the head of the Insurance Institute for Business and Home Safety commented: "We just have an incredible capacity for amnesia and denial in this country."[45]

Yet, amid all these backward steps, in some places, things move forward one small step at a time. As a case in point, in 2020,[46] Chicago adopted the International Building Code;[47] it was the city's first major building code revisions since 1949 (although, admittedly, multiple small modifications had occurred over that seventy-year period).[48]

In retrospect, it is instructive that in many cases—although not all of them—whenever a regulation has been enacted, it has often been an "after the fact" action intended to prevent recurrence of past problems. If implementation of a new rule is going to cost extra money, it takes a lot of indubitable, convincing evidence and sustained efforts to bring about its approval by consensus—a lot of hand waving and charisma alone will not cut it. In contrast, after a failure—or worse, a collapse—the time is ripe to fix the problem and move forward. Forward thinking does happen at times, but hindsight is always 20/20.

In short, it takes disasters to make progress. Fortunately, much progress will come because the future is pregnant with more disasters.

THE TINY CHRISTCHURCH EARTHQUAKE

On December 22, 2011, a magnitude 6.3 earthquake stuck Christchurch in New Zealand—a modern developed country that is part of the elite club of those few that can claim to have some of the most advanced design standards when it comes to providing earthquake-resistant buildings. A magnitude 6.3 earthquake is approximately one hundred times smaller than what is expected to occur on the San Andreas Fault in California, but it still has some punch. Two older reinforced concrete buildings that pre-dated modern seismic design standards collapsed, and a large number of unreinforced masonry buildings suffered major damage, but beyond that, the modern buildings in Christchurch's Central Business District (CBD) suffered damage but did not collapse, allowing the safe egress of their occupants. However, damage was so widespread throughout the CBD that the authorities decided to evacuate everybody and close off the entire area. At first, security forces controlled this limited access, then, a fence was built. The entire CBD was turned into a "red zone," allowing controlled access to professionals involved in authorized response and recovery activities, but keeping everybody else out. This created an eerie ghost town in which food was left to rot on restaurants counters, stores displayed their full inventory through broken windows, and wallets, purses, and passports accidentally left in hotel rooms during the rush of the evacuation could not be retrieved. These were tough measures to ensure safety and prevent looting, but harshly criticized by owners and residents. Eventually, limited access was granted and the fenced-off area was progressively reduced in size, but it still enclosed many city blocks by the second anniversary of the earthquake, with a new type of damage progressively taking root there while awaiting repairs/reconstruction—at least it was green damage, if any consolation to nature lovers.

By early 2015, more than twelve hundred of the CBD buildings had been demolished, with many more slated to suffer the same fate, turning the CBD into an oversized, unpaved parking lot. By 2018, although the

CBD felt like a massive construction site, rebuilding was barely halfway through.

Yet, structural engineers agreed that that the modern engineered buildings in the CBD behaved exactly as they were expected to do during the December 2011 earthquake.

What is wrong with that picture?

It has to do with the car crash analogy. This requires some explanations.

First, it must be acknowledged that, in light of all the barriers and hurdles met on the road to establishing building codes, the progress made on the development of building codes and standards over the past century is commendable. Even more impressive is the rapid evolution that has taken place over the last few decades. It has all been achieved through the relentless efforts of silent heroes who devoted themselves to enhancing the world's safety by minimizing the risk posed by many hazards, to possibly avoid some—but not all—future disasters. At the same time, all this important work was done by professionals from many disciplines who reached consensus and used judgment and experience to make educated guesses as to what the public would consider acceptable economically and politically.

Some would argue that the public does not know what it wants—or actually that the public knows that it wants everything, while wanting to pay for nothing. As a result, "what the market will bear" has been taken as a proxy for the voice of the public, and has been interpreted as the lowest possible cost—especially when that voice is expressed by developers driven by the goal of maximum return on investment.

Unfortunately, there is a problem with this approach.

What is problematic with it is the fact that the design philosophy embedded in building codes is one of "life safety," not "damage prevention." Life safety is as low as engineers are willing to lower the bar. This is because "engineers shall hold paramount the safety, health, and welfare of the public."[49] That is their top priority. It is the fundamental canon of the code of ethics of the National Society of Professional Engineers. In other words, save their lives—not their property. This "life-safety" objective has been adopted worldwide, but has typically not been explicitly communicated nor understood by all the public and building owners. In other words, most of them fail to recognize that building codes explicitly provide minimum standards—and many engineers firmly believe (and have vehemently argued) that this is exactly how it should be and that it should never change.

This philosophy is often justified by making an analogy, comparing the design of a building with that of a car. Much like compliance with the building code is required for construction to be legal, state vehicle laws define what makes a car "street legal." Most car manufacturers offer a basic entry-level model that meets these minimum requirements. It is allowed to circulate legally, but is typically a subcompact that does not have much—if any—luxury, gadgets, performance, space, comfort, and—sometimes—longevity. Although the analogy is somewhat imperfect here (because crash-worthiness is not a mandatory requirement for a car to be legal), when shopping for a car, most people nowadays assume that automobiles have at least a decent crash-test rating that will help them survive a major collision (and manufacturers with the best ratings make this distinction part of their marketing strategy). However, everybody understands that the car itself would be most likely "totaled" in a severe accident. If that is a concern to a specific customer, the option always exists to buy a Sherman tank instead. Although more costly, it would certainly allow one to drive through traffic oblivious to the risk of collisions. In fairness, the analogy is also imperfect because the percentage increase in cost to "upgrade" from a car to a Sherman tank grossly exceeds the more modest increase to go from a "life-safety" to a "damage-prevention" building design, but the fundamental concept is the same. It is the philosophy of providing a minimum design to achieve "life-safety" instead of damage prevention that has driven building codes—arguably, to make buildings more affordable.

The above car analogy is useful, but it has one major flaw. Most car collisions involve a couple of vehicles—sometimes a few more and sometimes only one. In contrast, if all buildings in a region struck by an earthquake have been designed following the life-safety perspective, a more appropriate analogy would be that of a massive car pile-up involving dozens of vehicles like those that happen due to icy roads or fog (Germany and Brazil being record holders in this regard, with more than 250 cars each in a single pile-up). In other words, during an earthquake, everybody ends up in the same "car-crash" at the same time. Such widespread building damage can paralyze a region, a city, or its core as happened following New Zealand's Christchurch earthquake. When the entire CBD was evacuated, cordoned, and then fenced-off for months, it affected everybody—even the owners of buildings that did not suffer any damage. In a pandemic, it is good but of limited benefit to be healthy when everybody else is sick; the same is true in a region devastated by any other hazard—to be the last one standing is of little consolation. It is fair to expect that a similar large-scale urban mayhem will happen in many other cities worldwide. Prominent structural engineers have already estimated that half of the buildings in San Francisco would have to be shut down following a major earthquake.[50]

A study conducted to investigate the effectiveness of investing in improving things before an earthquake found that each dollar invested to design to a level that exceeds the minimum code requirement provides a $4 benefit in return—in other words, repairs after an earthquake will be four times more expensive than investments before the fact.[51] The return on investment is even higher for floods and hurricanes. But that would require going above the minimum requirements of the building code— apparently not a popular idea. In California, in 2019, an attempt was made to pass a bill that would have mandated that all new buildings be designed to be fully functional after an earthquake. The substantially watered-down version that reached the governor's desk was vetoed,[52] even though it only called for the creation of a working group to study the option and make recommendations.[53] A subsequent bill followed, still asking for a similar workgroup; it died in the Senate.[54] A third one is under review, seeking to develop a standard that would optionally allow setting target times before a building regains its functionality following an earthquake, and setting the wheels in motion to discuss this idea with all stakeholders to determine the possible content of such a standard.[55] That still leaves a lot to discuss and resolve before action is taken, but that would be progress.

What are the chances of something similar to be even considered in the other western, central, and eastern states at highest risk of experiencing damaging earthquakes? For now, zero. This is because going above the existing life-safety design levels costs a bit more today—even though it can save a lot tomorrow.

In truth, costs matter. On a building project for which a developer is seeking rental income to produce a 7 percent return on investment, if extra measures increase the cost of the building by 5 percent, the return on investment drops to approximately 2 percent. The developer would be better off to invest in bonds that would provide a better and safer return. Unless, the cost can be passed on to the tenants—which brings it all back to what the public wants, or what the market will bear. In other words, by how many dollars per square foot can the lease rate of a building be increased for tenants to be excited and say that this is a worthwhile invest-ment? How about zero?

This is because, with some rare exceptions, nearly everybody is unaware of the fact that building codes are minimum design requirements only intended to provide "life safety." The public is usually surprised to learn that this is the case, which is usually after an earthquake has produced much structural and nonstructural damage. In fairness, more and more engineers nowadays, in some parts of the world, make it a priority to inform their clients of this situation and offer them options that could reduce or eliminate damage, but even when that happens, many owners choose the less expensive "life safety" option. This is what has happened in many cases even following the Christchurch earthquake, and even more so as the earthquake becomes more distant in the past. Except that, at least, in this case, with all information in hand and recognizing the consequences, the decision is effectively a deliberate bet that the earthquake will not happen soon—arguably a defendable position, possibly relying on insurance to cover the risk or gambling that the future cost will not be fatal.

Fundamentally, this is a pay-now versus pay-later decision, with all the potential pitfalls and perils.

ON THE DISASTER TRAIL

The Best of the Best

A few decades ago, as part of a political panel of top government officials invited to say a few words of welcome during the opening session of an earthquake engineering conference in Memphis, Tennessee, the representative from Arkansas made it a point to brag that in his state—across the river from Memphis—there were no building codes.

At the time, I had no idea that such a practice existed in some states, and I certainly did not understand why this was a matter of pride. When I caught up with him during the break, I asked, "How do engineers design buildings in Arkansas if there is no building code?" Without skipping a beat, he responded: "They don't need a code. Our state has the best engineers in the world."

There. Take note. If looking for engineers to design the next World Trade Center, look no further than Arkansas.

Fatalism: A Really Short Chapter

WHAT GOD WANTS

Typically, in most parts of the world, after buildings have collapsed and killed people during an earthquake, a window of opportunity opens, making it possible to change building codes to enhance the performance of buildings during future earthquakes and prevent the recurrence of similar disasters. That is good—at least, for those who survived.

Yet, in some parts of the world, even after an earthquake has killed thousands of poor souls when buildings of shoddy construction collapsed, nothing changes. During the reconstruction, sound advice that it might be time to do something different is met with a shrug. Probing further and asking why should people be dying all the time from one earthquake to the next leads to the answer: "If that is what God wants . . ." That is, fatalism—in its worst possible incarnation. This is not religious fanaticism, but rather unintentional obstruction.

Inevitably, after every disaster—be it an earthquake, a pandemic, or a locust infection—some people will interpret these disasters as expressions of the "righteous anger of God" invited by God-defying wicked souls. For those waiting for signs of the long-awaited apocalypse, the bigger the natural disaster, the more theologically titillating it gets. In some ways, this is religious activism, co-opting a disaster to promote a specific worldview. Fortunately, most religious folks are not so extreme; they take no joy in a disaster. Instead, like everybody else, they can see the opportunity to help build a better and safer world and can be proactive forces toward that goal—even if still promoting their specific worldview while doing so.

Fatalism is none of that. It is quite the opposite. Nothing is more frustrating for engineers—who, by their natural inclination, wish to improve things for the greater good—than hitting the wall of fatalism. Nothing hurts efforts to implement changes toward preventing catastrophes more than this frontal collision with a doctrine proclaiming that all events are predetermined and therefore inevitable.[1]

It is a personal choice to be religious—a constitutionally protected freedom in some countries—so that is fine in itself. However, being religiously fatalistic and ignorant is a total disgrace and an insult to humanity. Fatalism, in essence, implies that if people died during the earthquake, it is because God wanted them dead—as some sort of grand spring-cleaning event, weeding out the garden. If God wants something, then it will happen (as Roger Waters said in simpler terms in a song).[2] On that basis, any attempt to make buildings more earthquake resistant is pointless, as it would effectively be attempting to make buildings God resistant, an impossible endeavor from a fatalistic perspective.

With respect to the Three Little Pigs analogy, a fatalist is a brand apart. It is actually like a Fourth Little Pig looking at the straw, wood, and brick huts, who shrugs and rather pulls a reclining lawn chair to bask in the sun: what the Big Bad Wolf wants is for it to decide. Might as well provide the roasting spit and a bottle of BBQ sauce right next to the chair, for the world's benefit. If that is what God wants . . . or what fate has in store, for that matter, for those who subscribe to atheistic fatalism.

One could spend a lot of time here expanding on the nuances and variants of religious and nonreligious fatalism (such as determinism, fallibilism, nihilism, solipsism, defeatism, and other philosophical orgasms). However, for all practical purposes and the topic at hand, suffice to say that it is a massive waste of time to attempt convincing those who firmly believe otherwise that the outcomes of earthquakes are not inevitable and preordained by God (or destiny) and that human endeavor can change such outcomes—until they realize that they are actually the ones dying during an earthquake, which is a bit late to come to grips with reality.

Therefore, fatalists of this world will continue to die in one earthquake after another, maybe removing some fatalistic genes from the human genome in the process—although history has not shown this to have much of an impact in this regard.

FAKE FATALISM

A word of caution is in order. Fatalism is sometimes nothing more than politely expressed resignation in the face of abject poverty, ingrained government corruption, and other societal ailments that make it difficult—if not outright impossible in the short term—to improve the state of things.

In some pictures of collapsed buildings following the Haiti earthquake of 2010, one can see, among the rubble, pieces of columns without a single reinforcing bar. Plain concrete without steel bars is brittle, like chalk. These bars play such an essential role in reinforced concrete construction that an independent inspector is usually hired to verify that these reinforcing bars have been properly placed before the concrete is cast. In corrupt places where shoddy construction does not induce moral anxieties in contractors, because concrete is cheap compared to steel reinforcing bars, there is much temptation to pour the concrete into the formwork without all the required bars, or without any bars outright. Even if the inspector has integrity and cannot be bribed to look the other way, it might be possible to take the bars out of the formwork after the inspector has seen them and has left, pour the concrete into that empty formwork, and dump the bars into another column ready for the next inspection. Since the bars are not visible after the concrete is cast, who can tell? Until the earthquake destroys it all, that is. If that slimy trick can be played in countries where building codes exist and inspection is mandatory in an attempt to enforce them, imagine what happens where no such things exist and construction pretty much follows an "anything goes" mentality.[3] One should never forget that, in some places, shoddy construction practices are so widespread that buildings sometimes collapse on their own—without help from any earthquake.[4]

Suspicion of bribes and widespread government corruption are inescapable when brand-new construction collapses. For example, the brand-new $15 million wing of a hospital was severely damaged during the magnitude 7.3, November 12, 2017, earthquake in Iran,[5] roughly a year after its construction. Experts who assessed the structure in the days following the earthquake stated that the wing not only needed to be demolished, but that it would likely have entirely collapsed if not for the fact that it was braced by the barely damaged older wing of the hospital built forty years earlier.[6] Outraged citizens deplored the country's state of systemic corruption and angrily shouted at the president. Maybe the broadcast brutal punishment of one sacrificial goat will appease the public, but it remains that this could have been prevented—unless, of course, this is what God really wanted.

ON THE DISASTER TRAIL

Seeing It Is Believing It

I considered myself lucky one summer to have landed a job working at *La Grande-4*, in Northern Québec, for the Société de développement de la Baie James, an organization modeled on the Tennessee Valley Authority and tasked to develop the massive dams, dikes, power stations, and other infrastructure for one of the largest hydroelectric projects in the world.[7] In the summer between the sophomore and junior years, civil engineering students have little marketable technical skills to offer, so the opportunity to work as an inspector in such a massive project could not be turned down.

That far north, the boreal forest consists of hundred-years old black spruces that look like dwarf shrubs, the difference between mosquitos and B-52s is that the B-52s do not bite, and sunrise and sunset at the summer solstice occur at 4 a.m. and 10 p.m., respectively. The construction camp consisted of clusters of prefabricated bedroom units linked together around central restroom and shower utilities—all with the thin walls, bouncy floors, and other luxuries that come with temporary construction. Employees were flown there because the 1,550-mile-long road to get there from Montréal is best traveled only by large trucks and wildlife (particularly the last 300 miles on very rough gravel). The luggage loaded into the plane was carefully weighed because fully loaded planes were too heavy to land or take off from the short runway in La Grande-4—which made for some quite memorable acrobatic landings.

Arriving at the construction camp fresh off the plane, all new employees were directed to the check-in office to get their room assigned for the summer. The clerk who called my number looked like an archetypical member of the Hells Angels, in stature, girth, beard, and tattoos per square inch, but was impeccably clean and well mannered—a government employee after all. He told me: "During the summer, the camp is filled two persons per room. By the way, I have not been assigned a roommate yet, so if you want, you can share my room." I am not sure how others would react to that offer, but instinct (or selective exposure) told me it was best not to accept when a big bearded, tattooed dude offers to share his room—so I declined. "No problem," he said, as he assigned me another room. "Don't hesitate to come back and see me if you change your mind."

When I arrived at my assigned room—a bare ten-by-ten-foot box with two single beds and two small desks, the roommate was not there, but I could immediately guess his literary preferences from the colorful, hard-core centerfolds plastered on every single inch of walls on his side of the bedroom. An hour later, the door opened and my roommate arrived, in his dusty construction clothes. Without even removing his steel-toe boots, he crashed on his bed and was snoring profoundly within minutes.

I returned to the check-in office hoping the polite biker would still be there. Delighted to find that he was, I asked him why exactly he had offered to share his room. He said, "During the cold months, things slow down here, the number of workers drops, and I have my room to myself. During the summer, we have to double up. I am in charge of assigning rooms, so you can bet I have one of the largest and nicest ones—with a satellite TV and stereo too. By far, I prefer to have a clean university kid in there who will leave at the end of August than a construction worker that will dirty up everything. Besides, you would have my room to yourself because my girlfriend has also been assigned a room on her own and I am staying with her all the time."

"Deal!"

All of that is to say that making an enlightened decision is simply not possible when the consequences of the decision are unknown and preexisting beliefs stand in the way. With beliefs, as with first impressions or confirmation bias, perception and reality can be at odds and lead far from the most desirable outcome. At least, it almost did for me.

MEET THE FUTURE

Gloomy Predictions

THE "SLIPPERY SLOPE" DISCLAIMER

It has been established so far that it typically takes a disaster to trigger actions leading to major changes in how the world prepares for and deals with extreme events, such as earthquakes, hurricanes, floods, and other natural and anthropogenic hazards. Building on this fact, some extrapolation is performed here to get a glimpse of what the future might have in store when it comes to some existential threats, focusing specifically here on monetary instability, climate change, overpopulation, and nuclear holocaust, and—on the way there—putting to rest the myth that resilience will save the day.

That being said, one should approach the material presented in this third part of the book with a reasonable dose of skepticism, because a lot of the interpretations and inferences that follow are about the future. As such, they are—inescapably—predictions, which is the very thing that has been argued all along to be unreliable. The quote about the difficulty of prediction, especially about the future cited in an earlier chapter, is equally applicable here. Full disclosure: the author has no psychic power or connection with supreme powers, has had dismal success when playing the stock market, and—like most married people—has been reminded of past errors multiple times.

In addition, for those who have not noticed yet, it is emphasized that the author has slightly pessimistic inclinations—although some would argue it more correct to state that the author suffers from a *major* pessimistic bias. Arguably, that negative outlook is sometimes sprinkled with bits of optimism—but with much fewer sprinkles than typically found on a cupcake, for sure. In fact, the cupcakes in the following chapters are not only sprinkles-deficient, but they also do not have much glazing either.

Hence, before diving into these predictions, some concepts, problems, scenarios, and delusions—and, yes, predictions by others—will be examined, not necessarily in that order.

END OF THE WORLD SYNDROME

"End of the world" theories typically have legs because, as P. T. Barnum, founder of the Barnum & Bailey Circus, supposedly said, "there is a sucker born every minute."

First, the "end of the world" is apparently impossible because the universe, which consists of dust, gases, and other particles—sometimes lumped into planets where, in rare cases, life forms have evolved—has no clear end in sight. Hence, outside the realm of the Avengers universe (or Marvel's multiverse for the purists), nobody is likely to figure out how to wipe out the entire universe anytime soon. From a more focused navel-gazing perspective, what is more plausible is human extinction, or, in other words, the end of the world as far as humans are concerned—sometimes called "the end of the world as we know it." In this case, humans only leave behind traces of their past existence—like fossils—while the planet is left to continue circling the sun.

Scenarios describing possible paths toward that outcome abound. A not very original one predicts that this will happen due to a giant asteroid impact, like the one that is blamed for the extinction of dinosaurs—as if featuring humankind in a bad Hollywood remake called "Dinosaurs II, the Extinction Sequel." Other potential existential threats often mentioned include a massive solar flare creating a cosmic superstorm, a global pandemic due to a vicious virus emerging from the wild or from a biotech experiment gone wild, a string of supervolcano eruptions, global climatic changes, a collapse of the food chain due to the extinction of pollinating bees, global warfare with weapons of mass destruction (nuclear, chemical, or biological), or artificial intelligence logically concluding—*Terminator*-style—that eradicating the human race is the sensible thing to do.[1] Or—why not?—an extraterrestrial invasion by aliens having (spoiler alert) a better resistance to the earth's pathogens than those in *The War of the Worlds*,[2] because nobody listened to the prominent scientists who warned everybody that continuing to beam messages into the cosmos is a stupid idea that can only bring the earth to the attention of belligerent alien invaders.[3]

Second, by its very nature, the end of the world cannot be predicted, except for the fact that the sun will eventually turn into a red giant star on its 12 billionth birthday, expanding to swallow both Mercury, Venus, and possibly Earth[4]—not a pretty sight, and definitely not an event covered by fire insurance policies.

As such, any cataclysmic event—or definite "point-of-no-return" threshold—pegged to a specific date, or a specific year, can only be bogus. In particular, anything predicted to happen in exactly ten years is fundamentally suspicious. Any mention of a round number in a prediction of any kind is usually a sure giveaway that the statement is a marketing ploy more than a scientific fact. The ten-year window is frequently used by attention seekers as it has a nice ring to it, sure to attract disciples. There is nothing new to that strategy and it always works for a while, and then backfires spectacularly.

For instance, in 1926, Baily Willis, then president of the Seismological Society of America, frustrated by the lack of interest and inaction of Californians to take the earthquake threat seriously, built on the fact that some newspapers and magazines had misquoted him in stating that he had predicted that an earthquake producing widespread damage in Southern California was to happen within ten years. He co-opted these misquotes, nuancing them by adding *likely* and *probably* in his own statements.[5] Nobody would have raised an eyebrow if that kind of prediction had been made by an unemployed Joe Schmo who had dunked one too many beers, but this was the president of the Seismological Society of America speaking—a leading scientist from a cathedral of knowledge presumably making such a statement after having sifted through overwhelming evidence. Unfortunately, as it often happens when scientists inadvertently get entangled in politics—behaving like amateurs unskilled in the art of persuasion—that prediction was not anchored in solid scientific evidence.

While the Southern Californians that Willis wanted to wake up from their torpor essentially ignored his prediction, the insurance industry was jolted by it. The industry combined that prediction with its own previous studies to assess the seismic vulnerability of the insured portfolio, and decided to triple earthquake insurance premiums. This was on top of a previous doubling to tripling of premiums that had been spontaneously imposed a few months earlier as a "nervous" reaction to a damaging earthquake in Santa Barbara. Compounding these events, this resulted in earthquake insurance premiums jumping five to ten-fold in a very short time.[6] This infuriated many real estate promoters because banks required all commercial projects to have earthquake insurance before issuing a mortgage. Not only did the increased cost of insurance affect the economic viability of new projects, but it raised a concern among developers that it tarnished the reputation of Southern California, making it look like a seismically dangerous place and therefore affecting economic growth in the region. They used their collective political clout to discredit Willis and undermine his prediction, threatening to silence the local seismologists by lobbying politicians to cut off all earthquake research funding, and hiring other geologists to critically review the prediction and expose critical flaws in the seismological theories at the time. Unfortunately for Willis, a major error was found and acknowledged by the US Coast and Geodetic Survey. A preliminary survey of the state that had been used to calculate a relative displacement of twenty-four feet between both sides of the San Andreas Fault in Southern California from 1880 to 1923, when recalculated correctly, showed only a movement of five feet. Being off by a factor of five is not a small thing: who would refuse a fivefold increase in salary, or who would want a fivefold increase in taxes? Likewise, the end of the world in fifty years instead of ten years makes a huge difference. With the subtlety and effectiveness of a bull in a china shop, it did not take long for the business community to tarnish Willis's reputation and credibility, leading him to resign his presidency from the Seismological Society of America, and making other scientists think twice before entering into the political ring.[7]

On the positive side, Willis's brash prediction did prompt some members of the business community to recommend that engineers be required to certify the earthquake-resistance of buildings before a mortgage be approved, and swayed others to assemble a group of engineers and contractors tasked to develop over many years a building code intended to achieve earthquake-resistant construction.[8]

Some would argue that it might have been overly brash, given the state of science a century ago, to make doomsday predictions. Yet, even nowadays, it is hard to resist the temptation and easy to be blinded by results, particularly when the state-of-the-art tools are so powerful. Serious engineers and scientists expertly using supercomputers and highly sophisticated models are fully aware that results obtained are highly sensitive to small variations in the values set for some parameters—often parameters for which it is difficult, when not outright impossible, to obtain reliable values. As shown earlier, the path and point of landfall of a hurricane cannot be predicted with great accuracy, even only a few days in advance. As such, it is blind faith to believe doomsday studies that use even more complicated models—and considering even more input parameters—to pinpoint a D-day exactly ten years in the future. Or, if not blind nor naive, it is political maneuvering. Trends can be clear and undeniable, the need to take a problem seriously may be undeniable, the urgency to act to reduce a threat may be undeniable, but using hype, fear, and scary predictions to achieve results can be a turn-off and a flawed approach, in spite of best intentions.

History is full of flawed predictions and promises. Many people took to calling World War I "the war to end all wars," which it not only failed to do,[9] but is a striking contradiction in itself—as if claiming that the best way to fight diabetes is to ingest tons of sugar. Considering the scale of destruction wrought by WWI,[10] that was a lot of sugar, and—surprise—even more diabetics down the line. Or, for a more positive example, in 1954, the chair of the U.S. Atomic Energy Commission proclaimed that nuclear electricity would be so inexpensive to produce that it would be "too cheap to meter."[11] This meant that the cost of the energy itself would be so minuscule that an "all-you-can-eat" approach would be preferable to the bureaucratic costs of metering energy consumption.[12] In those days, that contagious giddiness with nuclear energy led other prestigious visionaries to predict that all of the world's electricity would be generated by nuclear reactors by 2000. Not quite: roughly 10 percent of the world's electricity nowadays comes from nuclear energy[13] and costs are not always competitive with other forms of power generation.[14]

It is part of the human experience to live through a fair number of doomsday predictions that never materialize and to be worried or scared for a while, only to find out that nothing happened in the end. Therefore, it is only natural and mathematical that younger individuals—due to their shorter life experience—are the ones most stressed and disturbed by doomsday predictions.

Most people thirty years or older will remember that computers across the world had been predicted to crash at midnight on New Year's Eve at the end of 1999, because software written years earlier only considered the last two digits of a year for every date—a practice that had been used to reduce the size (and cost) of file storage.[15] As a result, a global catastrophe was to be triggered when computer dates rolled from December 31, 1999, to January 1, 1900, instead of 2000. When that happened, power outages would occur, the banking system would collapse, medical devices would stop working, and planes would fall out of the sky.[16] The "millennium bug" was a legitimate computer glitch, and more than $600 billion was invested worldwide to patch all kinds of software prior to the deadline,[17] but many credible scientists and economists were convinced that it could not be fixed in time. Many people stocked up on food, leaving grocery store shelves empty.[18] By New Year's Eve, all governments were on stand-by and police officers worldwide were on duty.

And then, nothing. The New Year's Eve parties continued through the night, undisturbed, and the world lived on for another day—with the usual morning headache for those who drank too much.

The Y2K bug was not a hoax. It was a real software issue that had to be fixed and, fortunately, got fixed, but it received a tremendous amount of media coverage through 1999, making many people expect the worst.[19] Thanks to the hard work of computer programmers (the good ones) and engineers, all the major potential problems had been fixed ahead of time,[20] in spite of the fact that many firmly believed it could not be done. However, from there on, people who lived through the whole Y2K frenzy became a little bit more skeptical of doomsday predictions. And so it goes with age, one doomsday prediction at a time.

ON THE DISASTER TRAIL

Death of a Salesman

It is rare to meet someone who does not want to sell you a product offered by their company. "You don't need it" was the reason given to me.

We had purchased a new home in Ottawa, a region that the National Building Code of Canada clearly identifies as seismically active. Granted that compared to the West Coast, the earthquakes are far and few in between, and smaller, but our house sat on more than 100 feet of clay deposits reminiscent of the Mexico City lakebed and its propensity to amplify seismic waves as they travel to the surface (recall the Jell-O bowl effect mentioned earlier). The region's contractors that sold starter homes offered standard designs where the only options purchasers could pick were the colors of the carpets and of the outside cladding. It is also the only part of the world where I have seen regular homes (not log cabins) built using a chain-saw, giving a new flavor to the term "cookie-cutter" home. Whatever seismic resistance the house provided was incidental and unquantified.

Insurance was the only way to protect our investment. Careful reading of the small print revealed that earthquake insurance was not included as part of the standard policy, but that this coverage could be included for a modest additional premium of $15 per year. This is what the insurance salesman did not want to sell me.

"Why would you need that?" he questioned, as if it was surreal to have a customer asking to pay a higher premium.

"I need it."

"Why? The mortgage company does not require it."

"This has nothing to do with the bank," I replied, followed by a long explanation, as if I had to justify why this was important.

"You are not in California. This is Ottawa. There are no earthquakes here," he replied. Even though he was most likely not a seismologist, he talked with the assurance of a salesman (British English pun intended).

My response at that point, while not suitable for printing, was convincing enough. The extra $15 per year I paid was worth it, even if I am pleased that no claims related to earthquake-damage were made during that period.

The Silent Heroes

SILENT SUCCESSES

When it comes to making changes that will enhance the disaster preparedness and resilience of society, many professionals working in this field came to the conclusion, decades ago, that trying to engage the public to advance this cause was an exhausting and often futile pursuit, much like trying to push water uphill. Grassroots activists and movements have typically invested their limited energy pursuing other priorities. Young and old activists are eager to fight when it comes to immediate social and political matters of justice and rights, such as abortion, immigration, economic injustice, and many other legitimate hot topics, with passionate advocates on both sides of each issue. Likewise, many will rally to confront global, long-range, existential threats and environmental crises, such as climate change, nuclear energy, war. Multiple organizations exist to marshal this grassroots energy. In comparison, when it comes to lobbying for a safer infrastructure, good luck getting people excited—except when a disaster sparks public outcry, making it suddenly possible to pass measures that had long been waiting in the wings. Such windows of opportunity rarely open, but it pays to be ready.

Thousands of upset people protested and joined the Society for the Preservation of the Real Thing and Old Cola Drinkers of America[1]—and one went as far as launching a class action lawsuit[2]—when Coca Cola launched New Coke in 1985, apparently altering the very fabric of the universe by having added a bit more sugar to the recipe of its flagship soft drink.[3] These were passionate folks jolted by an overnight crisis. The American Society of Civil Engineers would drown in its own tears of joy if its annual call to invest in fixing the nation's crumbling infrastructure generated a smidgen of such passion from the general public, but watching a bridge rust over decades is not an overnight crisis. Evidently, they (and other similar groups) are not deluded by wishful hope. All the silent heroes that have relentlessly endeavored to make society safer against disasters have wised up early on to the fact that public opinion is fickle. As such, they have found ways to improve the state of things without having to wait for public outcry—not always successfully, often moving only a small step at a time, in a slow and sometimes frustrating process, but still—winning small battles, one after another.

To do so, they have sought advocates and supporters in key government agencies, pitching ideas or lobbying for the creation of organizations or commissions with budgets to empower activities toward disaster mitigation. For example, in the late 1960s, many seismologists, architects, engineers, and other experts in San Francisco were frustrated to see rampant development in some areas—such as reclaimed tidal areas along the San Francisco Bay shores—that they believed would make construction more vulnerable to earthquakes, while others argued that specific soil remediation measures, special design considerations, and limiting construction to low-rise buildings would provide adequate seismic safety. As this was being debated, some started to promote the idea of creating a public Bay Area Earthquake Commission that could make recommendations on zoning, land use, and other matters to reduce hazards. Coincidentally, this happened during a time (mentioned earlier) of profuse apocalyptic predictions by clairvoyants, hippy visionaries, and other eccentrics, all suggesting in different ways that a gigantic earthquake was soon to level California's cities, leaving millions dead—some of these visions possibly fueled by the consumption of certain chemicals of dubious legality. Amid this frenzy, State Senator Alquist supported a bill (drafted by one of his administrative assistants fascinated by earthquakes who had followed the previous discussions on the proposed Bay Area Earthquake Commission) to establish a State Earthquake Commission. Under the fiscally conservative regime of Governor Reagan, the bill was rapidly killed by the Senate Government Efficiency Committee. However, Alquist recast the bill as a resolution to establish a joint committee of the legislature, a procedural strategy that made it possible to escape the purview of the Senate Government Efficiency Committee as well as to not require the governor's signature. With a measly operating budget of $5,000, rather than the $100,000 originally intended, the Joint Committee on Seismic Safety was established.[4] Thus was born what would eventually become in 1975 the California Seismic Safety Commission, with a sizeable budget and authority, thanks to the 1971 San Fernando earthquake that damaged more than thirty thousand buildings across Los Angeles, disabled expressways due to collapsed bridges, and damaged the crest of a dam that could have flooded eighty thousand residents had the water level been four feet higher at the time of the earthquake.[5] Establishment of the Commission is but one example of how seismic safety advocates successfully found access to legislators and got dozens of seismic-safety bills through without having to convince and mobilize a public already tied up fighting other more visible or preoccupying causes.[6]

Silent heroes also include the hundreds of engineers who have served voluntarily on code and specifications committees to enhance practice—a slow process, but again winning many small battles against the status quo, advancing one step at a time toward structures and systems better performing during earthquakes and other hazards. As a result, much of the infrastructure that is designed and built today is substantially safer than it was in the past. Much like a new car today with seat belts and airbags is safer than a brand-new Ford Model T a century ago. In parallel, hundreds of scientists (including social scientists) have been proactive in educating the public and key stakeholders on hazards, risks, and ways to make the world safer.

While this book is not about all these silent heroes, as many books in various fields of expertise have been devoted to the topic, it is important to recognize their contributions. Far from working in the spotlight, like Greta Thunberg and other media darlings of this world, they are a community of individuals that moves forward in small steps, overcoming hurdles along a treacherous obstacle course, to get things done for the greater good.

The following section provides one example of some of the challenges that had to be overcome for a single, modest achievement toward the goal of enhancing public safety: preventing people from being killed by showers of bricks.

THE URM OBSTACLE COURSE

When silent heroes succeed in making a positive change happen, it is often only after having gone through an obstacle course of considerable length. The obstacles can be technical, political, economic, and much more.

It has been mentioned before that unreinforced masonry construction was banned in California following the 1933 Long Beach earthquake that damaged so many buildings of that type. Preventing the construction of damage-prone buildings was definitely a step in the right direction by not adding more problem buildings to the existing inventory, but that did not address the fact there already were thousands of vulnerable unreinforced masonry buildings across the state. There were fifteen thousand of them in Los Angeles County alone.[7] The necessity to strengthen these existing buildings was recognized and already expressed at the time.[8] For decades, cities were not sure that they had the authority to force private owners to fix these buildings. The problem was also compounded by extended discussions on whether these buildings should be condemned and demolished, or strengthened to some level (with various opinions as to what this level should be), and on how much of a cost burden this would impose on owners.

Then, in 1966, ending a seven-year court battle, the California Supreme Court ruled in a case involving the city of Bakersfield, which had condemned a hotel it considered to be a fire hazard, and the hotel owner who challenged on the basis that the hotel met all of the statutes in effect when it was constructed in 1929. The Supreme Court stated that a city has the right to eliminate an existing danger to the public, and that something considered a danger by current standards needs not be forever tolerated on the basis that it was code-compliant when originally constructed.[9] This empowered the city of Long Beach (in Los Angeles County) to move forward and condemn unreinforced masonry buildings, but resistance immediately ensued. Owners of such buildings organized themselves to stop the process, request financial assistance, and seek a more conciliatory approach. Possibly recognizing that demolishing three thousand buildings would also have an impact on the city's inflow of tax dollars, the city sought expert opinions from consultants, engaged in much deliberations, assessed the economic and financial consequences of various options, dealt with lawsuits, and established more refined approaches and priorities to determine which buildings should be evaluated and repaired or demolished. By 1976, Long Beach had an ordinance by which buildings started to be inspected and rated, albeit at somewhat of a slow pace.[10]

In nearby Los Angeles, which is a larger city with a more heterogeneous population, stronger special interest groups—and, according to some, a more disjointed decision-making process[11]—the process was even more convoluted. Shortly after the nearby wake-up call of the 1971 San Fernando earthquake, a 1973 City Council motion launched the process, requesting that the Department of Building and Safety investigate the feasibility of a program to deal with the problems posed by unreinforced masonry buildings, and determine if the Long Beach experience could be implemented in Los Angeles. Then, after years of revised draft ordinances, public hearings, legal advice, citizen protests objecting to proposals to post warning signs on seismically hazardous buildings, pushback from the motion picture industry accusing the city of using this as an excuse to close down theaters that showed sleazy movies, concerns for the loss of housing for the poor, financial planning, discussion on compliance deadlines, setting priorities for different types of building occupancy, research on possible retrofit methods, retrofit cost studies, environmental impact studies, and much more, the City of Los Angles passed a law in 1981 for mitigating the earthquake hazards posed by unreinforced masonry buildings. A mere forty-eight years after the Long Beach earthquake.[12] By 2004, of the 8,268 unreinforced masonry buildings targeted by that law, 194 had been exempted, 1,939 demolished, 6,124 retrofitted, and 6 remained to be demolished or retrofitted.[13]

In that context, it is important to mention that, as a concession to achieve some level of enhancements in public safety while making the cost of seismic retrofit work more palatable, unreinforced masonry buildings have typically been required to be strengthened to resist earthquake forces lower than what is considered the minimum for new buildings. In other words, because of that trade-off, these masonry buildings should be able to survive some earthquakes, but maybe not the larger ones typically considered by buildings codes for new construction. This could make it challenging after a large earthquake to explain to an owner who has invested lots of money to comply with the seismic retrofit law why the bricks that used to be in the building's walls are now scattered everywhere on the ground. Again, this brings everything back to the world of probabilities—an umbrella can be quite effective during a storm, but less so to protect against a nuclear blast.

ALL THAT FOR NOTHING?

The Swiss do not only like their cheese to be full of holes, but also their underground. At the height of the Cold War, in 1963, Switzerland enacted a law requiring each new home to include a nuclear shelter in its basement. The law specifies the required thickness of the shelter's reinforced concrete walls, that an armored door must be used, that the rooms must be equipped with a number of special air filters, dry toilets, and bunk beds, and that food supplies must be stocked (including bottled water and dried cheese[14]—most probably Swiss cheese, although not mandatory). Because the government promised to each citizen a place in an easily reachable shelter, building owners are informed by the law that their private shelters will be not-so-private; it must accommodate other citizens, including members of the Swiss civic protection forces.[15] To prevent nuclear-bunker slumlords, inspections of private shelters are conducted every five years to ensure that they are in working condition.[16]

As an alternative, homeowners were allowed to pay a fee to public authorities, which then embarked on the construction of public shelters on a grand scale. Some of these bunkers were designed as seven-story underground buildings and tunnels to accommodate twenty thousand people, although some of the older facilities were "downgraded" in 2006 to allow only two thousand people[17]—presumably to prevent the eruption of "internal wars" over the long run in overcrowded confinement conditions, as friendly neighbors would start to get on each other's nerves.

By 2006, Switzerland had more than three hundred thousand shelters with a capacity for 8.6 million individuals (equal to 114 percent of the country's population at the time).[18] Although the Cold War was long over by then, the government kept the program alive to provide protection in case of accidental radioactive release at one of the country's nuclear power plants.

No other country on earth has so heavily invested in nuclear shelters. Sweden and Finland are the runners up, but are only able to shelter 81 percent and 70 percent of their population, respectively[19]—interestingly, the Swedish government has published online the GIS locations of its public shelters, making the shelter assignment process less secretive (and possibly less orderly during a surprise nuclear attack) than the Swiss.[20] Beyond these three, when it comes to providing shelters that can completely protect against radioactive fallout, no other country can cover more than 50 percent of its population—most can hardly protect more than a few percent of their population, like Germany, at 3 percent.[21]

For the Swiss, constructing and maintaining these shelters has been an expensive endeavor over decades—in the billions of dollars. Fortunately, Switzerland has not been the target of a nuclear attack, nor has it had a nuclear power plant meltdown. Nobody will complain about that. However, given that the Big Bad Wolf never came, the price tag of the whole operation naturally started to draw attention. As such, in 2011, after nearly fifty years of investments in shelters that never got used, the Swiss parliament decided that it was time to repeal the law. Then, two days later, as the Fukushima nuclear power plant was drowned by a tsunami, the parliament reversed that decision.[22]

ON THE DISASTER TRAIL

Need to Know
Switzerland is a wonderful country. As the saying goes, the Alps are amazing, the country runs like clockwork, and the flag is a big plus. It was a pleasure and privilege to enjoy Swiss hospitality for an entire month as a guest of the Zürich Institute of Technology. We lived in a beautiful apartment in the upscale Zürichberg district overlooking Lake Zürich, a thirty-minute walk—and 450 feet change in elevation—from the office. Easier to stay slim in spite of all the Swiss chocolate and cheese when going anywhere always requires an uphill climb.

On our street, a few steps from our apartment, was a long narrow park where we never saw anybody play, walk, or picnic. We heard rumors that it might have been a nuclear shelter, but if it was one, it was a well-guarded secret. We never saw a door or a sign to that effect. We spent time online trying to find out where the famous Swiss nuclear shelters were located, but that also is apparently a well-guarded secret. In the event of a nuclear disaster, Switzerland tourists might find themselves being the only ones wandering the streets of Zürich, wondering where everybody else went. Alone above ground in Zürich and glowing-in-the-dark radioactive.

Truth and Lies about Resilience

RESILIENCE EVERYWHERE

In the previous millennium (only a few decades ago), "resilience" was an obscure term to most people, used by scientists, such as physicists,[1] ecologists,[2] psychologists, and psychiatrists,[3] to describe the ability to recover from deformations, trauma, or stress. Then, in less than a decade, it became a buzzword. In July 2016, Google searches on the word "resilience" returned 47 million "hits" on the internet, an 800 percent increase from six years earlier. In March 2020, it scored 211 million hits—another fivefold increase in less than four years. Eighteen months later, 303 million hits. There are pandemics that do not spread viruses that fast!—although a pandemic actually helped in this case, given that 100 million hits were obtained when searching for resilience and COVID "bundled" together. In fact, "resilience" has become more popular than "Beatles" (only 165 million hits in September 2021), and is catching up to "Christianity" (339 million hits that same month).

Hence, like "cool" and "you know," people nowadays seem to plug "resilience" into sentences whenever they are too lazy to open the dictionary or a thesaurus. This makes it possible for the Buffalo Bills to get "a resilient effort on defense and on offense" to overcome a fourteen-point deficit at half time and win a game (probably meaning "solid" instead of resilient),[4] for a food critic to describe the best chocolate truffles as having a resilient shell enclosing the ganache (possibly meaning "harder"),[5] for someone to buy "resilient flooring" (meaning "resistant and durable"),[6] and for a car to be called "the most resilient car in the history of sports cars" (if anybody knows what that means).[7] The market offers graphic novels, movies, and board games titled "Resilience," as well as a Resilient wine, a Resilient bourbon whiskey, a Resilient juice, and a Resilience dietary supplement. There is even a cruise ship named *Resilient Lady*,[8] for being launched after the rough seas of COVID. To put it mildly, the term *resilience* has evolved in an incredibly elastic manner. Even the US Department of State had a web page providing advice on "Ways to become more resilient,"[9] which recommended "laughing" as one item in a long bullet list. Quoting from the website: "Laugh: Even when things seem to be falling apart around you, try to find time to smile and laugh. It is very healing and it will help you forget your worries for a few moments. Rent a movie that makes you laugh or spend time with a friend with a good sense of humor." (Important note: It is absolutely true that laughing is healthy and has psychological and potential curative benefits, and this is why it is strongly recommended that everybody buy multiple copies of this book as gifts to heal friends and family while making them simultaneously more resilient.)

Evidently, there is a new Tower of Babel where everybody talks a different "resilience" language and nobody understands anybody else, but thankfully, there is still a rock-solid place where words have specific and clear meanings: the dictionary. Even though definitions vary slightly from one dictionary to the other, all of them agree that resilience is essentially and fundamentally the quality of being able to return quickly to a previous good condition after problems have occurred. To "spring back" and "recover" from all kinds of stuff (physical properties, illnesses, misfortune, etc.), back to "normal." The faster the recovery, the more resilient. Of course, deeper dives from the normal usually entail longer recoveries too, and thus less resilience.

As such, when referring to anything as being resilient, there first has to be a "baseline" that defines some original condition, as a starting point before—it being the topic of interest here—a disaster. Second, there has to be something measurable that is lost and to be recovered, or, in other words, a specific drop of "functionality" at the onset (which can be quite sudden in the case of an earthquake) gained back over time as things are repaired and return to normal. This functionality can be the number of customers having power, gas, or water; or the waiting time in emergency conditions compared to waiting time in normal conditions when dealing with transportation, distribution of goods, and emergency room operations; or any other measure. Therefore, quantification of resilience must be able to address both the *loss* of this functionality, and the path of this *recovery* of functionality both in time and space.[10] Without quantification, there is always what could be called wishful resilience, or aspirational resilience—meaning things done with the best of intentions to provide some tangible enhancement in resilience, but with a lot more hand waving than math.

THE "JUST IN TIME" STRAW HOUSE
OF MODERN COMMERCE

Like every other country, Canada has had its share of disasters. However, beyond floods, blizzards, wildfires, tsunamis, train wrecks, and other messes,[11] the most devastating national catastrophes occurred when the country failed to win the gold medal in hockey at the Winter Olympics. The women's team won gold in 2002, 2006, 2010, and 2014, and the men's team won in 2002, 2010, and 2014. In 2018, disaster struck: they only won silver and bronze, respectively.[12] The country raked in its largest number of Winter Olympics medal ever that year (29) only bested by Norway (39) and Germany (31), but the entire citizenry would have readily traded all those curling, bobsledding, snowboarding, luge, skating, and skiing medals for the only one that really counts in the country that claims to have invented hockey. The country where, in the process of creating the game, prosperous companies that manufactured hockey equipment were also established, including CCM (founded in 1899)[13] and Bauer (founded in 1927).[14]

In early 2020, beyond the Olympic losses, the national sport faced another major crisis: the players of the National Hockey League risked running out of hockey sticks because 75 percent of the sticks they use are manufactured in China, where everything had shut down in the early stages of the COVID-19 outbreak—including the CCM and Bauer Chinese factories.[15] The remaining 25 percent of hockey sticks used by NHL players are manufactured in Mexico.[16] Apparently, national pride only goes so far.

In the name of optimizing efficiency and profits, the trend over the past decades has been to configure operations to minimize costs. Inventories that, by definition, consist of stuff piled up in warehouses while awaiting to be used, were replaced by "just in time" manufacturing to eliminate the "waste from overproduction and inventory."[17] As in a well-timed choreography, suppliers scattered all over the map adjust their production and shipping schedules such that every part needed to assemble a product arrives at the plant—as the name says—just in time. Where some manufacturers used to keep enough material in warehouses to feed their assembly lines for one or two months, the norm nowadays is down to a few days. This has combined with globalization, which in essence has been a global search for the cheapest labor on the planet. As a result, much of the world's manufacturing has ended-up concentrated in a few countries, China and Mexico being two of the most popular among many.

In other words, in the span of a few decades, the world has reconfigured itself into "lean and mean" manufacturing supply chains to successfully maximize profits. At the same time, the most efficient systems are also often the most "fragile," as they lack the ability to adjust to sudden disturbances.

A world that depends on fragile systems may be in equilibrium and quite prosperous when everything runs like clockwork, but the day that something happens to disrupt the existing momentum, the systems break down. In other words, a highly efficient world that operates without redundancy is a more unstable world—a less resilient world.

A disaster striking an industrialized region not only can shut down its economic activity for a long time but also disable its shipping activities. As mentioned earlier, the 1995 Kobe earthquake destroyed the city's port, which at the time was the sixth-largest container port in the world and accounted for 39 percent of the city's industrial activity.[18] Other major industries affected by that earthquake included Kobe Steel Ltd (a major steel producer that lost roughly $1 billion in the process), shoe factories, and sake breweries, which suffered major damage. It took two years to rebuild the port and three years to remove all the earthquake debris from Kobe. Most industries eventually recovered, but some took longer than others to do so. By 2007, shoe production was still only 80 percent of what it was in 1994; the sake industry had recovered to 40 percent of its 1994 production.[19] However, because this all happened before the globalization frenzy had reached today's peaks, the international impact was not too significant.

Skipping ahead twenty-five years, when the COVID-19 pandemic wrecked the world's economy for a while, provides a glimpse of how a disaster can send ripples across the world's fragile supply chains, leaving many firms scrambling to find alternative suppliers. Only a glimpse, because much worse global disturbance is possible. During the COVID-19 pandemic, as the entire Western civilization shut down at the same time, the demand for many goods and products that are not essential in a public health emergency collapsed at the same time, reducing the pressure on the corresponding supply chains. There was a huge increase in demand for masks, ventilators, and—strangely—toilet paper, but few rushed to purchase SUVs, skis, sea-doos, jeans, cell phones, and so on. Imagine a different scenario, in which the world learned of the virus outbreak right from the start and decided to curtail all trade and travel with China to force the entire country into confinement—a China forced into quarantine to protect the rest of the world, leaving it to cough on itself without contaminating the rest of the planet. Imagine if the virus had overwhelmed China and spread over the entire country. It would not have been business as usual in the rest of the world, because China's shutdown would have cut the supply chains of most companies.

In 1914, Henry Ford shocked the world by doubling the salary of his autoworkers. While he did it as a strategy to reduce horrible problems of absenteeism and staff turnover on his assembly line, beyond the positive surge in productivity that it created overnight, it also made it possible for Ford's employees to purchase the very cars they were assembling.[20] As an accidental byproduct, it provided the middle-class with more disposable income, which in turn, boosted the economy as a whole.[21] A century later, as if getting dizzy at an open bar that served stock options, CEOs closed one plant after the other, moving production offshore to reduce production costs and sell their products cheaper to undercut the competition. As many of the laid-off people became poorer, all they could afford were cheaper goods, which in turn led to more plants moving offshore to produce the goods they could afford, and so on. All in a downward spiral decimating the manufacturing sector, trading the inconvenient lessons of Henry Ford for multimillion-dollar bonuses in the (possibly offshore) accounts of some CEOs.

Over the years, one iconic brand after another has moved its production lines offshore to cut production costs and sell at lower prices. Fisher Price toys, once produced in Buffalo, New York, are now made somewhere in China. Mattel's Barbie and Hasbro's G.I. Joe, Black & Decker's tools, Dell computers, the Etch A Sketch, Rawlings' baseballs (for America's very own pastime), Levi's jeans—all produced in China.[22] Certainly not because manufacturing was of better quality there. If anything, even though 80 percent of the world's toys were manufactured in China by 2007, millions of toys have been recalled over the years because they were dangerously malfunctioning, coated in lead paint, or defective for some other reason.[23] Hence, it had to be all about—and only about—cheap labor.

Today, China has a near total stranglehold on the production of some goods, including some strategic products. For example, the pharmaceutical ingredients used in thousands of generic medicines are only produced in China.[24] To list a few well-known products, China produces 95 percent of the ibuprofen (as in Advil), 91 percent of the hydrocortisone, 80 percent of the antibiotics, and 70 percent of the acetaminophen (as in Tylenol) used in the United States.[25]

When it comes to the rare earth minerals used to manufacture electronics, radars, and other high-tech gadgets that the military needs for warfare nowadays, a 2017 report from the USGS indicated that none were produced nationally, while China produced 81 percent of the world's supply.[26] To those paid to contemplate possible war scenarios, this does not bode well. Most significantly, nothing prevents the Chinese from using this manufacturing strength to their advantage to create disasters worldwide, which they could easily do by shipping defective hockey sticks to all other countries, propelling China's ice hockey team to Olympic gold. As past

Olympic Games have shown, there is no ethics when it comes to winning medals—or to winning any other war for that matter.

Today it is China, tomorrow it may be another more or less corrupt country. The point is, simply put, that as a result of globalization, most nations have lost their self-reliance. In the end, in the post COVID-19 world, after years or decades, organizations hopefully will find a new equilibrium—as it often takes a disaster to trigger changes. However, in that process, it is hoped that the legacy of the pandemic will be a renewed appreciation for the value of redundancy in supply chains, and for the tangible resilience that can be provided by strong and dependable domestic manufacturing capabilities.

RESILIENCE FOR ALL

Wanting resilience is all well and good, but there are some hurdles.

First, like the absence of consensus among the Three Little Pigs as to what type of hut it is best to build—at least until the Big Bad Wolf shows up—when it comes to earthquakes or any other hazard, most people do not care about resilience. Until the day a disaster occurs. This is largely because if someone spends dollars today to reduce damage later, therefore enhancing resilience by being better prepared, that person will only benefit from that investment if an extreme event occurs in his/her lifetime. In other words, even when fully aware of the risks of inaction, some folks will prefer to bet that the disaster will not occur in their lifetime, freeing that money for other pursuits that will provide immediate rewards instead. That being said, for the other folks ready to invest today to help them achieve disaster resilience, the inaction of others creates a problem: a resilient building surrounded by non-resilient buildings may not be resilient after all. Like nonsmokers who can get cancer sitting every day in smoke-filled restaurants, one cannot achieve resilience alone. Isolated islands of resilience are not enough to achieve success. Resilience is a team sport. Some stars may shine bright, but in most team sports—as accurately demonstrated by multiple Goofy cartoons—one or two highly paid stars surrounded by a bunch of losers does not make a championship team. Nor a resilient community.

Second, as mentioned earlier, building codes contain minimum design requirements only intended to ensure "life safety," which is good but not sufficient if the intent is to achieve resilience. The consequences of this design philosophy are generally not well communicated to the public—recall the car-crash versus Sherman tank analogy presented earlier.

When the Big Bad Wolf shows up, the balance tips for a while. As mentioned earlier, after the Christchurch earthquake, most of the buildings in the city's Central Business District were demolished and the area underwent a massive rebuilding effort that spanned more than a decade. Interestingly, among the ensemble of new buildings that have arisen as a result of the reconstruction process, most have very different structural systems than what was the norm prior to the earthquake. Partly consciously, partly unconsciously, much of Christchurch is being reconstructed with more resilient structural systems that will allow a more rapid return to functionality following future earthquakes, without necessarily incurring higher initial cost.[27]

Third, emotions can also drive decisions. For example, a major debate raged for years on what to do with the heavily damaged Christchurch cathedral. After the Canterbury bishop, with support from dozens of other local Christchurch churches and Christian groups,[28] announced that the badly damaged cathedral would be demolished, petitions and lawsuits by groups of parishioners stopped and stalled the process. Eventually, after more than seven years of back-and-forth arguments, it was decided to restore the cathedral, keeping the same stone masonry it was originally built from, because a big stone cathedral is also a big emotional symbol.[29] Some would have been happy to restore the cathedral to a lesser level of seismic resistance than required for new buildings[30] (to hell with resilience), recreating the grandeur but prone to damage in a future earthquake. In the end, common sense prevailed. The reinstated cathedral will be sitting on a new foundation designed to isolate it from ground shaking (using a technology known as base isolation[31]), and a reinforced concrete and steel framing system will be embedded into the stone walls that survived the earthquake.[32] The final product, at more than $150 million, and to be completed more than fifteen years after the earthquake,[33] will look almost like the original heritage building, but more resilient.[34]

BLIND RESILIENCE

As mentioned earlier, resilience involves both a loss and a recovery. Unfortunately, when seeking ways to enhance resilience, some folks tend to focus only on the recovery part. This is all well intentioned, but if the deficiencies that can cause losses in the first place are fixed before the extreme event, there will not be a disaster. Alternatively, if losses cannot be prevented but actions can be taken to reduce them significantly in a future disaster, recovery will be faster and things will also be more resilient. This is common sense. Yet, when it comes to proactively mitigating the vulnerability of the infrastructure, which means both enhancing design standards to make new construction more resilient from the starting point, or bringing the existing infrastructure up to the current standards (or close), it is often said to be "too expensive." This begs the question: "Too expensive compared to what?" To doing nothing instead ($0), or to the war in Afghanistan ($2.26 trillion)?[35] Or to paying the bill after the disaster happens? In the latter case, cost-benefit studies have repeatedly shown that investing $1 upfront saves on average $6 down the line ($4 for earthquakes, more than that for other hazards, and up to $13 in some circumstances).[36] The math is clear, but money upfront is always stressful. Furthermore, when money is tight, not everything can be made resilient—and since money is always very tight, this leaves people arguing on priorities for the little bit that can be done at any point in time. With odd arguments at times.

Yet, like it or not, while enhancing the capacity to respond and recover following a disaster is good, it is absolutely better to mitigate the vulnerability of the infrastructure in the first place, to reduce possible loss of functionality right at the outset. In other words, a fast response and recovery help make a nation more disaster resilient, but only mitigation can eliminate the massive initial losses that make it a disaster in the first place. Without mitigation, communities and countries can get stuck in an endless cycle of destruction and reconstruction.

What is somewhat unfortunate is that the resilience concept has appealed to many stakeholders who should have invested in mitigation but who did not want to do so, giving them a way to cop-out by investing instead on cheaper measures that are claimed to achieve a more rapid recovery following a disaster. Arguably, planning ahead of time to develop a network that will make it possible to expeditiously dispatch water bottles to a disaster area is a good thing, but nowhere as good as preventing that area's grocery store from collapsing during an extreme event in the first place. Preventing losses seems wiser, even if more expensive—particularly if the Big Bad Wolf does come.

RESILIENCE FROM SCRATCH

Given that it is expensive to strengthen the existing infrastructure and that mitigation dollars are limited, many researchers and policy makers have argued that efforts at making communities more resilient should focus on the critical infrastructure that would need to be operational after an extreme event, such as the water network, the power grid, and hospitals. At first look, it sounds logical. These are essential services that form the "backbone" of a functioning community. In particular, fully functional hospitals are needed after a disaster to provide emergency care for injured victims. Water is crucial, as humans typically get very low mileage without it. And almost everything in a modern society needs electricity to be functional. Therefore, the continued operation (or rapid restoration) of these services is a necessary condition for overall community resilience, and investments to upgrade their ability to remain operational through extreme events is wise. In fact, the general public is often shocked to discover that these critical facilities and networks often fail, making a disaster worse. Yet, this should not be surprising given that much of this critical infrastructure was built in an era when things were generally not designed to survive extreme events—meaning that, without upgrades, they will perform like the rest of the infrastructure (which means, not great). It will take many decades to make these critical infrastructures resilient, because costs need to be spread over time.

While making key lifelines more resilient is essential, and good, and valuable, expecting that this alone will make a community resilient is being overly optimistic. For example, after the Christchurch earthquake, the hospitals were functional, and the Central Business District remained serviced by water and power and was accessible even though a few damaged bridges were closed. Yet, given the fencing of the District, the demolition of more than a thousand buildings there, and reconstruction taking more than a decade, one can hardly say that Christchurch was resilient—at least, per the real dictionary definition. This suggests that for a community to be truly resilient, its buildings must also be resilient. Pursuing the goal of community resilience without doing anything about the inventory of existing buildings, while well intentioned, is like increasing the speed limit on a road full of gigantic potholes in hope of reducing the time it takes to reach destination in a world where people drive with a bag on their head.

Unfortunately, achieving resilient buildings is not easy. First, as mentioned earlier, for the foreseeable future, building codes are likely to remain minimum requirements intended to ensure life safety. Second, buildings in a community are owned by different owners, who can have different views and opinions on just about anything, from sports to politics, including preparedness to disasters—a problem compounded by the fact that most people do not know much about hazards, disasters, or why they should want resilient buildings in the first place ("Isn't the building code taking care of that already?"). In that perspective, it is commendable that some organizations have developed resiliency rating systems, similar to the rating systems that are being used to highlight the environmental and health impact for green buildings.[37] A resiliency rating can be established by assessing how much damage the building is likely to suffer during an extreme event, and the time it will take to repair it and return it to full functionality. Some critics may worry that awarding a building the top resilience rating may lead its owner to believe that an ironclad protection is guaranteed, missing the point that resilience is a complex issue broader than individual buildings, but anything that voluntarily engages owners in discussion on hazards and resilience can only be seen as a good initiative.

However, irrespective of whether resilience is to be achieved by optional ratings or mandated by future buildings codes, it remains that achieving community resilience "one building at a time" will take a long time—for new as well as for existing buildings. It may also be pointless until all buildings in the same district become equally resilient. For example, a highly resilient building may be destroyed if its roof is punched through by a massive shower of bricks from the not-so-resilient neighbor. Or, as in the case of Christchurch, a building surviving intact will still not bring any income if the neighborhood is shut down for months. Again, it will take many decades to meet this overwhelming challenge.

What else could be done to achieve resilient communities?

There is always the option of building new resilient cities, from scratch. Not only resilient buildings—because community resilience requires more than only resilient buildings—but entire self-contained resilient districts. New communities of resilient buildings, having their own independent emergency backup power generation, their own resilient bridges and resilient transportation network, their own waste-treatment and water-purification capabilities, their own hospital—and presumably their own security forces (moats and walls optional) to prevent looting by residents from the non-resilient neighboring towns after an extreme event, at least until all towns move into the resilient age.

The resilient city concept may seem as crazy today as the thought of sending someone to the moon was a century ago, but it may not be so crazy. The proposed "resilient building district" can be thought of as a "Resilient EPCOT," by analogy to Walt Disney's original Experimental Prototype Community of Tomorrow (EPCOT), referring here to EPCOT the living community, and not EPCOT the theme park. Not the Goofy kind of EPCOT—the real deal. Most people only know the EPCOT that got built in Florida, but Walt Disney's original plan was actually an entire urban planning concept.[38] It was a prototype community to be built from scratch, in a ring-like layout, with a central high-density urban center, surrounded by an industrial area, a residential green belt, and a commercial zone, with its own airport (for a future world filled with hundreds of small commercial airlines needing only short runways), and everything it needed to be self-reliant. Disney envisioned it to be the model "city of tomorrow" relying on (and showcasing) the country's latest technologies.[39] However, Disney died in 1966 and while some of the ideas from the original vision were implemented, the prototype city did not get built and EPCOT became something else.

A Resilient EPCOT would somewhat resurrect the idea, admittedly different but built to be highly resilient, meaning that on top of every new technology likely to be embedded in a new city from scratch, all of its systems and buildings would be able to most rapidly (or maybe even immediately) return to be fully functional after an extreme event. The flaw in this concept is that "living on an island" is not possible because a disaster in a foreign country can have ripple effects on the world's globally integrated economy, but a resilient island is still better than no resilient island.

Critics will argue that this unorthodox idea reeks of elitism, but at the same time, it could be viewed as a necessary step toward model cities of the future. Just like power windows, once only offered in luxury cars like Cadillacs and Lincolns, have become ubiquitous nowadays—very few automakers still offer models with hand crank windows, and possibly even fewer young drivers will lower their window when seeing the "classic" rotary hand-gesture requesting it.

In any event, those who do not like the Resilient-EPCOT idea can always follow the advice from the State Department to be more resilient. In the meantime, new cities are being built from the ground up across the world,[40] and China is already on track to build many of those to be as high-tech and futuristic as possible,[41] such as with Chengdu Future Science and Technology City.[42] It is unknown how many of these new cities of tomorrow will be disaster resilient.

THE FLAWED THEORY

Somewhat related to the concept of rebounding following a disaster, serious studies have shown that some business sectors thrive following a disaster and that the consequences of a disaster are therefore nowhere as severe as computation of losses leads us to believe. In other words, because the construction industry sees a boom due to the need to reconstruct and repair the infrastructure that has suffered damage, it sort of makes up for the fact that others have lost their home and possibly their job—if not their loved ones. The corollary to that observation is that governments should not worry about disasters because, from an actuarial perspective, it has little impact on the balance sheet, and it does not break the bank. It is like an infrastructure stimulus package, but unplanned.

Maybe true and logical to some degree. Yet, as titillating as this concept could be to those with an MBA or a PhD in economics, thinking in those terms is not like resilience on steroids, pumping up the economy. Rather, it simply amounts to robbing Paul to pay Peter—in polite terms, that is, although less polite terms would be also appropriate to describe this theory. If moving the dollars from one column to the other in the spreadsheet does not matter to an accountant, it makes a hell of a difference to those in the losing column.

The absurdity of the whole argument emergences when following the theory to its extreme. If there was any logic to the argument that a disaster has a positive impact on economic activity, then the next time the country enters in a recession, all that would be needed to restart the economy is to flatten some cities here and there. The next time the Dow Jones dives by 20 percent, order the Air Force to drop a few thousand bombs on Manhattan, and let the reconstruction frenzy push the stock market through the roof. Think of the many new jobs created during the rebuilding—and the squeaky clean, brand-new, modernized Manhattan that all will get to enjoy in the end. Sure, it would inconvenience those living in hotels or with friends and family, waiting for the insurance company (if they had insurance) to process the claim for their lost home and then for their new home to be rebuilt, but how could these folks not possibly say, "It was all worth it, we would do it all over again anytime"?

Nonsense.

It is true though that there will be winners after any disaster. It is true that disasters are good for some sectors of the economy, such as the construction industry and material suppliers—or the scoundrels who sell facemasks at exorbitant markups during pandemics—but it does not mean that nothing should be done to prevent disasters. Like it is true that oncologists make a fortune treating cancer patients, it does not mean that nothing should be done to prevent cancer. Along with the winners, there are the losers—and, in disasters as is any poker game, there tend to be more losers than winners, no matter how big the pot is for whoever cashes in.

Another flaw of this theory lies in the fact that, contrary to what could be intuitively expected at first, contractors do not necessarily flock to disaster areas. First, repairs and reconstruction require materials, and many suppliers cannot meet the surge in demand needed for these post-disaster activities. The rush on supplies can rapidly deplete local and regional inventories, and it takes a while for the industry to replenish the stocks. In fact, even in regions undergoing a construction boom—without help from any disaster—it already can take months for builders to get all the concrete blocks, tiles, timber, roof shingles, and other parts they need. Second, many subcontractors get their work on the strength of the relationships that they develop with a few prime contractors that are generally pleased with their work. A subcontractor that has worked hard to establish this network of solid relationships has little to gain to move overnight to another region only because some post-disaster opportunities have been created there—unless that subcontractor had, from the onset, a latent desire to relocate there in the first place. There may be massive needs for certain trades and skills in the disaster area, but rebuilding a network of relationships takes time. Selling services door to door may be lucrative for a while in the immediate post-disaster climate, but it cannot replace the long-term benefit of having a constant volume of work that comes with the trust of a prime contractor. Likewise, a prime contractor moving to the disaster area has to build from scratch new subcontractor relationships, and hiring subs by trial and error is fraught with risk.

Furthermore, things never move as fast as they should when it comes to massive reconstruction, particularly with respect to public-owned infrastructure. After a disaster—no differently than at any other times for that matter—everybody has an opinion. This means that in some democracies, the population—and every special interest group—has to be consulted to make sure that the reconstructed city will have all the state-of-the-art features that can be dreamed up. Bicycle paths everywhere, parks and flowers everywhere, free Wi-Fi for everybody, striking architecture, fresh-air and sunshine in all public places, world-class retail areas, a top-of-the-line transit system, new hospitals, a state-of-the-art convention center, and a performing arts center—oh, and a big sport stadium too.[43] Politicians eager to revive the region's economy will overpromise full reconstruction and a return to "normal" in record time, because it is urgently needed. Then, recovery funding does not quite add up to what was promised, budget constraints require adjustments, schedules slip, and what was to be accomplished in a few years ends up taking decades. Following that pattern, Christchurch's ambitious rebuilding plan after the 2011 earthquake was originally projected to be completed in five years, but some landmark pieces of the plan were not complete by 2021, and might not be completed by 2031 either—in spite of the fact that parts of the plan were scaled back.[44]

No turmoil happens without bruises. All of this adds up to an already severe emotional toll.

The stages of that stressful roller coaster ride are painful. The anxiety of watching a hurricane approach or floodwater rise, the guilt from not having heeded the warnings, the sense of vulnerability, the fear of the outcome. The shock facing the losses, the confusion, the disbelief, the anger. The temporary upswing and adrenaline buzz for those playing hero or helping their community. The optimism that all will return to normal soon, only to be crushed by reality that this will not be case. The disillusionment that follows and that translates in exhaustion and a feeling of abandonment—and substance abuse or suicide for some. And then—and only then—the long, long, long reconstruction phase, with all of its own emotional ups and downs through the process of rebuilding both the broken infrastructure and the broken lives, while adjusting to the new "normal."

Hence, as great as it might look on the spreadsheet of financial geniuses and economics experts, it is best not to drop bombs on Manhattan during the next recession. Even though generals might be delighted to help (to prove that the doctrine of solving problems by generous bombing campaigns is sound, even though that did not work for volcanos), there are better ways to stimulate the economy. Waiting for a disaster to happen is not one of them.

ON THE DISASTER TRAIL

Resilience Is Not a Joke

I will confess to having contributed to the spread of the resilience "virus," not by being the "patient zero" who unleashed the word "resilience" onto the world, but certainly for having coauthored a seminal paper that provided a framework to establish the field of disaster resilience.[45] Little did I know at the time that "resilience" would become like a pair of one-size-fits-all stretch pants that poorly fit most people. To some degree though, it should have been foreseeable. Researchers who "follow the money" are notorious for repackaging their work to make it fit the buzzword of the day, be it innovative, transformative, sustainable, resilient, collaborative, convergent, or "whatever-else" research. If the hot topic of the day for a granting agency becomes disaster resilience, a researcher who has spent a lifetime studying the sex life of ants will inevitably submit a proposal to study the effects of disasters on the sex-resilience of ants.

When it comes to qualifiers, those who dare call a spade a spade deserve the upmost respect.

During post-earthquake reconnaissance activities in Christchurch's Central Business District, which was an urban area closed to all unauthorized personnel, I was performing earthquake reconnaissance activities with colleagues from the University of Canterbury and professional engineers from Christchurch. An earthquake reconnaissance activity consists of surveying the state of damage to various structures—in this case, buildings—to identify conditions that have led to this damage. It a rewarding activity, but a grim one.

We were walking in the ghost town that Christchurch downtown had become, with debris in the street and surrounded by damaged buildings, when someone tapped us on the shoulder. Turning around, we found ourselves face-to-face with a clown—red nose, honker, and all, but in a doctor's white coat. We had run into a member of the Clown Doctors of New Zealand organization,[46] whose mission at the time was to relieve the stress of the individuals involved in performing building safety assessments and emergency stabilizations.

The organization website at the time explained that their silly clowning, throughout a depressing post-earthquake Christchurch filled with grief and sadness, provided relief and hope by giving people the permission to laugh in circumstances where it would otherwise not be possible. Given the recognized power of humor in healing, Clown Doctors endeavored to raise the community's morale.

Thank you, Clown Doctors, for being wise enough to call it "laughing," not "resilience."

Dollars Are Frequent Flyer Miles

IMAGINARY WORTH

A few decades ago, someone came up with the brilliant concept that if a butt is on the seat of a commercial airline, as long as this seat is airborne, the owner of that butt earns one point per mile traveled. The promise made was that a massive amount of accumulated points could eventually be exchanged against free plane tickets. The novelty in the idea lay essentially in adapting to the airline industry a structured marketing strategy that had existed for decades, going back as far as 1896[1] when supermarkets, gas stations, and stores started to issue trading stamps.[2] These stamps could be collected and glued on the pages of specially designed little booklets, up to the point when massive quantities of stamps could be redeemed in exchange for goods from the company that issued them. This became extremely popular in the 1960s when more trading stamps were printed in a given year than legitimate stamps by the U.S. Postal Service.[3] These were the first customer loyalty programs.

Nowadays, nearly every retail chain, movie theatre network, supermarket, pharmacy, hotel brand, car rental company, cruise ship line, gasoline company, credit card, and mom and pop operation, has some sort of loyalty program.[4] These programs are so common that merely listing them in bullet form would fill a book—albeit a boring one. Cutting and pasting all their rules and regulations would create an encyclopedia—an even more boring book. Not surprisingly, hardly anybody ever reads these rules and regulations, except evidently those who create them—and therein lies the catch.

Coming back to the airline industry, in the immediate aftermath of 9/11 when the fear of terrorists kept many people off airplane seats, many airlines bled billions of dollars. One after the other, from 2002 to 2005, United Airlines, US Airways, Northwest Airlines, and Delta Airlines were on the edge of bankruptcy and filed[5] under "Chapter 11" of the U.S. Bankruptcy Law to get protection from creditors while attempting to re-organize.[6] With layoffs and cuts to employee salaries and pension plans,[7] they returned to profit by 2006, in time to face the historical spike in oil prices and the global financial crisis of 2008.[8] Airlines returned in the red. Delta and Northwest Airlines merged in 2008, followed by United and Continental Airlines who did the same in 2010.[9] American Airlines filed for Chapter 11 in 2011, and then merged with US Airways in 2013. Which brings everything back to Loyalty Programs, as these had to be merged too.

When United and Continental Airlines merged their MileagePlus and OnePass programs, many frequent flyers felt that their benefits had been devaluated, in many ways.[10] A disgruntled MileagePlus million-miler went as far as suing the airline for breach of contract, stating in the submitted class-action court filing that, "in stark contrast to the gutting of promised and bargained-for benefits for Million Milers, the new MileagePlus Program perversely and arbitrarily rewards former members of Continental Airlines' former Continental One Pass frequent flier program."[11] The argument was that, prior to the merger, the MileagePlus program had pledged to all customers who had demonstrated their allegiance to United by having their sore *derrière* flown over a million miles in United's planes, that they would receive "guaranteed Lifetime Premier Executive status for life,"[12] which came with a bunch of neat perks. By contrast, customers with Continental had also amassed massive quantities of points that would count toward their frequent flyer status, but they had apparently been able to do so in a number of ways, including without actually flying those miles. It was also alleged that benefits had been debased to the disadvantage of MileagePlus members. A federal judge dismissed the class action, on the basis that the rules of the frequent flyer program allowed United "to terminate the Program, or to change the Program Rules, regulations, benefits, conditions of participation, or mileage levels, in whole or in part, at any time with or without notice" as well as to "withdraw, limit, modify or cancel any award."[13] In other words, promises made need not be kept when it comes to Loyalty Programs.[14] It is all there in the small print.

Of course, who is going to cry when elite members of the jet set lose some of their freebees and perks? Certainly not those who rarely set foot in an airplane, nor those who get shafted in ways that more substantially and dramatically affect their lives.[15] However, the crucial take-away message from the outcome of this class action lawsuit is that when it comes to currency—any currency—those who make the rules of the game control the game, and can change these rules as they wish.

Some would counter that points and miles from frequent flyer programs are arbitrary, hollow currencies dispensed at the will of the organizations that have created them, that they have no real value, and that they should not be considered a currency. Part of the confusion is created by the fact that the IRS has issued a statement that it "has not pursued a tax enforcement program with respect to promotional benefits such as frequent flyer miles."[16] This is partly because the IRS is at a loss to figure out what these miles are worth, which is a problem severely exacerbated by the fact that some people never cash in those miles, and the fact that the value of those miles depreciates nearly every time the airlines change the rules of their frequent-flyer programs.[17] In fact, if an airline decided to shut down its frequent flyer program overnight, the points would become worthless; in fact, on some airlines, the points that have been banked with their frequent flyer program are set to automatically expire unless at least one flight is taken over a certain time period. To the IRS, this makes them look more like a rebate that must be used by a certain date rather than a currency—which is a confusion appreciated by most business people who are prime bene-factors of these programs. However, the counterargument is that points and miles are a currency,[18] because they are a medium of exchange that has a specific value and unit system (accounting for relative worth) that stores value for a deferred payment. They are backed by the issuer of the currency, just like the bank notes that were considered legal tender in the era of "free banking" before central banks took over the system because too many banks failed due to lack of liquidity or due to the shoddy practices of "Wildcat" banks.[19]

The common denominator of all currencies is that it is nothing more than a promise made by the issuer of the currency. The stability of the currency rests entirely on the faith that when the piece of paper—or virtual number in today's digital age—will be redeemed, it will be worth its face value. The fool's gold in that precept is that the actual value of a currency rests entirely on the collective imaginary construct that the promise is believable, and thus reliable. As for all promises in life, short-term promises are always more credible than long-term ones. As far as immediate transactions are concerned, the currency serves its purpose, but when looking far on the horizon, those who make the rules of the game can easily change them by devaluing the currency. This is no different from what is commonly done with frequent-flyer miles. In the end, after devaluation, the number of points (or dollars) in everybody's account remains the same, but it buys a lot less. Everybody is upset, but those in charge can always blame someone else—and laugh all the way to the bank, as long as that bank uses a different currency.

Any currency is essentially a high-wire act where all the spectators have a stake in not seeing the performer fall and crash. Yet, statistically, over time, it happens. In plain view.

It is easy for anyone to write a bunch of IOUs when there is no genuine interest in repaying the debts. Arguably, that is why accumulating debt is not a concern for many governments: All that is needed to pay it back is to print massive amounts of paper money. It will not be worth much, but it will make it possible to pay back the amount promised at the currency face value. If that were to happen to the dollar, few US citizens would cry when the foreign interests who own 30 percent of the US Treasury bills, notes, bonds, and securities[20] found themselves owning worthless dollars. However, most of these same citizens would be devastated to find out that the part of their retirement funds invested in trust of the government has vanished.

Printing tons of money is certainly an effective way to erase all debts contracted in a certain currency—as the Argentinian government did in 1992 when it could not find any other lender willing to loan at reasonable interest rates, thereby devaluating its peso by a factor of 100 billion from its previous value.[21] With that kind of hyperinflation, a trillion-dollar IOU is effectively paid back with ten dollars of real value. However, to be clear, nothing pretty happens with this solution. While it can be thought of as the nuclear option of debt abatement, it should be clear that it is like dropping the bomb on one's own economic system. Those who decide to self-inflict such pain better have a sound reconstruction plan to create confidence in the new currency—and likely, in the new government—or have a plan to operate with someone else's currency while awaiting better days. The German papiermarks became worthless after the First World War, down to a low of 4.2 trillion German papiermarks per US dollar in 1923,[22] and everybody knows where that crisis eventually led.

It is dramatic and disheartening to witness the total collapse of a currency from a "shit-hole country"—to use a technical term coined by a famous president. It is an ugly situation that causes widespread misery. However, while it is possible for the International Monetary Fund to strong-arm countries plagued by hyperinflation, walking in after a mini bloodbath to clean up the mess, nobody has the capacity to come to the rescue of big players shooting themselves in a mega bloodbath. On the global scale, the total collapse of the Zimbabwe dollar due to an annual inflation of 89.7 sextillion percent (8.97×10^{22} percent),[23] or the Yugoslavian dinar (annual inflation of 5×10^{15} percent), or the Hungarian pengő (annual inflation of 41.9×10^{15} percent), were all small "earthquakes." The big earthquakes are the ones that will be devastating on a global scale.

If it takes a disaster for things to change, what are the blessings of disaster in this case? How can surviving a monetary disaster be good for anyone? Of course, being a survivor is already a small win in itself. Also, when almost everybody suffers losses, no matter how terrible the event that occurs, some will come through unscathed. Even the most horrible Ponzi schemes that have dispossessed hundreds of people of their lifetime savings have made a few winners who bailed out before the pyramids collapsed. However, for the unlucky survivors, when complex structures and highly fragile financial structures collapse, sucking hapless souls into the maelstrom, what remains is an opportunity to create a new order cleansed of deviant financial practices, and to enhance the robustness and resilience of the financial system (maybe even weeding out some compulsive greed, gambling, and arrogance in the process). That is, of course, as long as massive bailouts do not come to the rescue of political friends at the expense of victims, swallowing a "too big to fail" mantra—which, if repeated often enough, can become a dogma.

PREDICTING THE IMAGINARY

To predict the expected performance of retirement plans, financial analysis models often rely on Monte Carlo simulations—a methodology name quite revealing as to its possible reliability. The fundamental idea behind a Monte Carlo simulation is to assume that since the future is unknown, future annual earnings are predicted using statistical data of annual earnings from the past, and pulling a random number from that data to predict the return for each future year—as if in a Monte Carlo casino.

In a simple way, this is like having a die that shows returns of 1 percent to 6 percent painted on each of its faces, and repeatedly rolling it to predict what the interest rate for the return on investment will be for each and every year ahead—a very casino-like method indeed. To dress the gambler in a nicer suit, multiple dice can be painted with different numbers on their faces so that when thrown together at the same time the results replicate the fact that some percentage returns have happened more frequently than others in the past. Still a random pursuit, but more statistically convincing. Then, to take away the fact that any resulting single prediction of return over a period of thirty years has a low probability of being right, many thousands of thirty-year predictions are generated by Monte Carlo simulations, and statistics can be computed on what the chances of running out of money are before dying—usually a bad scenario when it comes to retirement. The results can even state the probabilities of doing so if, by luck of the draw, the simulation turns out to include a year of truly bad returns happening at the worst possible time, at retirement—often using the negative stock market returns from the Great Recession of 2008,[24] but without necessarily considering the ones from the Great Depression of the 1930s. Throw in annual returns from the Great Depression, and correlate the returns in successive years to match that period of history, and it is not clear if any of the Monte Carlo simulations can replicate that terrible scenario, or if it is an option that statistically disappears like the magnitude 7 vanishing act presented earlier. Hyperinflation and other black swan events that could wipe out all savings are also not included in the prediction, since the models use past data, and there is—by definition—no black swan in the data.

In other words, as sophisticated as this mathematical tool (and other ones) can be to calculate the funds that will be available at retirement based on returns and inflation over time, it remains that these are predictions based on statistics. Some will put great faith in those numbers, some will call it smoke and mirrors, and some, like Mark Twain, will say, "There are three kinds of lies: lies, damned lies, and statistics." If the future becomes an extension of the past, and the models assume the same, then the retirees will reach their promised retirement goals. If things get substantially more volatile, leading to a financial crash, the cakes served at retirement parties will be empty calories.

The survivors of a future financial meltdown might be those who did not put too much credence in financial models based on games that assume future returns dictated solely by the past—a very hard thing to do given the shortage of alternatives. After such a titanic event, those who had the wisdom to wear life jackets might prudently rebuild a less complex, less fragile, less speculative, and less gambler-focused financial system—at least for a few generations before history is forgotten and eventually repeats itself, as goes human nature.

IMAGINARY WEALTH

When placing a bet on a football game at a Las Vegas casino, one option (and the simplest one) is to bet that one of the two teams will win. Going a few years back to illustrate the concept (still valid today), for a game between the Buffalo Bills and the New England Patriots when Tom Brady was their quarterback, the casino used a "money line" approach to account for the fact that more people were likely to bet on the Patriots. (Love or hate Brady, the fact remains that the Bills had won three times and lost thirty-one times against him.)[25] For example, if the casino assumed that ten times more people would bet on the Patriots than on the Bills, those betting $100 on the Patriots could earn $104 if the Bills won, while those betting $100 on the Bills could get $1,110. Of course, no matter which team wins, the biggest winner remains the casino, which pockets a hefty percentage of the proceeds for providing this essential service called gambling. Before a 2018 judgment of the Supreme Court of the United States, a federal law decreed that betting on sports was illegal[26]—except in Nevada.

When someone on Wall Street places a bet on the price of crude oil by purchasing "futures," that trader speculates that the barrel of oil currently sold at $80 will be worth $100 six months later. The trader makes an $80,000 purchase for 1,000 barrels six month into the future at the current price. Move six months forward. The trader must now pay $80,000 for the barrels. If the guess was correct and market price at that time is indeed $100 per barrel, the 1,000 barrels purchase is made and resold immediately, for a net profit of $20 per barrel, or $20,000. If luck has it that the market took a dive to $40 per barrel, the purchase and resale translates into a loss of $40,000. Of course, irrespective of the price of oil, the broker's fees are charged in all transactions and the bank pockets a hefty percentage of the proceeds for providing this essential service called investing.[27]

There is more than one way to bet on a football game (covering the spread, betting on the total number of points, using parlay and teaser cards, etc.)[28] and more than one way to invest in futures and other derivatives (leveraging, hedging, etc.).[29] The big difference is that gamblers typically play their own money while traders gamble with other people's money—pension funds, mutual funds, and other savings. Also, while casinos and banks generally never lose, in the rare instances when banks do, the government can print more money and bail them out.

There are probably many disaster stories in gambling and investing. To add to the fun though, there is now a new investment tool called "Catastrophe Bonds" that is just as thrilling to play. The rules of the game are as follows. If, for example, the It Takes a Disaster Insurance Company (ITAD-IC) expects (and is solvent) to cover $500 million in losses due to natural disasters each year. Following the traditional route, ITAD-IC would purchase coverage from a re-insurance company, which is effectively a "mega-size" insurance company that is in the business of providing coverage to the smaller players like ITAD-IC.[30] The top two re-insurance companies (Swiss Re and Munich Re) collectively earn over $70 billion in insurance premiums annually.[31,32] If ITAD-IC's payouts in a given year were to exceed $500 million, it would turn to its re-insurance company to pay the difference. The premiums to re-insure are not cheap.

However, since 1997, insurance companies have created a clever tool called "Catastrophe Bonds." Instead of re-insuring itself, ITAD-IC could issue bonds to be purchased by any interested investors. To illustrate the process, imagine that ITAD-IC issues one hundred thousand $1,000 bonds good for three years, paying an interest rate of 7 percent per year. This would correspond to a total inflow of $100 million that ITAD-IC will keep in a special account, as a reserve. The rules issued with the bonds would stipulate that if the disaster payouts from ITAD-IC on a given year do not exceed $500 million, then all the investors would receive their 7 percent interest payment. If the same happens three years in a row, in addition to the three years of interest, the same investors would get their money back at the end of the three years. However, should it happen on any given year that the ITAD-IC disaster payouts need to exceed $500 million, then the $100 million stashed in the special reserve becomes entirely the property of ITAD-IC. This effectively makes the investors who purchased Catastrophe Bonds play the role of the re-insurance company and absorb the losses.[33]

No different than for the above games and investments, the Catastrophe Bonds can be purchased with many options, such as specifying that the investment would be lost only if the number of Category 4 hurricanes hitting a given state in a year exceeds four, or if the magnitude of the earthquake exceeds 7, and so on. They are also wonderful because they make it possible for Little Pigs who believe that the Big Bad Wolf is just a scary story and that disasters never happen, to put their money where their mouth is.

Nowadays, nearly $10 billion in Catastrophe Bonds are issued each year. As Dirty Harry (Clint Eastwood) replied to the punk who thought that the six bullets had already been fired from the .44 Magnum pointed at his head, "do you feel lucky?"[34]

ON THE DISASTER TRAIL

How Much Is a Ruble Worth?
The street vendor looked right, left, behind, right again, left again, and pulled out the matryoshka doll from his pocket. Tourists visiting Moscow inevitably buy one or more of these hand painted wooden dolls—also known as babushka dolls—that open up to reveal a smaller replica of itself inside that again opens to reveal a smaller one inside it, and so on and so on. For a few rubles, official government stores sell them in many different sizes, some with as many as twenty or more dolls nested inside—all with colorfully painted cartoonish images of Russian women in traditional dress and scarf.

On the Arbat pedestrian street, though, in the 1980s, where all the artists, protesters, and other nonconformists gathered to check if the tiny bits of freedom promised by General Secretary Gorbachev's perestroika were for real, the babushka dolls pulled out of the vendor's pocket stretched perestroika a bit more than was believed safe. The outer doll depicted Gorbachev, and in succession for each inside doll, Brezhnev, Khrushchev, Stalin, and Lenin—a bit of Russian humor apparently not shared by the authorities.

When I showed him a few rubles accompanied by hand gestures that I hoped made it clear I was asking how much it cost, the vendor said: "Dollars, not rubles. Real money only," in what was maybe the extent of his English.

In an attempt by the Communist government to prevent the inflow of foreign currencies, the customs officer at the Moscow airport had asked me (as he asked every arriving foreigner) how much foreign currency I was bringing into the country. He stood motionless, in silence, for a good minute, maybe as a challenge to see if I would dare try to bribe him, or maybe expecting a tip in dollars for his services. Then, as if an officially required "wait time" condition had been met, he had me fill a form declaring the amount of dollars I had with me, he stamped that form and my passport, and told me (as he told everybody else too) that I had better leave with the same amount of foreign currency when flying back.

As a result, "real money" was rarer than rubles, but standard currency on Arbat Street.

The street vendor had been wise to insist on being paid in dollars. In 1990, when I visited, the official rate was $1.62 per ruble, but the black-market rate was more like 10 rubles per dollar—before haggling. Then, in 1991, reality struck as the USSR dissolved, and a ruble bought $0.16—a new official exchange rate restricted to visiting tourists or those permitted to travel abroad.

We're All Cooked

GLOBAL WARMING

It is 2,800 miles south to north from the Canadian city of Windsor, Ontario (population: 217,000)[1] to the Canadian outpost of Alert, Nunavut (where the population jumped from 5 in 2011 to 62 in winter 2016 and 110 in summer 2016 to cater to tourists[2] suddenly attracted to the northernmost settlement in the world).[3] That is 1,000 miles more than the distance from Key West (population: 24,500)[4] to the northernmost tip of Maine (population: moose, loons, bald eagles, black bears, and other wildlife).[5]

In spite of its 2,800 miles north-south span, 90 percent of Canada's population live within one hundred miles of the United States. The fact that most Canadian cities are lined up along its southern border does not reflect a strategy to invade the Americans in an imminent surprise attack, contrary to what was alleged in *Canadian Bacon* (directed by Michael Moore and featuring John Candy in his last movie). Rather, it is because Canada's winters are brutally cold, so its citizens bundle up as far south as its borders allow. Do not tell Canadians that car emissions will accelerate global warming, because once made aware of this cause-effect relationship, some will leave their cars running all night in hopes of accelerating the process.

More pragmatically, like every other nation, Canadians are struggling to figure out how much climate will change, how to adjust to this change, and how to slow the process. Those involved in the maple syrup industry, whose main worries used to be acid rain and bug infestations, must now contend with the impact of a shorter transition period from winter to summer temperature, which reduces the production season, and with winters of less snow, which can affect the health of its trees.[6] The lobster industry also worries, as the lobster population is shifting in search of the colder waters it needs.[7] At the same time, Canada has the third-largest known oil reserves in the world,[8] estimated at 170 billion barrels in the tar sands of Alberta's Athabasca region—two orders of magnitude larger than the reserves being pumped out by the Hibernia offshore platform,[9] a hundred miles away from Newfoundland's lobster territory—and the country is cashing on that oil as fast as possible. Unfortunately, extracting oil from tar sands is an environmentally destructive activity[10] that creates considerable environmental pollution and produces more greenhouse gas emissions than conventional oil drilling.[11] Presumably to demonstrate that the country has a social conscience and is a good world citizen, Natural Resources Canada published a flyer acknowledging that the oil sands operations contributed to 7.8 percent of Canada's total greenhouse gas emissions in 2011, but that this only amounted to 0.1 percent of the entire world's global emissions. It added that Canada, in the totality of all its economic activities, produced only 2 percent of the world's greenhouse gases, displaying a beautiful pie chart showing that the United States, China, and Europe, respectively, contribute 22, 20, and 17 percent.[12] Yes, Canadians are pros too, when it comes to packaging facts in favorable statistics, because rearranging the numbers to instead plot greenhouse gas emissions per capita would have shown that Canadians are on par with all these other emitting countries.

In fact, Canada is no different than many other countries that proclaim to be leaders in reducing greenhouse gas emissions and that promise to achieve net-zero emissions decades into the future.[13] They may have to use clever accounting tricks to meet these goals down the line—such as getting credits for the reduction in emissions that will occur if China builds power plants fueled by natural gas (exported by Canada) instead of coal. However, while Canada does not appear to be in a position to lecture others when it comes to climate change, others are not in a position to lecture Canada either. In fact, most countries seem to suffer from split personality disorder—or even schizophrenia—when it comes to climate change.

Climate change disasters maybe be disasters in slow motion—because it will take a while for the palm tree to replace the maple leaf on Canada's flag—but they are disasters, nonetheless. Therefore, everything presented in earlier chapters about disasters due to natural hazards is equally applicable to climate change. Problems related to time scale, predictions, human behavior, politics, and the rest, all remain true, but all of that is complicated by one compounding factor: the behavior of the messengers.

SOPHISTS

In the 1970s and 1980s, Jim and Tammy Bakker hosted *The PTL Club*, which was a highly successful evangelist Christian television program (PTL = Praise the Lord). The show was reported to raise in excess of $1 million per week in donations,[14] which were used in part to create a religious theme park that attracted five million people per year[15]—making it the third largest theme park in the U.S. at the time[16]—and in part to further the PTL mission at the Bakker's discretion. The Bakkers also reportedly lived a flamboyant lifestyle[17] referred to as prosperity gospel[18]—which is pretty much what it sounds like.[19] The husband-and-wife televangelist team promoted the tenets of evangelical Protestantism, a denomination that embraces the authority of the Bible as a strict "gospel truth" that dictates all actions and beliefs in life.[20] Sexual encounters with a secretary behind the wife's back, paying hundreds of thousands of dollars to purchase her silence, and a forced resignation all interfered with delivery of the message.[21] It also suggested that there might be lots of room for interpretation in the Good Book. But it is interpretation of the "good book" of the Internal Revenue Service that somehow got Jim Bakker in trouble with US Government law, as he was found guilty and sentenced to forty-five years in a federal prison on counts of fraud and conspiracy.[22] Eventually, the sentence was reduced to eight years. He was paroled and released after less than five years and has since launched the *Jim Bakker Show* to spread biblical revelations that the end of the world is near,[23] collect donations to help in broadcasting that message, and sometimes sell generators and special food supplies to survive the Apocalypse.[24]

Some people have no problem with that career path and generously support him financially. Less forgiving, others find it contemptible that someone who preached high moral values found it perfectly acceptable to partake in adultery and fraud. Yet, conflicted preachers[25] are not an exception when it comes to contradictions. Countless politicians claiming to be standard-bearers for traditional values based on religious beliefs have been caught red-handed doing some serious hanky-panky with secretaries, lobbyists, reporters, strippers, prostitutes—or even enjoying secret homosexual encounters of the type they vehemently condemned in speeches.[26]

Rare seem to be those who can actually walk the talk—those that do often become the cornerstone of new religions. Throughout history, there have been a lot of well-intentioned folks eager to share with everyone their version of wisdom, their understanding of the truth, and their perfect solution to all problems; unfortunately, sometimes blatantly, sometimes unintentionally, this amounted essentially to "do what I say, not what I do." Messengers that contradict what they profess tend to kill the message. That clumsiness is at the root of many tensions when it comes to climate change. When someone calls wolf, that someone better be credible because asking everybody to drop everything and take immediate action is a serious matter. For a messenger, being genuinely scared does not count as a credential, and having solid credentials is also irrelevant if the solutions advocated are full of loopholes.

Scientists generally agree that global warming is taking place. A NASA website refers to this consensus by stating that this is the opinion of more than 97 percent of climate scientists who are active researchers, and that this is most likely due to human activities[27]—referring to eighteen national scientific organizations and agencies supporting this statement and more than two hundred similar groups worldwide.[28] Data collected worldwide indicates that the global temperature is rising, that oceans are becoming warmer, that glaciers are receding, that the Arctic and Antarctic ice masses are shrinking, and that water levels are rising. Even climate change deniers generally do not deny that global warming is occurring; rather, they claim to be skeptical that this is attributable mainly to human activity.

Scientists are usually credible people. Their job is to question, probe, observe, study, investigate, and hopefully make new discoveries about how the world works—from the infinitely small to the infinitely large, and everything in between. Discovery is a serious endeavor, and—nowadays—a most expensive one.

Take astrophysics, for instance. In 1990, the Hubble Space Telescope was brought into orbit by the space shuttle *Discovery*. Original cost estimate for the project was $400 million, but by the time it was finally launched in 2010, the bill had reached $4.7 billion[29]—clearly, great astrophysicists do not make great accountants. Unfortunately, once images from space started to be beamed down to earth, some were blurry and it was found that the mirror had been incorrectly polished. No sweat; a repair crew flew up to fix it in 1994. Thereafter, four other space shuttle missions followed, up to until 2009, to replace old instruments that had reached their lifespan—sometimes much sooner than anticipated. The total aggregated cost related to the Hubble Telescope construction, launch, and five servicing missions is estimated to be on the order of $10 billion.[30] Yet the telescope is aging, so the James Webb Space Telescope (launched December 25, 2021) now is complementing and extending the discoveries of the Hubble Space Telescope, with a telescope mirror three times larger than Hubble, packed with newer and better toys, and carrying a bigger overall price tag—originally estimated to cost $1.6 billion, but ending up at more than $10 billion at launch time. With an orbit higher than Hubble, it will be too high for servicing; it also will have a planned service life of only five to ten years. This makes it imperative to already start planning for its replacement. Therefore, astrophysicists are already dreaming of a (more expensive) space telescope with a forty-five-foot-wide mirror to be launched in space sometime around 2030, to search for evidence of life on the universe's distant exoplanets or dig deeper into space to pry from the oldest black holes information on how galaxies form and evolve[31] so that we can better predict our destiny, either when the Milky Way (home) will collide with the Andromeda galaxy four billion years from now[32] or when the sun will turn into a giant red star that will engulf the earth maybe slightly sooner.[33]

One would be hard pressed to find a scientist saying that this valuable research is a boondoggle. To them, this is top science and worth every dollar of it. Yet one launch of a SpaceX Falcon Heavy emits more carbon dioxide in a couple of minutes than a car does in two hundred years[34] (the Ariane 5 rocket that launched the James Webb Space Telescope burns about 25 percent more fuel than that).[35] Vaporization of space junk reentering the atmosphere also produces pollution, although that has not been quantified.[36] Every launch also spews iron, lithium, nickel, mercury, and other metals into the lagoons around Cape Canaveral that are home to a rich wildlife.[37]

To some folks, scientists calling wolf with respect to climate change but giving the quest for the stars a special free pass to pollute sounds like preachers sleeping with porn stars. Scientists will respond that the sum of all rocket launches amounts to less than one-hundred-thousandth of all global CO_2 emissions, compared to the airline industry that contributes nearly 3 percent of all emissions (a great line that the new space tourists will remember to use).[38] This is somewhat like the obnoxious neighbor with the junk cars on the driveway and garbage littering the untended lawn asking to be left alone because this is nothing compared to the gigantic city dump where all the owners of pristine homes on manicured lots send their garbage.

However, scientists are unfettered by such frivolous and sophistic accusations. The math is definitely on their side when it comes to rockets. As for the airplane emissions, it is a serious problem, but exceptions are fully justified when it comes to big scientific powwows where important matters are discussed. The American Association for the Advancement of Sciences regroups 273 societies and academies of science, for more than ten million members. That makes for hundreds of annual meetings, conferences, and other junkets in attractive locations for which these members—probably feeling very guilty—jumped in an airplane. Annual meetings of the American Association of Geographers typically attract roughly nine thousand people.[39] The national conferences of the American Chemical Society (one of the world's largest scientific societies) attracts between twelve thousand and eighteen thousand participants—twice each year.[40] The Society for Neuroscience annual meeting typically attracts between twenty-five thousand and thirty-five thousand attendees. It is fair to assume that few, if any, of these scientists sailed—Greta-style—to any of these conferences, as serious scientists typically find the airlines more accommodating of their busy schedule and have little time for publicity stunts.

Not to be outdone, 26,706 participants registered for the United Nations Climate Change Conference in 2019—a tad shy of the 28,141 that participated in 2018.[41] Although, in this last case, the conference organizers made it clear that lots of trees were planted to offset the carbon emissions related to the conference. This is commendable, as few other conferences bother with such forestry projects.

Planting trees is definitely convenient to purchase peace of mind—somewhat like donating a few dollars to a hospital foundation each time one binges through an ice cream bucket. Even more convenient, as a bonus to those who purchase carbon credits from firms catering to this market, the carbon offsets are credited on the day of the purchase—apparently, in some cases, the benefits corresponding to the full one hundred years of a tree's life can be accounted for when calculating credits.[42] Unfortunately, a planted tree does not achieve maturity for twenty years,[43] and many of those traveling scientists are categorical that it will be too late if emissions are not reduced to specific caps within ten years. In some other cases, instead of planting new trees (as those who purchased the carbon offsets possibly expected), what has been sold is the right to claim credits for the carbon absorbed by existing trees—arguably because they have not been cut.[44] This is a bit like promising a romantic dinner and a sunny vacation to a loved one, and delivering on the promise by pulling a lawn chair in the backyard and ordering pizza when the sun shows up between two clouds.

With the carbon offset world still in its infancy, it remains largely unregulated in many ways,[45] which apparently makes it possible to be highly creative in calculating what a carbon credit is, how many credits a specific activity is worth, and how many times the same credit can be sold to multiple buyers—to the point of where some have called carbon offset credits an "imaginary commodity,"[46] which is an interesting perspective given that it is an idea that has been embraced by investment firms and politicians.

Arguably, buying carbon credits is somewhat like shoveling "doggy doo" into the neighbor's backyard.[47] If the goal is to reduce carbon emissions, the most effective actions might be those taken at the source. Everybody can participate in reducing greenhouse gas emissions and everybody has a number of solutions ready to be applied to others, but, there again, the messengers tend to preach abstinence while behaving like sailors on a shore leave.

There are those who advocate greater use of public transportation but have never set foot in a transit bus or have not done so since they were students[48]—which is even more damning when these are managers working at public transportation authorities.

There are those who advocate curtailing air travel. Dominic Champagne wrote and directed the show *Love*, jointly produced by the Cirque du Soleil and Apple Corp, featuring the music of the Beatles.[49] More than eight million people, since 2006, have flown to Las Vegas and paid hundreds of dollars to see it.[50] The same Dominic launched the "Pact for Transition" initiative in 2019,[51] whereby people could sign an online pledge to reduce their personal greenhouse gas emissions by, among many things, minimizing air travel. The list of those who took the pledge includes the Canadian billionaire Guy Laliberté,[52] former street performer and founder of the Cirque du Soleil,[53] also known as Canada's first space tourist for having paid $35 million for a personal twelve-day trip to the International Space Station from where he participated in a webcast to create awareness about the importance of water conservation.[54]

There are those who advocate that people should move into smaller homes to save the planet, recommending to cut today's average 2,600-square-foot home in half.[55] Those who adhere to the tiny house movement promote even more simplicity, squeezing down to 400 square feet.[56] At the same time, to enjoy post-presidential life, President Obama purchased an $11.75 million, 6,892-square-foot house in Martha's Vineyard.[57] Given that President Obama's 2015 action plan to address the threat of climate change emphasized the use of clean energy and decreasing carbon emissions,[58] the mansion's occupants will likely heat the place by cuddling together or by installing the more than one hundred solar panels needed to power a house of that size on a sunny day.[59] The sizeable percentage of its 29.3 acres that is covered by manicured lawns will also likely be cut using scissors instead of a fleet of lawnmowers.

Maybe house size is a moot point. As clarified by Al Gore's communication director when it was revealed that the author of *An Inconvenient Truth* owned a 10,000-square-foot home near Nashville, Tennessee[60] that used twelve times the energy of a typical Nashville home, the former vice president makes up for it by purchasing green energy and carbon offsets.[61] This is also probably the case for Gore's 6,500-square-foot, six-fireplace second home in Montecito, California.[62]

There are those who advocate only eating locally produced food—incidentally, achieving food self-sufficiency regionally would not only reduce greenhouse gas emissions, but also reduce vulnerability to natural disasters and events that happen elsewhere and that can disrupt the global food supply chain (a benefit only if the local food sources are not themselves vulnerable to a local disaster). Saying goodbye to bananas and encouraging the local economy makes sense, although "local" is an elastic concept given that most developed places are not self-sufficient when it comes to food. The 1.4 million residents of Hawaii,[63] who currently import more than 85 percent of their food, might be able to adapt to some degree, as stated in the State's Increased Food Security and Food Self-Sufficiency Strategy.[64] The 730,000 residents of Alaska may be more challenged to do so.[65] Incidentally, self-sufficiency by itself is not synonymous with a healthy diet. For example, a 2006 study by the British Columbia Ministry of Agriculture and Lands reported that the province's farmers only produced 48 percent of the food consumed by the 4.1 million[66] residents of the province at the time. This self-reliance measure would have dropped to 34 percent if the good people of British Columbia had put in their plate as many fruits and vegetables as recommended by Canada's Food Guide to Healthy Eating— and possibly even lower percentages when considering that the province's population increased since that study, to 5.1 million in 2020[67]—because the province imports a lot more fruits than it produces.[68]

Then again, eating local is not necessarily a sure bet to reduce carbon emissions. Farming is an activity that can also produce carbon emissions when all is taken into account. A study calculated that by the time it ends in grocery stores, New Zealand lamb shipped from halfway around the globe to the United Kingdom has produced only a quarter of the greenhouse gases emitted for local British lamb.[69] This is apparently because New Zealanders let their sheep roam around in wide meadows to graze on naturally growing grass and clover, without the need to use additives or hormones[70] whereas British farmers feed them grain that requires a substantial amount of energy to grow.[71]

At the same time, to remain energized through all these activities, many climate change advocates and activists presumably drink coffee. Does their credibility erode with each cup? While California, Hawaii, and Puerto Rico produce some coffee,[72] more than $4 billion worth of coffee is imported from far greater distances (that do not qualify as "local") to quench the country's addiction to caffeine[73]—which is less than half of what Europe imports.[74] It has been estimated that 140 billion pounds of CO_2 is produced to make it possible for the world to drink its five hundred billion pounds of coffee every year,[75] which is as much emissions as sixty-four million passengers flying from London to New York.[76] For comparison, the major US airlines collectively transported twenty-six million passengers across the Atlantic in 2018.[77] Not bad for a product nonessential to life and that has a near-zero nutritional value[78]—before it gets loaded-up with milk, sugar, whipped cream, and other trimmings that make calories skyrocket.

To avoid being perceived as individuals who see the straw in the others' eye but not the beam in their own, trading the coffee drinking habit for the healthier one of drinking water (directly from the faucet, without plastic bottles) would go a long way. That might be quite a challenge though, given that, for comparison, when asked to choose between a month without coffee or without cell phone, 49 percent of coffee drinkers answered they would prefer ditching the phone.[79]

Actually, for those messengers eager to live by their message, ditching both coffee and the phone might even be better. All the emails, chats, photos, and videos that end up posted on climate change websites or simply posted on the cloud, consume energy—90 billion kilowatt-hours in the United States alone, 416 terawatts worldwide, and growing.[80] For comparison, a large coal powered plant produces approximately five hundred megawatts. This power-hunger is because the cloud is not in the sky, but rather scattered all over the world, in thousands of data centers that are essentially large warehouses (of up to half a million square feet) filled with storage technology heated/cooled to maintain optimal operational temperature for the hardware. Fortunately, some of the more modern ones use solar energy with a few hours of battery backup (and maybe someday enough battery backup to run 24/7 on solar energy).[81] All of that to make sure the World Wide Web is ready to instantaneously satisfy anybody's urge to access all of the planet's data—or watch cat videos on YouTube.

CLIMATE CHANGE IS AN EARTHQUAKE

The awful (and inconvenient) truth is that all that clowning around does not make climate change less of a reality.

Data on annual global temperature since 1880, as reported by NOAA, indicates a slight decrease of 0.35°F from 1880 to 1910, and from then an increase of 1.4°F from 1910 to 2020—at an accelerating rate of 0.32°F per decade since 1981.[82] Evidently, that is an average using temperatures measured across the planet; variations are significant, even across the same country, as the rate has been three times slower than average in Florida, but two times faster than average in the Northeastern United States. This warming has left its imprint in many ways, such as seen through the melting of glaciers. Photographic records as well as on-the-ground markers have clearly documented the retreat of 90 percent of the world's glaciers—by more than a mile in some cases.[83] Although a few have expanded (such as the Jokobshavn glacier in Greenland that has grown 100 feet thicker per year from 2016 to 2019),[84] the net total globally is a significant loss. These are facts, based on tangible data. What the future holds falls within the realm of predictions, some more pessimistic than others, some with more sophisticated models than others, with all the caveats and uncertainties that this entails.[85] Alternative scenarios predict a further global warming of 2.1°C to 3.9°C (of 3.8°F to 7.0°F) by the year 2100 if continuing current practices.[86]

Nobody deliberately wishes global warming to occur. Owners of ski resorts are seriously concerned that their business is melting away, but, so far, none of their clients have used sailboats to travel uphill to the slopes. The 130 million skiers worldwide have typically used cars, buses, trains, or airplanes to reach one of the six thousand ski resorts located in sixty-seven countries, and use some of their twenty-three thousand lifts to reach the top of the mountains.[87]

In parallel, the average global sea level has increased by roughly nine inches since 1880—again, non-uniformly across the planet when it comes to impact on shores, due to variable local ground settlement, ongoing slow "rebounding" of the continent from the previous Ice Age, regional currents, and many other factors.[88] Two of the factors that have caused this increase in sea level are the thermal expansion of seawater as it gets warmer (everything expands with increases in temperature) and the added meltwaters from receding glaciers located on land (and only those on land, because a glacier that floats displaces its own volume in water, resulting in a zero net change in water level when it melts). Again, these are facts, based on tangible data. Again (and it is worth repeating), what the future holds falls within the realm of predictions, some more pessimistic than others, some with more sophisticated models than others, with all the caveats and uncertainties that this entails. Alternative scenarios predict a further global sea level increase ranging from ten to one hundred inches above current levels by the year 2100.

Nobody deliberately wishes oceans to rise and coastlines to flood. Owners of beach resorts are seriously concerned that the beaches are being eroded away. In Florida alone, beach tourism brings nearly $50 billion per year to the state's economy.[89]

Yet, facts are facts. The data shows a clear and definite trend in everything that relates to global warming—something that everybody can see, in spite of the flaws, contradictions, and goofiness of the messengers. Not surprisingly, surveys conducted worldwide reveal that, depending on the country, 75 to 97 percent of people believe that climate change is occurring[90] (the number in the United States hovers at 75 to 80 percent, depending on the survey).[91] Furthermore, between 50 and 70 percent of Americans believe global warming is human-caused, the actual number again depending on how the question is asked[92]—as is the case with statistics, surveys can be structured (some would say "gamed") to give different answers. More than two-thirds of Americans indicate that they are worried about global warming and think that it will harm the country (50 percent think it will harm them personally).[93]

Yet, while the automobile industry saw an overall 1.3 percent decline in the number of vehicles sold in 2019, sales of pickup trucks and SUVs increased by 2.6 percent,[94] in spite of the fact that passenger cars generally have a 50 to 100 percent superior fuel efficiency.[95] Apparently, worries do not always translate into action.

All the above is perfectly understandable when recognizing that climate change is nothing more than an earthquake—albeit in slow motion. As such, it is possible that marches, protests, sit-ins, bed-ins, and other mass mobilizations will have an impact and "move the needle" to some degree, one nudge at the time—as all the silent heroes have done for decades when it comes to earthquakes and other hazards. However, as for all hazards, nothing is as effective as a disaster by itself: it not only moves the needle, but it also puts jet engines on it. A disaster is a truthful and reliable messenger.

Sadly, when low elevation atolls and coral reefs are swallowed by the rising sea, few industrialized countries will pay attention. An earthquake far away is generally ignored—beyond well-wishing thoughts, donations to humanitarian relief, and a few benefit concerts.

However, when downtown Miami and Manhattan find themselves flooded weeks at a time, damaging trillions of dollars in real estate and creating massive economic losses, the calls for a national state of emergency will be heeded. An earthquake at home can have an impact.

Therefore, it is the redeeming feature of global warming that it is a relatively slow process, not a sudden one. Its ravages will be felt over decades, but the problem is more akin to a cargo ship running into a shallow sand bar than to the *Titanic* hitting an iceberg.

Climate scientists may be calling for a global state of emergency to enact dramatic reductions in greenhouse gas emissions before the doomsday beyond which damage will be irreversible, but all this wisdom, all this science, and all these predictions will likely continue to fall—for the most part—on deaf ears until real pain from a few major disasters related to global warming are felt locally. At home. Where it hurts. In ways that are unambiguously clear.

The "unambiguously clear" part is critical. There have been hurricanes, floods, heat waves, hail, and other disasters before. Attributing every future such disaster to global warming is not necessarily convincing—particularly when well-intentioned eco-alarmists argue that every time someone sneezes it is because of global warming. Dozens of cyclones have developed year after year over the oceans, but not all of those are large, and not all make it to shore. Florida was hit by five Category 4 hurricanes from 1945 to 1950, but none struck the state in the thirty-year span from 1961 to 1991.[96] Hurricane landfall is a hit and miss game, so when two will hit Florida in a given year, those attributing this catastrophe to global warming will get a lot of media attention but possibly will achieve nothing else—whether global warming increases the number of hurricanes per year or not.

Attributing every wildfire to global warming is not convincing either. On a good year, most of California sees less than an inch of rain per month for five to seven consecutive months.[97] Months with zero precipitations are frequent. In fact, most of California would still be a desert if not for the fact that water has been channeled to its cities from hundreds of miles away (in some of the most geographical and political tortuous ways).[98] As a result, much of California's hills are most of the time covered by dried grass ready to be ignited at the first spark. Equipment owned by the Pacific Gas & Electric company (PG&E)—which provides power to more than half of California—has apparently generously provided such sparks. The California courts declared the utility company responsible for no less than seventeen wildfires in 2017—out of a total of twenty-one that year. Since the company has been deemed negligent in adequately trimming the vegetation around its power lines and equipment, bringing combustible material closer to sparks,[99] California courts approved a $25.5 billion[100] settlement by which PG&E compensated the victims of the wildfires it caused in 2017 and 2018.[101] For good measure, the California Public Utilities Commission tacked an additional $1.9 billion fine on top.[102] As part of the process, PG&E also pleaded guilty to eighty-four counts of involuntary manslaughter.[103] Somehow, the litigating parties apparently did not think of suing the millions of licensed drivers in California, who produced more than a third of the state's greenhouse gases[104] that contributed to the same global warming blamed to be responsible for increasing the number of wildfires per year.

None of the above are unambiguously clear events—whether or not global warming contributes to an increase in their frequency.

An unambiguously clear event is something new and measurable that has never happened before and that can be directly attributed to a specific cause, without any reasonable doubt. This is because it takes more than massive consensus to move the needle; it takes an emotional earthquake. If it were easy to move any needle, it would not have taken until 2017 for the citizens of Boulder, Colorado—known to be so environmentally conscious and avant-gardist in promoting sustainability that detractors and supporters alike refer to the city as the "People's Republic of Boulder"[105]—to have their local electric company shut down its coal-fired power plant and transition to natural gas.[106]

PHLEGETHON

It is pleasurable nowadays to take a romantic stroll along the Seine in Paris. People living along its banks during antiquity felt similarly blessed, not so much for the romance in that case but rather for the abundance of pure drinking water and fishes it provided. More problematic, for the centuries between these two extremes, someone on a date would have made a serious strategic mistake to suggest a walk along the river, for the same reason that a guided tour of the sewers is generally not propitious to triggering a lasting romance. As it grew to become the biggest city in Europe in the Middle Ages,[107] everything was conveniently dumped into the Seine: sewers, animal carcasses, industrial waste, tanners' dye, and more. As early as the twelfth century, the king preferred to keep his windows closed to avoid smelling the stench from the river.[108] As industry grew, so did pollution, up through the twentieth century when heavy metals, chemicals, fertilizers, antibiotics, and other poisons were added to the mix.

This was not a unique situation. In the 1950s, the Thames through London was said to be a foul-smelling sewer of dead water for miles downstream of the city.[109] The rest of Europe did not do much better, from the Danube[110] to the Elbe.[111] The European practice of dumping everything and the kitchen sink into nearby waters traveled with the explorers to the new world where polluting rivers was accepted practice up to the late 1960s.

In Greek mythology, Phlegethon was a stream of fire—probably a river of lava. Not to be outdone, the United States created its own burning river. The Cuyahoga River runs through Cleveland, which was for decades one of the major industrial centers of the Great Lakes region. As such, it collected pretty much everything these industries along its shore could dump into it, as a standard operating procedure. In those days, trashing a river was considered a normal tradeoff for prosperity.[112] However, the Cuyahoga was not just polluted, it was the most polluted in country. In fact, so polluted that it frequently caught fire: thirteen times from 1868 to 1969. In 1912, a fire on the river killed five people. In 1952, another river fire caused $1.3 million in damages. Nothing changed then.

However, in 1969, when oil-soaked floating debris ignited, the fire did not kill anyone, and it only inflicted $100,000 in damage to two nearby bridges, but it was the right earthquake at the right time. It is hard to find a better poster child for stopping pollution than a burning river. Something is seriously wrong when water, beyond not being safe to drink anymore, catches fire. Through the fifties and sixties, the population was progressively becoming more aware and upset at the growing pollution problems, most notably due to the smog visible in big cities, the recognized dangers of pesticides (such as DDT), endangered species, major oil spills washing ashore, and many more.[113] Images of the blue planet taken by the budding space program provided snapshots that made it clear that the earth was of a finite size. The burning Cuyahoga River topped it off. It was the earthquake moment that led to the National Environment Policy Act (NEPA), which itself led to the establishment of the US Environmental Protection Agency (EPA). One of the first successes of the EPA was the Clean Water Act of 1972, which had as a goal that all of the nation's rivers be cleaned to allow the safe return of swimmers and fishes by 1983.[114] Nowadays, people are swimming in the Cuyahoga River (except after heavy rains that raise the concentrations of E. coli in water—a different problem) and are catching fish that are safe to eat.[115] Concerns nowadays are that the river, like many others,[116] may be filled with viruses, bacteria, and microparasites,[117] but this is a different problem that will need its own different earthquake.

Similar earthquakes are needed—and will happen—to tackle most modern problems, and not only global warming. For example, the plastic crisis will take more than a ban on plastic forks to be resolved. The fact that the pile-up of plastic garbage is invisible to North Americans makes the crisis intangible and could possibly allow it to continue until the much-needed earthquake happens.

Collecting plastic out of the Great Pacific Garbage Patch,[118] by deploying an array of floaters that drag a debris-collecting skirt using the ocean's energy itself, is an unprecedented undertaking that is receiving worldwide acclaim. A possible danger is that, by demonstrating that "oceanic garbage collection" is an achievable technology, it could become a regular garbage collection service billed to users like any other business and thus allow polluters—with a clear conscience—to think of plastic as "not so harmful to the ocean after all."

Biodegradable plastics that break down into smaller bits have also been reported to be problematic, first because these microplastics can accumulate in the ecosystem and the food chain with some continued damaging effects, and second because it makes the problems less visible, giving the impression that it has vanished.[119] Perfectly biodegradable plastics would need to have molecules that completely disassemble themselves into carbon atoms. Typically, microorganisms are needed to "compost" these biodegradable plastics into water, carbon dioxide, and biomass.[120] Various chemical companies claim to be working in that direction or to have products already capable of such a feat.[121] There will be resistance to adoption if it costs more than the existing plastics of the planet-polluting kind. However, when a window of opportunity is opened up by the much-needed "plastic earthquake" of the future, the extra cost of plastics that can biodegrade to the carbon atomic level will be deemed fully worth paying.

Until then, many folks will continue trying to figure out how to deal with the contradictions of messengers. This will leave many to ponder whether it is better for the planet to buy six green peppers that have not been grown organically but that can be taken with bare hands directly from the vegetable counter, or to pay more for six green peppers that are certified organic but that have been individually wrapped in plastic for some perplexing reason.

ON THE DISASTER TRAIL

Where the Buffalo Roam

On April 20, 2002, an early morning magnitude 5.1 earthquake with its epicenter fifteen miles southwest of Plattsburgh, New York, woke up a lot of people across the state. Earthquake waves travel well and far in eastern North America because the underlying bedrock there is less fractured than on the western part of the continent. People reportedly felt the earthquake from Buffalo to Boston, and as far south as Baltimore.[122] For an earthquake of that size, damaged chimneys, settlement of road embankments, and other minor damage would be expected in close proximity to the epicenter area, and not much else—and that is exactly what happened. Nevertheless, it made for a busy day at the Multidisciplinary Earthquake Engineering Research Center, as we fielded media calls while trying to put together a small team to travel to Plattsburgh and document the damage, however small it might be—after all, it is not every day that an earthquake happens in our home state, so it deserved some attention even though it was insignificant on the grand scale of things.

One local TV station insisted on coming to campus for an interview to be aired on the 11 o'clock evening news. I was quite busy but agreed to meet them at 6 p.m. in our earthquake engineering laboratory. The lab was filled with advanced equipment to load structures statically and dynamically with hundreds of thousands of pounds of loading, as well as, on any given day, some rather interesting test specimens. Not surprisingly, many groups typically ask to visit the lab during the year, and we have hosted many tours of the facility.

When I arrived at 6 p.m., spotlights were on, the camera was ready to roll, and the airbrushed news reporter was pumped up and ready. Her first question was: "This was quite an earthquake for New York State. How much damage should we expect to find in Buffalo?" My answer—and the correct one—was, "None. Given the size of this earthquake, and the fact that Plattsburgh is three hundred miles from Buffalo, it is most unlikely that it could have produced any damage in Buffalo."

As soon she heard the word "none," she made a face and told her crew to start packing up. There would be no damage in Buffalo, so the whole thing became of no interest—instantly so. My offer to give a personal tour of the lab to showcase the research conducted at the university by many professors to prevent damage from future earthquakes anywhere worldwide was met with an empty stare. Her body language made it clear that she felt she had wasted time driving with her crew to our lab with no sensational story to bring back—something she could have avoided with a simple question during the original phone call she had made to set up the interview.

This was a fantastic example of "if it didn't happen here, who cares?" Incidentally, it was also a fantastic example of "reporters in search of infotainment are maybe not real journalists" but that is a different story.

Elbow Room

A NUMBER GAME

Imagine a single termite in a house. It could blissfully chew up a bit of the house every day without having to think much about the future. Even if there was only a single wood house within reach of that termite, that resource would seem inexhaustible, which would be a reasonable assumption considering that it would apparently take more than three thousand years for a termite to completely digest a 1,000-square-foot home,[1] and that the life expectancy of a termite is less than two years. That termite would not have any reasons to worry about starving. Chomping an infinite supply of food and basking in the sun the rest of the day makes a termite's life seem idyllic—except that it is not, because termites rarely dine alone. If a thousand termites attacked the same home at the same time, the original blissful termite could still see life as an endless feast, unless it started to develop a conscience when looking at the rate at which the food disappeared, wondering what would happen to future generations when the bountiful meal was all consumed. In the spirit of preserving resources, it might go on a diet, cutting its consumption of cellulose by half, to "save the world" so to speak. That would gain some time if, indeed, only a thousand termites were taking the house down bite by bite. Unfortunately, termite queens are prolific and pop-out thirty thousand eggs on a good day.[2] Not all will hatch, but a termite colony of one thousand can multiply into three hundred thousand in less than five years.[3] With such a massive invasion, no matter how much the original blissful termite curtails its intake of wood, that house is doomed; and if no other wood source is within reach, so is the colony. All that because of an overpopulation problem. In other words, overpopulation is also a disaster. A massive "earthquake." In slow-motion.

This analogy scales up pretty much the same to any human-created problem. Like cows. Except for cows who have the good fortune of being born in India, they are pretty much considered to be not-so-sacred dumb

318

animals. Or rather, whether cows have an IQ greater than dogs or not, they have had the bad luck of being milked or ground into burgers for centuries without anyone giving much thought about it. However, now, with the cow population on the planet approaching two billion, methane emissions from cows alone have been singled out as one more significant contributor to global climate change. It has been calculated that each pound of beef produces as much carbon emission as a car driven over 150 miles.[4] Considering that a well-fed steer can yield five hundred pounds of beef, that is equivalent to driving thirty cars from New York to Los Angeles. Multiply that cross-country trek by two billion, and that adds up to a lot of carbon. As a consequence, many are advocating a vegetarian diet as part of the strategy to resolve the climate change problem.

Certainly, if the entire developed world decided overnight to adopt a voluntary simplicity lifestyle, to downsize living space to a hut, and to shift to a diet consisting exclusively of Brussels sprouts, beans, and beer, it would significantly reduce the size of herds worldwide. Were such a dramatic diet shift to happen, cows around the globe (and not only those that bumped their snout on the closed gate when the last slaughterhouse shut down) would erect shrines to the cow-Gods who have freed them forever to graze in the wild. Celebrating even more, wolves and other predators would pop out the champagne to welcome this sudden miracle-release of a once unattainable source of proteins that, after centuries of docile captivity, might have evolved into a defenseless running and mooing herd. Nonetheless, having the entire human population sworn off red meat would only provide Mother Earth a momentary respite from her rowdy kids. Once the earth's population grows to fifteen billion humans, that will add up to an awful lot of Brussel sprouts, beans, and beer consumption, requiring massive deforestation to fulfill the agricultural demand, gazillion gallons of water to irrigate all that land, and a colossal fleet of donkeys to cart all this to the huts where it will be gulped—not to forget the impact on the ozone layer of fifteen billion methane-emitting bean eaters.

It goes without saying that those huts, if located in cold climate, would have to be heated. Some partial to word-burning stoves[5] will argue that burning wood is practically carbon neutral[6] and that modern, high-efficiency stoves, built with the latest technology, can meet and even exceed the EPA's stringent 2020 emission guidelines.[7] Others will object, arguing that wood-burning stoves emit particles ("particulate matter") of a size equal to about 3 percent of the diameter of a human hair[8] and that these particles have been identified by the World Health Organization, the European Environment Agency, and by the US EPA to be a health-hazardous air pollutant,[9] and that billions of stoves emitting these particles together with tons of carbon monoxide would foul the air and accelerate global warning. They would instead advocate for efficient electric heat pumps connected to solar panels. Yet others would raise concerns that the manufacturing of solar panels, just like the semiconductor industry, requires the use of toxic chemicals,[10] such as hydrochloric acid, arsenic, nitric acid, hydrogen fluoride, gallium arsenide, cadmium-telluride, and a bunch of other products that could create an environmental or public health hazard if not handled and disposed of properly,[11] and that manufacturing, transportation, installation, maintenance, decommissioning, and dismantlement of solar panels also has a carbon footprint, just like all of human activity.[12] These are great debates, and things evidently become more complex when considering all factors that come into the equation to calculate carbon emissions of a given product over its lifetime. Nonetheless, as wonderful as it may be if one method of energy generation emits three times or ten times less carbon than another, the point remains that, over time, there will be no net gain in the reduction of carbon emissions if the population of the planet triples, or grows by a factor of 10—unless everybody moves their hut to warmer climates and forgoes air conditioning.

In other words, reducing one's consumption of plastic and nonrecyclable waste, minimizing carbon footprint, and doing other planet-conscientious actions are all commendable. Anybody who deliberately and painstakingly makes energy choices and succeeds in reducing their personal carbon footprint from 20 to 8.5 metric tons a year should be proud of this achievement. Even more so given the fact that, as calculated by students at MIT, in the United States, a homeless person who survives on soup kitchens alone emits 8.5 tons a year—a target that might be unattractive to even the most environmentally conscious soccer moms and football dads.[13] Admittedly, the MIT class project that came up with those numbers did not consider whether adopting an Amish lifestyle could be an even better alternative to help bring down further the total carbon emissions per person, but, if that born-again Amish population were to increase tenfold over time, there would again be no net global gain in lowering carbon emissions. Each born-again Amish would be self-sacrificing like the termite in the above example, to no avail.

One of the reasons focusing on reducing carbon footprint is becoming popular might be that it is a "low-hanging fruit," meaning that it is achievable because it is sellable. Some have inferred that it is a market potentially worth trillions of dollars that will make billionaires out of those corporations who have invested in carbon trading[14]—and, simultaneously, out of politicians who will help make this happen. Unfortunately, it is focusing on treating a symptom of the disease, not on the cure itself, because it is easier to do so. The "earthquake" has not happened yet, so a lot of efforts are spent on making the planet more energy efficient, rather than on controlling population growth. Saving the planet might actually require both.

It could be argued that carbon emissions are the symptoms, whereas the cause of the problem is the continuously increasing population. Those focusing on the symptoms will lobby for laws and regulations leading to lifestyle changes. All good and all valuable actions in their own right. However, a challenge lies in the fact that, as argued in previous chapters, human nature makes it easier to impose changes to others than to oneself. Tribalism woven into human DNA over millennia sustains the natural tendency to huddle in groups that love to clash across political, cultural, and religious divides. As a result, the problem is always "them," not "us," which greatly and conveniently simplifies things. However, seriously tackling the problem of population growth makes everyone the problem—a far less popular idea.

When the problem is "them," it is fully justified (or deemed to be a necessary evil) for someone to burn jet fuel over six thousand miles for the sole purpose of disrupting presentations by the US government promoting fossil fuels and nuclear energy at the United Nation Climate Talks.[15]

This can be seen as a crusade to save the planet,[16] or challenged and depicted as mere "political exhibitionism,"[17] depending on the point of view. It sure makes for entertaining debates, as someone must be right and someone else wrong, but in the meantime, it does nothing to control population growth.

Attempts at controlling other people's minds often happen by laws or by wars. Focusing on the former, at first, is always a good start—at least for democracies. Some options are possible in this regard. If one agrees that global reduction of carbon emissions will only ultimately happen when reducing the number of emitters—that is, living beings—this can be done by targeting two different groups.

First, any person asking humans to sacrifice anything to save the planet cannot righteously do so without first contemplating the carbon footprint of their own Fluffy, Kitty, Bubbles, Nibbles, Spidey, and other pets (pet rocks being the exception). The United States is home to over 160 million dogs and cats—by far, more than any other country—and these furry freeloaders alone consume more dietary energy than sixty million people do.[18] Even considering that some of that food does not compete with what is on the dinner table—although a lot of homeless people would beg to disagree—one must add to the energy equation the carbon footprint required to process that food, package it, and ship it to grocery stores where much heated or air-conditioned real-estate space is used to stock pet food (often an entire aisle). Furthermore, all that pet food is eventually converted into feces that pollute the earth, and do even more so when picked up in a plastic bag by owners who do not want to be shot for leaving unwanted fertilizer on their neighbors' lawn. Unfortunately, to make matters worse, along with an increase in prosperity in some other countries (such as China) comes the wish to emulate the United States when it comes to pet ownership.

Therefore, now more than later, the first and simplest measure to implement to achieve population reduction would apparently be to rid the planet of all domestic pets, at once.

Simplest?

Not so. It is outrageous. "Fluffy is part of the family."

Really? If there were any pride in sharing DNA with a dog, then being called a bitch would be a compliment. Besides, all scientific evidence unambiguously points to the fact that dogs are animals, not family members—although scientific evidence is often trumped by unscientific conjectures found on websites nowadays.

One would think it a relatively easy thing to do to let go of domestic animals, but not so. Many of these animals, who never asked to be taken away from their natural environment millennia ago, were domesticated either to serve a purpose (like sheep dogs) or to entertain and comfort (like cats and goldfishes). Studies have highlighted the perceived benefits of pets in terms of improving mental health and social status—just like a five-thousand-pound 16 mpg Hummer H3 can provide the same benefits to its proud owners.

Evidently, exceptions are likely to be negotiated. Special permissions could be granted to K-9 units, guide dogs, and other animals performing critical duties that cannot be fulfilled by Chihuahuas and Dachshunds, while waiting for better technology to replace them. Nonetheless, the concept is clear: no more resources needlessly wasted on pets.

The above is an important first step because if population control cannot be achieved for pets, the next step might be outright impossible. This is because the second group of carbon emitters whose population must be controlled is the humans, and reducing the number of humans on the planet will not be a small task. In other words, as difficult as it may appear to be, eliminating pets is the least difficult of the population reduction strategies. Relatively speaking, at least.

Pet lovers will argue that it is more important to reduce the world's human population, given that humans are numbering in the billions, whereas the energy consumption and carbon footprint of domestic animals in the United States is nothing more than a drop in the bucket. Arguably, the argument may sound logical to pet lovers, but it raises an interesting ethical dilemma: How can anyone ask Dick and Jane (or Mohammad and Fatima) to refrain from having another kid on the grounds that it will have a negative impact on the planet, while someone else will be allowed to own a 120-pound Rottweiler? Attempting to control the world's population while searching for contrived arguments to justify the parallel existence of domestic animals is tantamount to admitting that, in the minds of pet lovers, cats and dogs matter more than people do—or, at least, matter more than some people. This reasoning does not even need to be verbalized in some cases. Beyond it being an expression of frivolous wealth, there is no denying that burying the remains of Fluffy in an expensive casket at the Pet Cemetery, in a plot adorned by a tombstone, is paying last respects to a four-legged being who was, after all—there it is again—a family member. Given that the burial of dead is a practice known to have existed for more than a hundred thousand years, performed out of concerns for the afterlife, this is effectively saying that Fluffy has more soul to preserve than Joe Homeless who was found dead in a dumpster, after a night of subzero temperature. In a world assaulted by population growth, that will not fly. At least, not without serious sparks (more on that later).

Nonetheless, irrespective of what is done with pets, it remains that there will be a need to do some serious human population planning—in other words, there will be a need to control the rate of growth of the human population before it reaches unsustainable levels. One problem, as for all other hazards reviewed earlier, is that action toward that goal is unlikely to start before the "earthquake" itself has happened, and since the overpopulation earthquake has not happened yet, possible solutions will be debated within some circles but not embraced and implemented to make a difference. In essence, there are only a limited number of ways to stop the growth of, and possibly reduce, the human population, and these are reviewed below.

POPULATION ZERO

Population growth can be expressed by a simple mathematical function. Less rigorously but just as effectively, it can also be illustrated with the following example.

Rabbits arrived in Australia with the first fleet from England in 1788.[19] Rabbit meat was considered a convenient food source, so the rabbits were bred accordingly, kept in cages or contained within closed walls. Until circa 1859, when a gentleman got the brilliant idea of releasing some of those imported European rabbits in the wild for the sake of sport hunting—in an era where rabbit shooting was considered trendy (like modern "shoot 'em up" video games, but with real blood). Apparently, the gentleman was not a good shot, or ran out of ammunitions, because some rabbits definitely survived. At least one male and one female. In a new continent with no natural predator, this couple of rabbits sure had a good time. If setting the nest far enough from hunters, it is possible to estimate population growth as follows. It is known that a single female rabbit can have, on average, 5 bunnies per litter and 5 litters per year, and that bunnies become fertile adults in only four months. Assuming 3 males and 3 females per litter, for the sake of argument, if an original batch of 6 bunnies is born in January, then mother and her three daughters will produce 24 bunnies in May, for a total of 30 rabbits. By September, mama with daughters and grand-daughters will together produce 96 offspring, growing the rabbit population to 120. By the end of the year, the total will be 510 rabbits. And so on. By January of the next year, there will be 37,566 rabbits, and by the end of the third year, over a million, eating all the vegetation in sight. Such a population growth rate is staggering. That is, in essence, what is called exponential growth (basically the same function as for virus spread). Unless rabbits serve as fast food to a lot of hungry carnivores, the population will grow at an accelerating rate—and it did so in Australia, where they have no natural predators. Even though the longevity of rabbits is only six years[20] (who could live longer with so many kids, grand kids, great-grand kids, etc.), within ten years (that is, by 1869), Australian hunters killed approximately 2 million rabbits each year, without making a dent in reducing the infestation and the ecological disaster that came with that invasive species. Nowadays, Australia's rabbit population is estimated at over 200 million[21]—and is a major problem.

The same is happening with humans—albeit slower, thankfully.

Arguably, a highly effective way to eliminate the rabbit problem in Australia would be to prevent the females and/or males from being fertile. No new bunnies, no rabbits left within six years. That is why Australian scientists are tinkering with ideas on how to render all females or males sterile, as one possible way to eradicate the problem.[22]

Interestingly, species eradication as the ultimate solution to save the planet is exactly what the Voluntary Human Extinction Movement (VHEMT)[23] is advocating for humans.[24] If the entire world buys into the program and abstains from reproduction, within roughly one hundred years, there will be no more humans, and thus no more human-created environmental problems. Simple mathematics. The VHEMT has a live-and-let-live philosophy, but one with a "do not reproduce" caveat. In other words, the VHEMT philosophy, at its core, does not advocate global mass suicide (only destructive cults do that, as in 1978 when hundreds of members of the Peoples Temple ended it all in Jonestown, Guyana, by drinking cyanide—while others refusing to drink were outright murdered).[25] Rather, it advocates a slow-motion suicide of the human race, not even asking the last one leaving to close the lights. As such, VHEMT advocates are a little bit more people-friendly than the Church of Euthanasia, which was recognized and received tax-exempt status from the state of Delaware,[26] and whose slogan was "Save the Planet, Kill Yourself." Embodying what could be classified as an extremist faction of the VHEMT, the "four main pillars" of that unique Church's faith were suicide, abortion, cannibalism of the already dead, and sodomy (defined as "any sexual act not intended for procreation"). One may question the viability of any organization that encourages its patrons to suicide, but the Church of Euthanasia's website made it clear that those who killed themselves after having joined the church would automatically attain sainthood status, and reminded its saints to feel free to donate their estate to the Church.[27] Apparently, as in many other religions, the head of the church does not necessarily have to follow the very tenets it preaches, as somebody has to stay behind and tend to the finances.

For all of its good intensions, the VHEMT has not exactly caught on like wildfire. In fact, while the goal of altogether eradicating the human population is an entertaining abstraction, it would be fair to guess that the chances for the VHEMT to succeed in its crusade are nil. This is in part what happens after millennia of evolution. Given that any species that loses its will to reproduce will effectively cease to exist, it is fair to assume that the DNA of all species that exist today (including humans) is packed with genes that seek transmission to future generations. Maybe some humans could convince themselves that civilization has run its course and that it is time to discontinue the lineage, but this would be a marginal group—an insignificant minority. Fertility clinics charging tens of thousands of dollars per treatment are the living proof that the human DNA is geared toward procreation of the species. People want kids for all kinds of reasons—reasons that often cannot be verbalized but that are no less tangible at some emotional level. DNA is much like a dictator with an urge to be remembered beyond death, and it will not listen to any dissenting views.

In short, if the human spirit were inherently not inclined to favor reproduction, humans would have long vanished from the universe. Furthermore, if an entire segment of the population managed to overcome the evolutionary drive and to successfully resist the procreation urge, this would achieve nothing other than some sort of artificial natural selection, whereby those who bought the arguments against reproduction cleansed themselves from the gene pool. Not unlike what happened in the movie Idiocracy,[28] the dystopian satire in which a time traveler wakes up in 2505, only to discover that, over five centuries of evolution, the elites who ratio-nalized against having kids had wiped themselves from existence, leaving the earth to be ruled by prolifically procreating idiots—a society where fart jokes were supreme.

Yet the leaders of one country have achieved the impossible, and have reined-in population growth already,[29] with its population projected to be the same in 2050 as in 1950, and all without any evidence of the kingdom's citizens degenerating into glorifiers of flatulence humor. The country achieved this population equilibrium not by allegiance to the VHEMT doctrine, but rather by devotion to other beliefs that, incidentally, have also resulted in making it the country with the highest wine consumption per capita in the world[30] (double what the French drink, which is not a small thing). Indeed, Vatican City might be the only country where the birth rate is zero, except for the occasional accident,[31] although, unfortu-nately for these surprise babies, Vatican citizenship is not granted by birth, but rather only on a temporary basis to those at the pope's service. Yet, in spite of this display of procreative self-control, VHEMT militants likely will be booing when the pope encourages increasing birth rates in parts of the world (outside of Vatican City, of course).[32]

POPULATION GROWTH ZERO

Short of a zero population, zero population growth might be the next best thing. This implies an equilibrium in which the number of births is equal to the number of deaths. It also implies that, to be helpful, this equilibrium is achieved at a total world population number that is sustainable. Nothing indicates that this optimum number has not been exceeded already. Some have estimated that the optimum world population should not exceed two billion if widespread wealth and human rights are to exist without depleting the planet's resources and biodiversity.[33] Most unfortunately, the two-billion threshold was crossed in 1927. The United Nations predicted (in 2017) that the world's population will reach a plateau of 11.2 billion in 2100,[34] and start to drop thereafter. That is the mean of all predictions, with the predicted population value reaching over 16 billion if the actual fertility rate turns out to be 0.5 children per family greater than what was assumed in this calculation.[35] Needless to say, most of the 16 billion people in this scenario eke out a living, malnourished or outright starving, dirt poor or unemployed; they do not live in a 2,500-square-foot home, use eighty gal-lons of potable water per day,[36] eat two hundred pounds of red meat and poultry per year,[37] and drive a $35,000 car.[38] Although they may aspire too, if they all did, the earth's resources would already have been depleted—for example, the United States alone consumes 20 percent of the world's motor fuel,[39] with roughly 4.5 percent of the total world's population, while China consumes 12.6 percent of it (and growing) with nearly 20 percent of the world's population.

While the standards of living are generally increasing worldwide and the world population continues to increase, some countries will see their population shrink. Japan is already on the down slope; from a peak population of 127 million, it is project to have only 85 million in 2100.[40] Like Japan, a number of countries will struggle with the inverted population pyramid problem—fewer young workers to support a massive number of retirees—which will put significant financial strains on the economic well-being of those nations,[41] and compound the global geopolitical stress in an environment of global massive population growth. Nothing to help resolve the global population problem, particularly if the countries suffering from population drops react by implementing incentives to families to have more children, as currently done in South Korea, Japan, Turkey, France, and others.[42]

REDUCING POPULATION

Whether the objective is to stop population growth or reduce the total population, many agree that there are basically three ways this can be achieved: voluntarily, coercively, or brutally.[43]

Voluntary Actions

Since a voluntary action is, by definition, self-imposed, it requires women to have a say in how many kids they wish to have. Obviously, voluntary decisions do not occur in a vacuum. The socioeconomic environment typically influences free will. A couple of young urban professionals with pressure-cooker jobs in a city where a $2,000/month rent includes a prime eye-level view on a six-lane freeway is hard pressed to figure out how to juggle life with one kid—let alone a flock of them.[44] A couple whose strong religious beliefs exhort them to be fruitful and multiply—which is standard party line for all religions that compete for worldwide faith-supremacy—is more likely to have busloads of kids.[45]

Data from many studies shows that in most nations where women are more educated and become active contributors to the economic growth, the birth rate has dropped.[46] If that is indeed the primary driver, given that half of the world is still living on roughly $5 a day,[47] the tipping point where population growth will stop would still seem to be quite far in the future. Furthermore, new data seems to indicate that when economic wealth passes a certain threshold and families accumulate more financial security, the birth rate increases again.[48] Wealthy folks are becoming big wealthy families[49]—which does not help the planet either, since wealthy families are heavy consumers of resources.

The counterpart and balancing factor to fertility is the death rate. Given the undeniable sanctity of human life, anything that deliberately tilts the balance to fewer lives could raise interesting questions that are often too touchy to ask.

Actions from charitable organizations are generally always well intentioned. For instance, the Bill and Melinda Gates Foundation aims to provide access "to safe and affordable food" to stop the death of millions of kids from malnutrition before they reach the age of two.[50] By all means and measures, a laudable goal that one should never criticize. However, in many of the sub-Saharan countries where malnutrition occurs, many adolescent girls are denied access to information on family planning and reproductive health—when not outright denied access to education and forced to marry in their teenage years.[51] Eliminating ignorance and oppressive social constructs would seem to be an effective and imperative prerequisite to eliminating child malnutrition, but where religious or societal traditions consider women to be chattels, pregnancies are out of control—and hence indirectly contribute to an uncontrolled population growth. In these regions, in spite of the millions who die from malnutrition every year, population has grown exponentially; it will therefore grow even faster as malnutrition is progressively eradicated, unless oppressive religious or social pressures are eliminated.

Finally, the challenge with voluntary actions is that everybody must buy into the concept for it to work. One large group who believes in popping out kids like a production line undoes the efforts of many more who believe in controlling and limiting pregnancies.

Coercive Actions

Since a zero-child policy is not likely to succeed, it being apparently counter to millions of years of evolution and thus against the human instinct to reproduce, one-child and two-child policies have been attempted in the past.

China's one-child policy was one such experiment. First, enacted in 1979, it rapidly had to be modified into a two-child limit in rural China because parents there badly wanted a second child if the first one—mother of all misfortunes in their eyes—turned out to be a girl.[52] Not surprisingly, with such a mindset, within a short time, the country ended up with more men than women, which comes with its share of societal problems, including the human trafficking of "brides-to-be" from other countries.[53] Furthermore, as is the case with any law, a series of sticks and stones measures were put in place to encourage compliance and penalize violators. Those who obeyed the law received a "one-child glory certificate" illustrated in a way that would have made Mao proud, and that blessed the proud couple with extra money, better housing, better schools, and a slew of other entitlements. Those who did not comply with the law had to cope with the darker side of the communist ideal, including forced abortions, "remedial surgery" (namely, sterilization), jail time, and a good deal of public shaming in the great proletarian tradition.[54] Then, in 2016, to cope with the problems of an aging population and labor shortages, China changed to an official two-child policy instead of a one-child policy. As these problems worsened and birth rate did not improve, it became an official three-child policy in 2021, although some predict this pronatalist policy will not increase the birth rate unless the country eliminates pregnancy-based discrimination by employers,[55] changes its culture of long working hours, and reins in the rising cost of living.[56]

A number of other countries have a well-intentioned two-child policy.[57] These are voluntary, for the most part, meaning that some achieve the goal and some do not. Irrespective of this, some studies indicate that stringent fertility control techniques, such as a worldwide one-child policy, will not stop population growth during this century.[58] In essence, it all depends on how "stringent" the measures are, and how stringently they are enforced. On a topic so emotional, it is difficult to achieve anything by coercion.

Disastrous Actions

"Disastrous actions" is that special category of events that regroup all large-scales horrors created by human activity. The list includes mass-scale starvation, pandemic, ethnic cleansing, and wars—from conventional to nuclear. However, for any one of those events to make a serious dent in future population growth, it would have to be disastrous on a scale never seen before. This is because, no matter how terrible and traumatic, none of the mass-scale disasters of the past have stopped population growth. At best, they have temporarily stalled it.

At first glance, since war essentially turns humans into cannon fodder, it would be natural to expect that global-scale butchery could significantly reduce population. After all, according to some estimates, 70 to 85 million people died during World War II, which at the time represented approximately 3 percent of the 2.3 billion alive in 1940. Roughly two-thirds died on battlefields, the rest from famine and diseases directly attributable to dire living conditions created by the conflict—not to forget the horrors of Nazi death camps. Some sources estimate the death toll to be closer to 118 million.[59] World War I produced about half as many casualties, out of a world population of roughly 1.7 billion.[60] Many other wars have been equally devastating. The Chinese civil war (1850–1864) killed about 100 million, and high estimates for the genocide of the North American first nations is 138 million.[61]

Yet, in spite of all those anthropogenic disasters, from a long-term perspective, population numbers have continued to surge forward. In fact, World War II was followed by a baby boom in North America, France, and even Japan.[62] In terms of the impact on global population, WWII was almost "one step back, two steps forward." Furthermore, a study considering the impact of various hypothetical disaster scenarios on population growth found that a five-year World War III conflict killing people in the same proportion as the previous two world wars combined would have no significant impact on that growth trajectory for the rest of the century.[63] Therefore, if relying on armed conflict to reduce the world's population in a major way, nothing short of Armageddon will do.

So if population growth cannot be bombed away, how about relying on bugs to do the job instead? When it comes to filling graveyards to capacity, cholera, smallpox, HIV, influenza, COVID-19, and all kinds of bacteria and viruses have proven their effectiveness, so one may be justified to believe that a pandemic will cleanse the planet from the human race.

Not quite.

With respect to pandemics, the Plague (also known as the Black Death) has been the most effective in reducing population throughout recorded history. It is estimated that it killed up to two hundred million people in Europe and Asia from 1347 to 1351, and twice that many worldwide during the rest of the century.[64] It is estimated by some sources that 60 percent of Europe's population died of the disease.[65] Yet, in spite of these substantial losses, after two hundred years, population levels had recovered and exponential population growth resumed. Granted it took a while to recover from that hit, but one must keep in mind that all this happened at a time when medical doctors believed that all diseases were due to excess blood and that draining blood out of the body, using cuts or leeches depending on the amount of blood deemed necessary to extract, was the obvious cure—a procedure most health insurance plans do not cover nowadays. As part of that great medical wisdom, depending on the various schools of thought on which that eminent medical training was founded, the cause of the disease was either determined to be due to divine forces, earthquakes,[66] or the presence of all kinds of foreigners. Jewish communities were annihilated in many European cities between 1349 and 1351 because of the misguided belief of hysterical mobs that the Plague emerged because Jews had poisoned the wells.[67] Note that, in that frenzy, even those with any type of skin diseases, such as acne, were exterminated—just in case.[68]

Resurgence of the disease continued throughout Europe and worldwide, at times with extensive but proportionally less numerous casualties. Note that the Plague bacterium has never been eradicated and a few people still die of it every now and then, but a vaccine exists,[69] so a Plague pandemic is unlikely nowadays—except among anti-vaxxers if it came to that. However, there exists no cure for many other deadly diseases. The world's population could be decimated by outbreaks of Ebola, multiple highly pathogenic strains of influenza, Severe Acute Respiratory Syndrome (SARS), Middle East Respiratory Syndrome-Coronavirus (MERS-CoV), Marburg, and Nipah, to name a few.[70] Yes, not only YouTube videos can go viral.

There is something humbling in the thought of dying because someone sneezed too close. However, with all international disease tracking and control measures in place, in spite of the COVID-19 debacle, a repeat of the devastation wrought by the Black Death pandemic of the fourteenth century is not foreseen. Before the COVID-19 pandemic, a team of economists had estimated that a moderate to severe world pandemic would produce annual losses of approximately $500 billion, corresponding to 0.6 percent of global income[71]—staggering numbers, but not wiping civilization altogether. Other studies predicted from 100 million[72] to 150 million deaths.[73] Out of 7.7 billion people, that is less than 2 percent of the total number of human beings on the planet. By all measures, this is still far from the kind of drop in population that would help the planet. The Dow Jones Industrial Index plunged by more than 50 percent during the fifteen months of the 2008 financial crisis, but it had fully recovered six years later[74]—in fact, the Index even increased a further 85 percent in subsequent years. In 2020, it plunged by 30 percent in the early stages of the COVID-19 crisis, and surprisingly recovered most of that in three months. From that perspective, while 150 million deaths is tragic and nobody wants to be counted as part of that statistic, the reality is that the remaining 7.55 billion will not stop procreating.

WHO GOES FIRST?

So, if none of the previous calamities, none of the previous policies, and none of the voluntary actions have succeeded in reducing the world's population, what will? Unfortunately, as demonstrated up to now, simply because of human nature, it will take a disaster. Impactful changes likely will not happen unless a critical threshold is crossed. In this case, that threshold is scary.

ON THE DISASTER TRAIL

Replacement Therapy

As a kid, my first pet was a two-inch-long tortoise. Not sure why. Maybe my parents saw benefits in the fact that it was quiet and did not need to be walked twice a day. In any event, it was a lovable animal that was showered with all the necessary care and attention it deserved. Unfortunately, the pet shop owner must have sold my parents the wrong kind of equipment because, one day, it mysteriously disappeared from its terrarium, only to be found weeks later all dried up under a sofa.

The next members of the animal kingdom to take residence in my bedroom were colorful fishes. Again, quiet, this time no legs—maybe furthering a pattern. This was a wonderful and soothing aquatic microcosm, perfectly contained. Unfortunately, the pet-shop owner had omitted telling me that a gorgeous angelfish will eat all the guppy fishes, until it owns the aquarium—a learned lesson to beware of beauty. Then, after getting rid of angelfishes, I discovered that mother guppies ate their own babies—which thereafter leaves quite an impression when hearing mothers tell their baby, "You're so cute I could eat you up." Then, disaster struck: Our neighborhood lost power for a week in the middle of a sub-zero winter, which taught me that tropical fishes do not know how to navigate around icebergs and will end-up floating belly up in cold water.

With that kind of track record, even though I get along splendidly with cats, I prefer to interact with animals in their natural environment. I am guilty of having fed peanuts to suburban chipmunks and I have tamed pond fishes to eat bread from my hands, but I prefer simply contemplating the wildlife. Ducklings furiously paddling to follow mama duck across the pond, a frog hanging by the suction-cups of its fingers on my office window, a lizard changing color as it sees me approaching, a deer jumping over a bush, rabbits hopping across meadows, and other fantastic sightings revealing the beautiful diversity of the world—all happier (and safer) there than in my house.

Nuclear Holocaust

BEFORE BIRTH

"Have you noticed," said one university professor, "every year, the freshmen are one year younger than the previous batch." That is evidently all a matter of perspective.

One of the challenges professors face as they grow older (other than the challenge of growing old) is to be able to keep track of the mutating social mindset of the entering class. Not that professors necessarily need to be attuned to what gets the young bloods motivated (that is their problem) or excited (that is even more so their problem). Not that distinguished faculty members need to understand the lingo of the day, "dig" the latest fad, or know the YouTube viral hit of the hour. No. What is truly important is that a witty cultural reference that could elicit a roaring response a decade ago might now result in blank stares—or worse. If upon hearing Yoko Ono–like erratic shrieking in the corridor outside the classroom a professor quipped, "That's what inspired the famous Barenaked Ladies success!"[1] in 1993, it might have triggered a few giggles. Decades later, in the #MeToo movement era, with the platinum-selling rock band unknown to the frosh, the same comment might lead to immediate suspension for inappropriate language (and possibly even dismissal if the comment should spread through social media).

Knowing that nothing is more disheartening to a professor than wasting a perfectly good joke on a herd of ignorami, in 1998 a small liberal arts college in Wisconsin started circulating the annual Beloit College Mindset list.[2] The sole purpose of the list throughout the years has been to remind faculty members that the life experience of the entering class only spans, on average, a meager eighteen years, which is significantly less than theirs—particularly for those faculty members long in the tooth and determined to not retire as long as they have a pulse. Hence, irrespective of its intelligence, the consciousness of the cohort of teenagers entering higher education on year "x" and expected to be the graduating class of year "x + 4" is defined by its limited world experience. This means that the class of 2002, since it was born with CDs, most probably had no idea what sounding "like a broken record" meant, whereas the class of 2022 does not know what kind of pyromaniac ritual could possibly be implied by "burning a DVD." Those born with the internet and Skype would not know the difference between a floppy disk and a coaster (of the DVD type). Today's freshmen have never licked a postage stamp, and the World Trade Center is, like Pearl Harbor, something described in history books—or, for those who feel no shame in admitting that they do not read books, something found in the history section of YouTube or mentioned on Twitter once every five years. Therefore, and sadly, everything that happened before a person's birth is considered by that person to be prehistorical.

It is not surprising, then, that for those in their twenties and thirties today, the most pressing social problem—if not the only one—is global warming, and nothing is more dangerous than a politician who fails to acknowledge the facts or to take actions to mitigate the crisis.

Unfortunately, those who were in their twenties and thirties in the 1980s will remember when the most threatening danger was not global warming but global annihilation by nuclear war. In other words, a nuclear holocaust. Those who are now the parents or grandparents of today's youth will remember that the end of the world was not expected to happen in slow motion because the United States elected a president who did not care if glaciers melted. Rather, it was expected to happen in a blinding flash because the country elected one who had proclaimed that a nuclear war again the USSR could be won by pushing the big red button faster than the enemy. (Note to this year's entering class: The Union of Soviet Socialist Republics (USSR), or Soviet Union, is a prehistorical political construct that was created in 1921 and folded in 1991—its ability to wreak havoc across the world during those seventy years is described at great length in multiple textbooks, but thankfully, those proudly allergic to books can learn all about it in a six-minute animated cartoon on YouTube.)[3]

During Ronald Reagan's tenure as president, the world teetered on the brink of disaster. The Soviet Union and the United States were both convinced that the other had developed a first-strike doctrine.[4] A first strike consisted of a preemptive surprise nuclear attack of massive intensity intended to annihilate the enemy's arsenal. The belief was that the attacked country's ability to retaliate would be so weakened that the aggressor would sustain only limited and survivable damage, while the country attacked would be unable to continue the war. Adding fuel to the fire were President Reagan's comment that he could envision a nuclear war limited to Europe,[5] and his branding of the USSR as the "evil empire"—an appropriation from Star Wars that seemed natural from a president who formerly was a movie actor (whose most memorable performance took place in real life actually, and consisted of denouncing to the FBI fellow actors believed to be communist sympathizers).[6]

From the sidelines, the idea that there could be a winner in any nuclear exchange—and especially in one of the intensity hinted in the first-strike doctrine—was seen as ludicrous. Absolute madness! In 1980, the United States had 23,368 nuclear warheads and the USSR 30,062; the United Kingdom had 498, France 188, China 180, and the rest of the world, maybe one (in India). These nuclear warheads were not of the prototype size first used in 1945. The two uranium bombs dropped on Japan in 1945 had a power equivalent to roughly 0.02 megatons of TNT and flattened nearly all construction within a 1-mile radius,[7] whereas the B53 thermonuclear devices in the US arsenal in the 1980s delivered a 9-megaton blast—nearly five hundred times more powerful—that would produce a 3.4-mile fireball, flatten most construction within a 9-mile radius (254 square miles of destruction) and burn to death anyone within a 20-mile radius (1,250 square miles).[8] Given that 29 percent of the earth's 198,000,000 square miles is land—the rest being water—it would take only 46,000 of these B53 bombs to fry everything living outside of water. Truly the kind of stuff that gets the military industrial complex to pop open the champagne. Not to be left behind, and never shy when it came to mass killing, the USSR had built up its arsenal not to play tit for tat but to win the death game. For show, in the 1960s, it detonated the largest nuclear bomb ever tested: a 50-megaton baby 2,500 times more powerful than the 1945 uranium bombs.[9] The Tsar Bomba, as it was named, had to be equipped with a parachute to slow its descent so that the airplane that dropped it would be able to fly 28 miles away from the detonation point, giving it a 50 percent chance of surviving the blast.[10]

So, with over twenty thousand nuclear devices at the ready in each country's arsenal, if one country truly had wanted to disable the retaliation ability of its enemy, thousands of nuclear warheads would have had to be launched at the same time—from both countries if the attacked country responded faster than anticipated by believers of the first-strike doctrine. Who, in their right mind, figured out that somebody could be declared the winner if forty thousand nuclear weapons blew up within minutes across the planet?

A step back is in order to understand what is at stake, not only because it is estimated that the United States and Russia still have roughly seven thousand nuclear warheads each,[11] but because many other players are racing to join the nuclear club, and not only to play defense.

OFFERING MORE THAN ONE WAY TO DIE

While the option was provided in earlier chapters to skip descriptions of the physical mechanisms that create earthquakes, tornadoes, hurricanes, and other hazards, no such dispensation is offered here. It is imperative that everybody understands in its spine-chilling details the phenomena created by a nuclear explosion within the scope of a nuclear war; otherwise, the world is doomed, and not just beyond the end of civilization.

From the penny firecracker to the megaton nuclear bomb, there are a few common attributes shared by every explosion. First, at the point of ignition, there is a temperature spike. In the case of a small explosive device, the temperature may spike over 1,000°F, but the surfaces of materials can survive the explosion without being charred because this high peak of temperature only lasts a few thousandths of a seconds. This is not unlike the party trick of running one's finger left and right through the flame of a candle without consequences. There is no doubt that keeping the finger static above a flame will inflict a burn, but for that to happen, the high temperature has to be kept constant long enough for the heat to be transferred to the finger, which might be a second or two in this case. By running the finger through the core of the flame for only a fraction of a second, there is not any significant heat transferred to the finger to burn it. This is also, in part (because thermal conductivity also matters), why walking barefoot over a bed of hot coals (whose temperature ranges from 1,000 to 1,800°F) is possible—having sweaty feet also helps in this case.[12] (Disclaimer: Do not try any of this at home—no liability is assumed here in case of failed experiments.)

Therefore, if a bird lands next to a small bomb before it explodes, it will definitely die, but not because it is turned into a roasted chicken. It might even keep all of its feathers, unburned. Rather, birdy will be killed by the second physical phenomenon created by the explosion, namely, the pressure wave. The shock wave is what will blow eardrums and shred things to pieces. It is a supersonically expansive pressure wall at the front of the expanding gases created by the explosion. In a nutshell, the bigger the bomb, the bigger the peak pressure, the bigger the damage, and the bigger the casualties and injuries.[13] To make matters worse, the debris carried away by the pressure wave become projectiles that can also be lethal on their own.

In addition to the above two phenomena, when a bomb is a nuclear device, there are a few additional attractions to the show.[14] The main events that unfold after detonation are in the following order.[15] First, the explosion itself produces a—literally—blinding flash that can permanently burn the retina of the eyes.[16] Then, the heat release mentioned above becomes massive—intense enough in temperature and duration to set things on fire miles away. Also lethal are X-ray pulses released by the explosion. All that before the pressure wave even arrives. When it does arrive, it is also on a scale of its own, literally flattening the landscape for miles. However, like a wind gust on a campfire, the pressure wave will not extinguish the fires triggered by the initial heat radiation; rather, these fires will grow into a firestorm up to soaring temperatures, sucking up all the oxygen at ground level in the process. Then, the radioactive material that rose up into the nuclear mushroom cloud eventually rains down on the survivors at the periphery of the devastated area, the resulting fallout creating radiation-induced cancer and other deadly problems.[17] Depending on their ulterior motives, experts can pick through the existing data with a fine-toothed comb to argue at length on how more or less severe each of these effects will be for various types of urban construction, depending on the type of bomb, altitude and time of day at which it will be detonated, and the age of the president. Contrary to the 1950s videos of nuclear tests in the Nevada desert produced to show that well-kept and clean homes might have their roof blown away by the nuclear blast but won't be set on fire,[18] and to Bert the Turtle teaching school kids that they can survive a nuclear blast if they duck under their desks,[19] a nuclear blast is pretty in the eyes of the beholder only when that beholder is a four-star general standing many miles away—and even then . . .

Adding up to the nuclear sting of a single explosion, the outbreak of a nuclear war would imply contemplating scenarios with multiple nuclear explosions scattered over continents, with consequences never seen before. Imperfect as they might be, computer models were created to assess possible outcomes. As a result, the concept of a "nuclear winter" following a massive number of nuclear explosions—as in an all-out nuclear war—was presented as another possible way to die in the 1980s. The theory went that, in such a war, the massive number of firestorms ignited by hundreds of nuclear bombs detonated at the same time would fill the stratosphere with soot that, like the ash from massive volcanic explosions, would block sunlight and plunge the world into an artificial winter that would last years, with catastrophic outcomes. That certainly raised the stakes of the mutual destruction game. Nuclear winter may have served as a deterrent to nuclear war for a while, but at the end of the Iraq war, the retreating Iraqi soldiers set fire to six hundred oil wells that burned for up to eight months. The nuclear winter models had assumed fires burning for days after nuclear attacks—not months—but in spite of all the oil well fires in Iraq, a noticeable cooling of the planet did not occur.[20] More studies have been conducted since to either reaffirm or discredit the nuclear winter concept, and controversy remains.[21]

However, whether or not a nuclear winter arises following an all-out nuclear war is actually irrelevant. In a first-strike model, the objective is to disable the enemy's nuclear warhead arsenal. It is fair to assume that there is no missile silo in Central Park, Disney World, Berkeley, Mall of America, and the Dallas Cowboy stadium. However, these would be perfectly viable targets for nuclear attacks when the objective is instead to kill civilians. Every metropolitan city and every landmark worldwide will become "fair game" when ideology sits in the driver's seat of future war games. So it is helpful to remember what a nuclear weapon can do to a city.

THE ORIGINAL NUCLEAR SIN

While there might be an initial aggressor in any war, it does not take much time for everything to escalate to the point where everybody is aggressing everybody. With respect to the Japanese, there is pretty much consensus in the Western world that they were the aggressors in many ways. Pearl Harbor is a great example of that, and an example well known to Western civilization—if only because Hollywood has made it a tradition to remind everybody periodically. The Japanese were probably even more aggressive in the Nanjing massacre,[22] where soldiers raped twenty thousand women, killed sixty thousand civilians, and executed prisoners of war by making them walk on landmines, setting them aflame, beheading them, or executing them in other presumably equally entertaining ways—but no Hollywood movie has ever covered this, because it pretty much happened to others (refer to prior chapter for description of why people take little interest in what happens to others). Incidentally, Japan's apology for the Nanjing massacre is still regarded as unsatisfactory in some quarters.[23]

However, when it comes to nuclear devices, Japan is the only country that has been at the receiving end of a nuclear attack—to date, and hopefully for many more decades. Although still a topic of debate among historians,[24] history books have emphasized that the use of nuclear bombs on Hiroshima and Nagasaki, by killing roughly two hundred thousand civilians, possibly saved millions of lives, because of the perception that both Japanese troops and civilians alike were willing to fight to the end, even preferring suicide over surrender. Some even apparently witnessed Japanese mothers grabbing their children and jumping off cliffs to evade capture.[25] Although they dropped sixty-three million leaflets over Japan warning the population of massive air raids intended to destroy cities, the United States never informed the Japanese government that they were about to use a devastating new weapon—apparently because they were not certain that the bomb would work as intended[26]—but even if they had done so, one can speculate that the official response from Japan would have been

"Damare kusotare" (Japanese is a polite language, so this does not quite translate as "go fuck yourselves," but the intent is the same). Given that the Japanese government ordered that anyone in possession of one of the sixty-three million leaflets be arrested, it probably would not have made much of a difference.[27]

As recorded in the annals of history, the *Enola Gay* dropped the bomb on August 6, 1945. It exploded 1,900 feet above Hiroshima's city hall.

After witnessing the devastation of Hiroshima and getting confirmation from their own atomic physicists that they had been the target of a nuclear bomb, the Japanese military brass, guessing that there might be only a couple more such bombs available in the American arsenal, assessed that they could survive a couple more such brutal hits and continued to fight with the tools of conventional warfare. In a radio broadcast picked up by the Japanese news agencies, President Truman warned of other impeding similar attacks. More leaflets were dropped from the air over Japan, informing the population that the United States had dropped a nuclear bomb on Hiroshima, and encouraging the population to convince Emperor Hirohito to surrender, for their own sake. Yet, the Japanese prime minister met the press and reiterated the country's resolve to fight on.

On August 9, 1945, three days after Hiroshima, a second nuclear bomb was dropped on Nagasaki. Combined with the fact that the Soviet Union had reneged on the Soviet-Japanese Neutrality Pact and declared war with Japan the night before, and had invaded the state of Manchukuo, which was under Japanese control, to make its intentions clear, this became a lot of war activity for Japan to handle. Emperor Hirohito likely thought "Watashitachi wa okasa re" (which, if Google Translate is to be trusted, possibly means, "We are fucked"), and ordered his army to accept the terms of surrender.[28] On August 15, the day after a failed coup d'état by rebelling generals who wanted to continue fighting forever,[29] the emperor announced to his nation, via broadcast radio, the official surrender of Japan to the Allies. All this to show that nuclear bombs, by themselves, might not be much of a deterrent at all to those who are, by ideology or training, hellbent on fighting—contrary to what some might wish to believe.

Now, as the victim of a nuclear weapon, the Japanese have kept alive the memory of what a nuclear attack can do to a civilian population. A pilgrimage to Hiroshima should be mandatory to anyone who believes a "limited nuclear war" is not a big deal. Not only have the ruins of the City Hall at ground zero (directly below the bomb's detonation point) been kept as a monument to symbolize the first footprint of nuclear power unleashed on a civilization, but a Peace Memorial Museum has been built next to it to remind the world of what happened to the city the moment the bomb exploded and throughout the years that followed. Being that the museum is regularly assaulted by busloads of kids on school trips, and that these kids will march in single file through the exhibits without stopping to read anything—out of necessity, to not clog the flow—multiple exhibits graphically depict the horrors created by a nuclear attack, using wax figures of children walking through rubble, flesh dripping from their extended arms, bricks partially melted by the heat, stone walls indelibly marked by the shadows of nuclear victims, and photos of burned victims.

THE FORMERLY HOT TOPIC OF COLD WAR

To pile up the clichés, given that the war ended with a bang, it did not fall on deaf ears. Anyone that was not buddy-buddy with the United States had reason to be concerned about the fact that Uncle Sam alone could wield a weapon of unmatched massive destruction, particularly Mother Russia, who was already not fond of that creepy Uncle. Although the USSR had teamed-up with the United States to fight the Nazis, it was a marriage of convenience that did not last beyond World War II. The postwar Soviet grab of Eastern Europe solidified the American conviction that communism was an existential threat to the planet,[30] and the USSR interpreted the US postwar military spending (at more than ten times the prewar levels)[31] and its expressed goal to contain communism as a clear sign that it planned to dominate the world.[32] This mutual mistrust and antagonism launched the Cold War and the arms race. The United States was busy blowing up plutonium bombs on decommissioned warships, turning the idyllic Bikini Atoll into a South Pacific radioactive wasteland,[33] when—surprise—the Soviets detonated their first atomic bomb in 1949. From there, there was no turning back. The United States exploded its first hydrogen bomb in 1952 (five hundred times more powerful than previous atomic bombs),[34] and the Soviets did the same in 1955.[35] Then, instead of being dropped from airplanes, nuclear warheads were mounted on intermediate-range ballistic missiles launched from ground bases, and then from submarines to make it harder for the enemy to pinpoint the launch locations.

By the mid-1960s, both sides had amassed enough nuclear firepower to achieve mutually assured destruction (MAD),[36] but the USSR was behind in the number of warheads, so the race continued. By 1980, the countries were even, having stockpiled roughly twenty-five thousand warheads each.[37]

While this madness went on, the rest of the world was horrified. Europe, squeezed in-between the super-powers, as a doormat to the USSR, was extremely worried, but most of the entire world also planned for the worst. Alert sirens to warn the population of an oncoming nuclear attack were installed in every major city and tested periodically, sometimes as part of

nuclear drills. Some countries embarked on programs to b
shelters. Beyond that, other than watching the two bulls gettin
lock horns and fight to death, all that most could do is hope that
sense would prevail—maybe helped by international diplomac
aware that a single spark could end it all.[38] When fear of mutual an
tion is the only thing preventing mutual annihilation, any small glitcl
derail this fragile equilibrium and launch an apocalypse—and docume
declassified over time have revealed that this launch almost happened.[39]

From the 1960s to the 1980s, people worldwide lived with the lingering
specter of an all-out end-of-the-world nuclear war. Those who remember
these unsettling times when everybody could have been instantly turned
into ashes in a nuclear flash will be forgiven for having had no sympathy
for teenagers moaning that they had been robbed of their life when their
high-school graduation dance was canceled due to COVID-19.

Then, things got a little bit better. Discussions to start reducing the
number of warheads started in the mid-1980s, culminating with the signature of the START I (Strategic Arms Reduction Treaty) treaty in 1991[40]—
in part made possible by the progressive collapse of the USSR. The treaty
engaged its signatories to reduce their total number of nuclear warheads
to no more than six thousand, down from the forty thousand and twenty-
three thousand that the USSR and United States had respectively stockpiled in 1991. Both superpowers abided by the terms of the treaty, and the
world felt safer, although six thousand warheads is still a massive nuclear
arsenal—in fact, more than what both countries had when the threshold of
mutually assured destruction threshold was crossed in the 1960s.

Then, in 2001, the United States withdrew from the treaty, alleging that it
had to rebuild its nuclear arsenal to protect itself against rogue states—like
Iran—that intended to build up nuclear capabilities.[41] Russia responded by
embarking on its own program to develop new nuclear weapons.[42] By 2014,
the rhetoric from both sides indicated that a new nuclear arms race was
already underway, with Russia working on unstoppable hypersonic nuclear
weapons,[43] and the United States boasting that it would match and exceed
these capabilities,[44] with the Pentagon not only already working on it,[45] but
also claiming that it could win limited nuclear conflicts[46] or even a nuclear
war.[47] Making the scenarios more convoluted, China may also wish to flex
its nuclear muscles and even consider a first strike. Even though it currently
has only a few hundred nuclear warheads, it has enough enriched uranium
to match the US arsenal within a few years[48]—and is already sending signals
that it may be willing to disregard its 1964 promise to not be the first to
use nuclear weapons in a conflict. To further complicate matters, in 2018,
Russia revealed its plans for an unlimited-range nuclear-powered cruise

ıt necessarily carrying a nuclear warhead—would
if it accidentally crashed.[49] Given that it was
ılready during early tests, some international
veapon the "flying Chernobyl."[50]

ـ current era, with all of its present and looming
.y people, too much pollution, too much plastic, too
emission, and global warming. Arguing whether global
ın be slowed or stopped is somewhat irrelevant. If it is real, then
ρroblem. If it is not real but perceived by most people to be a real
ᴊolem, then it is a problem just the same, because it can create similar
social conflicts. Humans are the only species that can believe in imaginary
constructs of the mind, to a point that can be devastating, and has been,
throughout history.[51]

While solutions to these problems are possible, none of the solutions
may be sustainable until the root cause of all these problems is addressed,
which is the population crisis mentioned earlier—the elephant in the room
that is likely to remain ignored until the day when technological solutions
will not be able to keep pace. With or without climate change, at some
point, with overpopulation, scarcity of resources will drive the day and the
problems will appear intractable. There will be too much of every problem,
with no relief in sight when relying solely on the goodwill of nations and
organizations to self-impose restraints. All of the creativity, ingenuity, and
financial motivation of free-market thinkers will not be enough to solve
the problems. As mentioned previously, reducing the consumption and
emissions of individuals by 50 percent will not make a difference if the
number of people consuming and emitting doubles at the same time—and
if the third world's consumption and emissions increase to match that of
developed countries.

At that point, coercive actions will become inescapable and international coalitions will call for their legislation, but implementation and
enforcement, as always, will be a problem. Some groups will rebel. When
entire nations go rogue, tensions will rise globally.

It is foreseeable that all these coercive actions will directly focus on measures to reduce and modify consumption habits. It would be surprising if
actions to reduce the population were directly contemplated—even more
surprising if any were enacted. This is because when attempting to reduce
the population, there always will be "not enough of us and too many of
them" and never "too many of us and not enough of them." That is what
makes the problem nearly intractable.

However, population reduction will happen. Indirectly. As a result of the massive "earthquake" that will get triggered.

Polemology[52] is the name given to the scholarly study of war. Given that war has been an engrained part of human activities for all of history, it makes sense that this topic has been studied by experts from many disciplines. These scholars have described the many possible root causes of wars and the evolutionary, economic, political, ideological, rational, and psychoanalytical motivations for it. Like marital conflicts, world conflicts are complex matters—but arguably simpler, less emotionally charged, and less costly, some divorcees would venture—so anyone trying to summarize the driving motivations for war games into a few categories incurs the great risk of oversimplification. Yet, this is exactly what will be done here: in a nutshell, the two main drivers of wars have been the need to acquire resources or the need to win some ideological argument—sometimes both at the same time. Wars for resources include those conducted over territory, such as: (i) the Anglo-Indian Wars (1766–1849), (ii) the Mexican-American War (1846–1848), and (iii) World War II (1939–1945). Ideological wars include: (i) the American Civil War (1861–1865), (ii) the Crusades (1095–1291), (iii) the Thirty Years' War (1618–1648), (iv) the Korean War (1950–1953), (v) the Vietnam War (1955–1975), (vi) the American Revolution (1775–1783), and (vii) the French Revolution (1789–1799), to name only a few. The Cold War (1947–1991) falls into both categories. Both resources-driven and ideology-driven wars have been, in some cases, viciously bloody, which is expected when both sides are convinced they are morally and absolutely right, as is the case in most armed conflicts, given that reasonable people open to compromise usually prefer to avoid war.

What will happen and how it will happen is anybody's guess. Many intertwined factors will conspire and the few surviving historians will strive to provide the global picture. History will likely focus on the straw that "broke the back of the camel" and triggered the global conflict. However, it is reasonable to expect that groups butting heads will include countries that seek green technologies versus those that could not care less and behave like Canadian Geese on golf courses—hint: an adult bird produces two pounds of poop per day and urban geese colonies that can count in the hundreds are nearly impossible to scare away. Ideological conflict will also arise where groups seeking to control demands on the environment will take exception to the large family pressures of some religious groups. Currently, these conflicts exist locally, but resources are still believed to be plentiful. Once some countries run out of water, oil, gas, uranium, and food, these conflicts will elevate to national levels. The urgency to provide for the electorate or directorate—depending on the political system—will take over the reign of policy and it will be decided that it is time to evict the planet's obnoxious tenants, blasting their souls into the ether. Many players will have a nuclear arsenal at hand and will not hesitate to use it, defensively or offensively. Some of it may be rinky-dink, black-market, hit-or-miss missiles, but much of it will be sure-fire, surgically precise, and deadly.

In a war where the objective is to take over resources, dropping nuclear warheads on cities will be a viable strategy, to kill the maximum number of potential combatants and demoralize the population to surrender, while keeping radioactivity away from the rural areas where resources are located. Whether wiping out hundreds or thousands of cities will be sufficient to win that war remains to be seen, and it may have to be decided that some resources with some level of radioactivity is better than no resources at all.

When such an all-out nuclear war occurs, in the heat of battle and considering the motivations that will have triggered it, it is conceivable that it will not stop before it is too late. Just like it is possible to survive a magnitude 8 earthquake, by being located hundreds of miles away from the epicenter, it is possible to survive a nuclear holocaust if one happens to live on the international space station, on the moon, or on a faraway planet, and does not need to rely on Earth for the resupplying of oxygen and food. That option, however, is available to pretty much nobody at this time.

In that scenario, total annihilation is a real possibility, but if, in a positive outlook, only 50 percent to 80 percent of the world's population is exterminated, that will still leave between one to three billion people on earth. As is the case after all disasters, those who will survive this cataclysm will have hopefully gained wisdom and use it to rebuild a somewhat better world. That, sadly, is the blessings of disaster: it takes a disaster for things to change for the better.

Some will argue that this apocalyptic view of the future is unnecessarily alarmist. However, when panic breaks out and everybody is running for the exits, there usually is a reason for it, and pretending it will all be fine in the end is somewhat misguided. Everybody knows that the security guard in *Jurassic Park* who goes around telling everybody to stay calm and not panic is the one that ends-up in T-Rex's stomach.

ON THE DISASTER TRAIL

Sergeant Pepper and Salt Treaties
It was nearing the end of our first day after moving into our western New York house when we heard the alert siren blast—the very same eerie sound

that we had heard as kids in Canada during the annual test of the civil protection warning system, as a reminder that nuclear war was always a real threat. The shortest distance between the Soviet Union and the United States being over the North Pole, it was clear that thousands of warheads would fly in both directions over the Great White North should a nuclear conflict erupt. Statistically, out of tens of thousands of nuke-carrying missiles, it was possible that some of them would develop engine problems and crash on us along the way. Furthermore, some Russian rockets might be intentionally aimed at Canadian cities in retaliation for Team Canada having won the hockey series of the century against the USSR's Red Army in 1972.

But now, decades later, we were hearing that long-forgotten siren for the first time in decades. In western New York. There had been no warning of an upcoming test of the alarm, so it had to be the real deal. A nuclear attack on the first day after having signed the mortgage. How unlucky can you be?

Yet, something was odd, as none of the neighbors seemed worried. It turned out that every day at 6 p.m., the nearby fire station marked the time by running its siren, like clockwork. It was 6 p.m. There would be no attack. All was safe.

It is fascinating to notice that the experiences of a lifetime remain engraved in memory forever. Likewise, things that have not been experienced are abstractions that have to be explained.

As happens to most faculty members, I sometimes arrive so early to a room where my class is scheduled that I get to catch a few minutes of the preceding lecture. In 1990, one of those serendipitously caught exposés was on the cultural and historical significance of *Sgt. Pepper's Lonely Hearts Club Band*, with the professor talking at length about "Lucy in the Sky with Diamonds," Henry the Horse, and "Fixing a Hole." At first, it struck me as odd that anybody could get academic credits to listen to and then discuss the content of a rock album—even though it was a mighty good one—but then it made sense. Most of the students in that classroom were born in the 1970s. To them, the Beatles were a boy band that their parents liked. The Beatles might have been cultural icons of the 1960s—and an earthquake in their own world—but they never existed as far as these students were concerned. They were prehistoric.

The Beatles have absolutely nothing to do with nuclear holocaust. However, their music is universally embraced for the uplifting and positive emotions it generates. Hopefully, the world will prefer to enjoy their music rather than play war games, and will make it possible for all to put on the *Sgt. Pepper* album and sing along: "It's getting better all the time."

Act V

Ending on the thought that nuclear annihilation is in the cards is quite a downer. It is also in direct violation of a fundamental story telling principle that requires the denouement to resolve the conflicts in a manner that creates a sense of catharsis. Translation: Ten minutes before the end of any movie (which can be easily verified, stopwatch in hand), something devastating happens to the main character that makes the situation worse than at the start of the story. At the bottom of the abyss, everything could be lost, and the possibility of a satisfactory outcome is jeopardized. Then Act V kicks in. Something clever happens that saves the day, which resolves the entire story and, more often than not, produces a happy ending, because everybody loves a happy ending—particularly book publishers. However, what specifically happens in Act V varies depending on the type of story.

Therefore, out of necessity, to provide due closure to so many pages full of disasters, some parting thoughts are in order. To accommodate the fact that different readers have different expectations about what should happen in Act V to bring the story to a satisfactory conclusion, these parting thoughts are provided in three subsections below; only one of these needs to be read, depending on what the readers believe this whole book was about.

ACT V IN COMEDY

This book has been a practical joke. There is really no reason to worry about disasters.

Governmental institutions are well staffed, well funded, and ready to respond to all emergencies. Future disasters will be inconvenient, for a while, but in due time, after the government bails everybody out, all will return to normal.

Do not worry. Be happy. Enjoy cat videos.

ACT V IN TRAGEDY

Yes, it will all end in a nuclear holocaust, on a Metallica soundtrack.

Humanity is screwed.

It will not be pretty.

(Wink wink: Act V is sometimes called "the catastrophe" because, in a tragedy, this is where the hero usually dies or meets complete destruction.)

ACT V IN A DOCUMENTARY

Anybody who could accurately predict the future would be busy getting rich with the stock market, not writing books. Therefore, all bets are off as to whether anything presented in this book will actually happen.

Different people reading the same book—any book—will understand it differently, based on their own experiences, culture, beliefs, and outlook on life. That diversity enriches life. As such, this book never intended to moralize—that would have been ill advised. However, it is hoped that the ideas presented here have intrigued or challenged readers, or have at least brought them an appreciation for the complex relations humans entertain with natural and anthropogenic disasters. By better understanding what drives and motivates certain actions or inactions, maybe something good will come out of it—but only maybe, because the author has serious "True Pessimist" tendencies.

THE TAKE AWAY

So, in retrospect, which Little Pig are you?

That probably can be determined based on what you remember the most from the book.

If you are remembering that:

- hazards are everywhere, but it is possible to live with them (and survive them) by taking action,
- since the infrastructure is not designed to provide protection against every hazard, it is advisable to plan ahead on what to do when a disaster strikes,
- one will need to be self-reliant for a while during a disaster,
- deficient infrastructure should be fixed before it leads to catastrophic consequences,
- it is best to be proactive to prevent disasters,

then you are the Little Pig busy building the brick house, where all the other pigs will come begging for safe haven when the Big Bad Wolf comes.

Nothing can be taken for granted in life, but you agree, as wise souls have always advised, that an ounce of prevention is always worth more than a pound of cure. Life is one hell of a roller coaster ride, but it is a single-ride ticket. Best to make sure that the coaster does not derail along the way. Simple inexpensive measures that can prevent that catastrophe should be implemented. As for the more costly measures, it is a bit blurry from there on.

If what left the biggest impression is that

- hazards are everywhere and buying a house on a seismic fault or on the slope of a volcano is possibly not the best of investments,
- the infrastructure is not designed to provide protection against every hazard but that is the engineers' problem and responsibility to deal with it,
- the governmental organizations will come to the rescue to rapidly provide assistance,
- the deficient infrastructure will be fixed when budget surpluses will make it possible,
- there will be disasters but this bridge will be crossed when everybody gets there,

then you are the Little Pig playing violin in front of your wooden house. You have a good conscience because you followed the building code, but your house will probably be blown away if reality hits harder than anticipated. There is a need to juggle many priorities in life competing for time and money, and striking a balance is the goal. Finding that balance point is a personal matter and there is no cookbook solution to help decide. There are fifty shades of earthquakes.

If you are remembering that

- hazards are everywhere but they are low probability events,
- the infrastructure is not designed to provide protection against every hazard but everybody will be in the same boat and the time to figure it out will be when disasters strike,
- once the disaster happens it will be possible to hire lawyers and sue everybody,
- the deficient infrastructure is not broken yet and so there's no need to fix it,
- and anyway there will probably not be disasters,

then, you are the Little Pig playing flute in front of the straw hut. You are having a hell of a good life, enjoying every minute of it, and if you die of old age, you will have a smile on your face. The last bank check you will sign before dying will bounce because you will have successfully accomplished the feat of spending all of your savings to enjoy life to the fullest before your last breath. If you make it there safely.

Finally, for those who remember nothing from the previous pages, that is perfectly fine too. Many read as a tool to help get to sleep. Not remembering a thing implies that you must have been dozing off before reaching any meaningful point in any chapter, snoring with the book spread open on you face. That strategy has been a total success. Even an earthquake could not wake you up.

Addendum

When the final draft of this manuscript was submitted to the editor at the end of November 2021, the daily number of new COVID cases in the United States was reported to be its lowest in four months, giving hope that the end of supply-chain shortages was in sight, and the thought of a nuclear war was far from everyone's mind.

Not even four months later, when the copyedited manuscript was being reviewed (mid-March 2022), things had taken a turn for the worse.

Supply-chain problems were nowhere near being resolved, to the point where cars were being sold with missing parts to be delivered later.[1] Not only were semi-conductors, cotton, shoes, toys, and tons of other stuff in short supply, but so were the truckers needed to haul it all.[2] Even the publishing industry suffered, impacted by a paper shortage delaying publication of books (including this one) by months.

To make it worse, to prove that the world has no supply-chain problems when it comes to despots, Russian troops invaded Ukraine. As a result, the price of gas at the pump was reaching new record highs worldwide, the value of the Russian ruble collapsed, Russia was on the verge of defaulting on its debt, and Vladimir Putin launched an attack on the Chernobyl nuclear power plant[3] and announced that he had put his nuclear arsenal on high alert, just in case he felt like pushing the button to launch World War III.[4]

What will be the state of the world when this book is finally published? Unpredictable, as all the disasters mentioned earlier. However, it is hoped that, as stated at the beginning of the preface, "If you are reading these words right now, you are most probably not dead. At least, not yet."

Notes

PREFACE

1. "History & Rules," Darwin Awards, http://www.darwinawards.com/rules/. Granted, YouTube is full of videos of folks who severely hurt themselves by ignoring various basic laws of physics (apparently, some have not quite understood yet how gravity works), but focus here is not on these outliers.

2. Michel Bruneau, *Shaken Allegiances* (CePages Press, 2009).

WHICH LITTLE PIG ARE YOU?

1. Reasons vary on why this happens, and some may speculate that this could include a showboating desire to demonstrate a deep understanding of key issues of some social or technical problem, or that it might occur impulsively, in panic, to respond to an unexpected question on national television (neuroanatomy might someday demonstrate striking similitudes between the brains of politicians and those of professional jugglers).

2. Sharon Begley, "Lessons of the Kobe Earthquake," *Newsweek,* January 29, 1995, http://www.newsweek.com/lessons-kobe-earthquake-182190.

3. Nora Zamichow and Virginia Ellis, "Santa Monica Freeway to Reopen on Tuesday; Recovery: The Contractor Will Get a $14.5-Million Bonus for Finishing Earthquake Repairs 74 Days Early," *Los Angeles Times,* April 6, 1994, https://www.latimes.com/archives/la-xpm-1994-04-06-mn-42778-story.html.

4. M. Bruneau, J. W. Wilson, and R. Tremblay, "Performance of Steel Bridges during the 1995 Hyogo-ken Nanbu (Kobe, Japan) Earthquake," *Canadian Journal of Civil Engineering* 23, no. 3 (1996): 678–713.

5. Louise Comfort, *Self Organization in Disaster Response: The Great Hanshin Earthquake of January 17, 1995* (Pittsburg, PA: Graduate School of Public and International Affairs, 1995), 12, http://cidbimena.desastres.hn/pdf/eng/doc8691/doc8691.htm.

6. "Kobe Earthquake of 1995," Facts and Details, http://factsanddetails.com/japan/cat26/sub160/item863.html.

7. There is also an emperor in Japan, but while this monarch was once considered a deity and referred to as the "heavenly sovereign," his role post–World War II has been limited to the less godly duties of serving as a figurehead—albeit a figure whose head is

never shown on bank notes or postage stamps anymore, contrary to the highly popular "Hello Kitty" cartoon character, who is featured on special-issue stamps almost every year. The emperor is empowered to perform important ceremonial functions, such as officially appointing the prime minister who has been democratically elected by the Japanese legislature (known as the Diet), but not much more, so it is OK if the emperor gets his news from the television.

8. Eric Johnston, "Lessons Learned in Kobe Aid Relief Effort," *Japan Times*, March 15, 2011, https://www.japantimes.co.jp/news/2011/03/15/national/lessons-learned-in -kobe-aid-relief-effort/#.W29jGLgnZ9A.

9. "Japan Deputy Mayor Commits Suicide," UPI, March 15, 1996, https://www.upi .com/Archives/1996/03/15/Japan-deputy-mayor-commits-suicide/4439826866000/.

10. Elizabeth Ferris and Mireya Solís, "Earthquake, Tsunami, Meltdown: The Triple Disaster's Impact on Japan, Impact on the World," Brookings, March 11, 2013, https:// www.brookings.edu/blog/up-front/2013/03/11/earthquake-tsunami-meltdown-the -triple-disasters-impact-on-japan-impact-on-the-world/.

11. Julian Ryall, "Mount Fuji Eruption Fears Prompt Japan to Draw Up Disaster Plans," *Telegraph*, July 31, 2018, https://www.telegraph.co.uk/news/2018/07/31/mount-fuji -eruption-fears-prompt-japan-draw-disaster-plans/; "Japanese Government Begins Discussion on Contingency Plans in Event of Mount Fuji Eruption," *Japan Times*, September 11, 2018; "Mount Fuji Eruption Could Paralyse Tokyo: Report," *France-Presse*, May 1, 2018, https://www.ndtv.com/world-news/mount-fuji-eruption-could-paralyse -tokyo-report-1845251.

12. Parenthetically, engineers would advise the third Little Pig to at least use reinforced masonry (that is, masonry that includes steel reinforcing bars), instead of unreinforced masonry, to minimize the risk of damage in future earthquakes, but that goes beyond the scope of the nursery rhyme.

EARTHQUAKES HAPPEN BECAUSE . . .

1. "Namazu," Wikipedia, https://en.wikipedia.org/wiki/Namazu.

2. David Bressan, "Namazu the Earthshaker," *Scientific American*, March 10, 2012, https://blogs.scientificamerican.com/history-of-geology/namazu-the-earthshaker/.

3. Mary Winkler, "What Exactly Is This Japanese Trend Known as 'Kawaii' All About?," Envato tuts+, Oct 2, 2013, https://design.tutsplus.com/articles/what-exactly -is-this-japanese-trend-known-as-kawaii-all-about--vector-15984.

4. http://www.17wcee.jp/.

5. Earthquake Early Warning, The Japan Meteorological Agency (JMA), https:// www.jma.go.jp/jma/en/Activities/eew.html.

6. http://chara.web-mk.net/articles/business/public/%E3%82%86%E3%82%8C%E3 %82%8B%E3%82%93.html.

7. Daniel McCoy, "Loki Bound," Norse Mythology for Smart People, https://norse -mythology.org/tales/loki-bound/.

8. Eileen McSaveney, "Historic Earthquakes—Earthquakes in Māori tradition," *Te Ara: The Encyclopedia of New Zealand*, http://www.TeAra.govt.nz/en/historic -earthquakes/page-1.

9. Malcolm Johnston, "Is There Earthquake Weather?," USGS, https://www.usgs.gov/faqs/there-earthquake-weather?qt-news_science_products=0#qt-news_science_products.

10. Carl A. Hanson, *Economy and Society in Baroque Portugal, 1668–1703* (Minneapolis: University of Minnesota Press, 1981), 30.

11. "Demographics of Portugal," Wikipedia, https://en.wikipedia.org/wiki/Demographics_of_Portugal.

12. "Lisbon earthquake," Wikipedia, https://en.wikipedia.org/wiki/1755_Lisbon_earthquake.

13. The Free Dictionary, s.v. "optimism," https://www.thefreedictionary.com/Optimistic+philosophy; Michael J. Murray and Sean Greenberg, "Leibniz on the Problem of Evil," in *The Stanford Encyclopedia of Philosophy* (Winter 2016 Edition), ed. Edward N. Zalta, https://plato.stanford.edu/archives/win2016/entries/leibniz-evil.

14. Tom Drake, English 258: Literature of Western Civilization, https://www.webpages.uidaho.edu/engl_258/lecture%20notes/leibniz.htm.

15. "Gotttied Wilhelm Leibniz," Wikipedia, https://en.wikipedia.org/wiki/Gottfried_Wilhelm_Leibniz.

16. "Seismology," Encyclopedia.com, https://www.encyclopedia.com/science/news-wires-white-papers-and-books/seismology.

17. "Robert Mallet," Wikipedia, https://en.wikipedia.org/wiki/Robert_Mallet; Jeremy Norman, "Robert Mallet Founds the Science of Seismology, 1850 to 1858," Jeremy Norman's HistoryofInformation.com, https://www.historyofinformation.com/detail.php?id=4185.

18. "Supercontinent," Wikipedia, https://en.wikipedia.org/wiki/Supercontinent; R. N. Mitchell, N. Zhang, J. Salminen, et al., "The Supercontinent Cycle," *Nat Rev Earth Environ* 2 (2021): 358–74, https://doi.org/10.1038/s43017-021-00160-0.

19. "Historical Perspective," USGS, https://pubs.usgs.gov/gip/dynamic/historical.html.

20. "Continental Drift and Plate Tectonics," Let's Talk Sciences, January 22, 2020, https://letstalkscience.ca/educational-resources/backgrounders/continental-drift-and-plate-tectonics.

21. "Historical Perspective," USGS.

22. "Timeline of the development of tectonophysics (before 1954)," Wikipedia, https://en.wikipedia.org/wiki/Timeline_of_the_development_of_tectonophysics_(before_1954).

23. "Difference between Meteorologist and Weatherman," Study.com, May 30, 2020, https://study.com/articles/difference_between_meteorologist_weatherman.html.

24. "Developing the Theory," USGS, https://pubs.usgs.gov/gip/dynamic/developing.html.

25. "Plate Tectonics," Wikipedia, https://en.wikipedia.org/wiki/Plate_tectonics.

26. "Plate Tectonics Revolution," Wikipedia, https://en.wikipedia.org/wiki/Plate_Tectonics_Revolution.

27. "GPS—Measuring Plate Motion:, How Fast Are the Tectonic Plates Moving?," Incorporated Research Institutions for Seismology, https://www.iris.edu/hq/files/programs/education_and_outreach/aotm/14/1.GPS_Background.pdf—a document summarizing information from www.gps.gov/, www.usgs.gov and www.faa.gov.

28. Grove Karl Gilbert, "The Investigation of the San Francisco Earthquake," *Popular Science Monthly*, August 1906, available from Wikisource: https://en.wikisource.org/wiki/Popular_Science_Monthly/Volume_69/August_1906/The_Investigation_of_the_San_Francisco_Earthquake.

29. "Geologic Activity," Point Reyes National Seashore California, US National Park Services, accessed May 8, 2020, https://www.nps.gov/pore/learn/nature/geologicactivity.htm.

30. "San Andreas Fault," Wikipedia, https://en.wikipedia.org/wiki/San_Andreas_Fault.

31. "TNT equivalent," Wikipedia, https://en.wikipedia.org/wiki/TNT_equivalent.

32. Lisa Wald, Kate Scharer, and Carol Prentice, "Back to the Future on the San Andreas Fault," United States Geological Survey (USGS), June 1, 2017, https://www.usgs.gov/natural-hazards/earthquake-hazards/science/back-future-san-andreas-fault?qt-science_center_objects=0#qt-science_center_objects.

33. San Andreas Fault," Wikipedia.

34. B. T. Aagaard, J. L. Blair, J. Boatwright, S. H. Garcia, R. A. Harris, A. J. Michael, D. P. Schwartz, and J. S. DiLeo, "Earthquake Outlook for the San Francisco Bay Region 2014–2043 (ver. 1.1, August 2016): U.S. Geological Survey Fact Sheet 2016–3020," https://pubs.usgs.gov/fs/2016/3020/fs20163020.pdf.

35. USGS, Frequently Asked Questions, What is the difference between earthquake early warning, earthquake forecasts, earthquake probabilities, and earthquake prediction?, https://www.usgs.gov/faqs/what-difference-between-earthquake-early-warning-earthquake-forecasts-earthquake-probabilities#:~:text=Early%20warning%20is%20a%20notification,not%20yet%20possible%20for%20earthquakes.

36. Evah Kungu, "Difference Between Forecasting and Prediction," Difference Between.net, accessed July 23, 2018, http://www.differencebetween.net/science/difference-between-forecasting-and-prediction.

37. Susan E. Hough, "Earthquakes: Predicting the Unpredictable?," *Geotimes*, March 2005, http://www.geotimes.org/mar05/feature_eqprediction.html.

38. "Parkfield, California," Wikipedia, https://en.wikipedia.org/wiki/Parkfield,_California.

39. S. J. Titus, C. Demets, and B. Tikoff, "Thirty-Five-Year Creep Rates for the Creeping Segment of the San Andreas Fault and the Effects of the 2004 M 6.0 Parkfield Earthquake: Constraints from Alignment Arrays, Continuous GPS, and Creepmeters." *Bulletin of the Seismological Society of America* 96 (2006): 5250–68.

40. See Google Maps: https://www.google.com/maps/place/Parkfield,+CA+93451/@35.8951744,-120.4345956,139m/data=!3m1!1e3!4m5!3m4!1s0x809336801bebd28b:0xb6a0ce739b80736c!8m2!3d35.8996856!4d-120.4326542.

41. Dan Robinson, "San Andreas Fault, Part 1: Parkfield, Cholame & Annette, California, California's San Andreas Fault," Storm Highway, https://stormhighway.com/san-andreas-fault/parkfield-cholame.php.

42. See home page of United States Geological Survey, https://www.usgs.gov/.

43. W. H. Bakun and A. G. Lindh, "The Parkfield, California, Earthquake Prediction Experiment," *Science*, 229, no. 4714 (August 16, 1985): 619–24, doi: 10.1126/science.229.4714.619 https://science.sciencemag.org/content/229/4714/619.

44. Kenneth Reich, "Lessons of a Quake That Didn't Happen: Geology: The Long-Predicted Parkfield Temblor Failed to Occur in 8-Year Period," *Los Angeles Times*, January 4, 1993, https://www.latimes.com/archives/la-xpm-1993-01-04-mn-908-story .html.

45. Steve Hickman, John Langbein, and Peter H. Stauffer, "The Parkfield Experiment: Capturing What Happens in an Earthquake," USGS, Fact Sheet 049-02, July 2004, https://doi.org/10.3133/fs04902 or https://pubs.usgs.gov/fs/2002/fs049-02/fs049-02 .pdf.

46. Naomi Lubick, "Parkfield Finally Quakes," *Geotimes*, September 2004, http:// www.geotimes.org/sept04/WebExtra092804.html.

47. See https://www.tripadvisor.com/Restaurant_Review-g1568653-d832575-Reviews -Parkfield_Cafe-Parkfield_California.html#photos;aggregationId=&albumid=101& filter=7&ff=292720110.

48. Cinna Lomnitz, *Fundamentals of Earthquake Prediction* (New York: John Wiley & Sons, 1994), 262.

49. C.-H. Geschwind, *California Earthquakes: Science, Risk, and the Politics of Hazard Mitigation* (Baltimore: Johns Hopkins University Press, 2001), 192–212.

50. Lomnitz, *Fundamentals of Earthquake Prediction*, 24–25.

51. See Google maps: https://www.google.com/maps/place/Liaoning,+China/@41.0 18199,117.8288578,6z/data=!3m1!4b1!4m5!3m4!1s0x358f5e2ea41fd385:0x7b5a542ee 2054d00!8m2!3d41.9436543!4d122.5290376.

52. Lomnitz, *Fundamentals of Earthquake Prediction*.

53. Christian Kallen, "Earthquake Prognosticator, Glen Ellen Raconteur Jim Berk-land Dies at 85," *Sonoma News*, July 28, 2016, https://www.sonomanews.com/news/ 5903516-181/earthquake-prognosticator-and-glen-ellen.

54. Donald Prothero, "Quacks and Quakes," *Skeptic* 16, no 4 (2011), http://www.donald prothero.com/files/92369329.pdf.

55. "Infotainment," Wikipedia, https://en.wikipedia.org/wiki/Infotainment.

56. "'Virgin Mary Toast Fetches $28,000," BBC News, November 23, 2004, http:// news.bbc.co.uk/2/hi/4034787.stm.

57. Joe Nickell, "Grilled-Cheese Madonna, Skeptical Briefs," *Skeptical Enquirer* 18, no. 3 (September 1, 2008), https://skepticalinquirer.org/newsletter/grilled-cheese _madonna/?/sb/show/grilled-cheese_madonna.

58. Brendan Koerner, "The $28K Sandwich That Grew No Mold: How the Virgin Mary's Grilled Cheese Stayed Mold-Free for 10 years," *Slate*, November 23, 2004, https://slate.com/news-and-politics/2004/11/how-the-28000-sandwich-stayed-mold -free-for-10-years.html.

59. Nickell, "Grilled-Cheese Madonna."

60. "Virgin Mary Sandwich," Museum of Hoaxes, November 16, 2004, http://hoaxes .org/weblog/comments/vmary_sandwich.

61. "Jesus Toaster," MustHaveStuff.com, https://musthavestuff.com/jesus-toaster/.

62. "Dutch Man Predicted 8.8 Magnitude Earthquake Would Hit California May 28," Q13 Fox Seattle, May 28, 2015, https://q13fox.com/2015/05/28/dutch-man-predicts-8 -8-magnitude-earthquake-will-hit-california-thursday/.

63. "'Quake Mystic' Frank Hoogerbeets Warns Cosmic Event Could 'Trigger Mega-quake' over Christmas," news.com.au, December 7, 2018, https://www.news.com.au/technology/environment/quake-mystic-frank-hoogerbeets-warns-cosmic-event-could-trigger-megaquake-over-christmas/news-story/8d302c7dc630923b524c11585449ee7e.

64. "How Often Do Earthquakes Occur?," Incorporated Research Institutions for Seismology, https://www.iris.edu/hq/inclass/fact-sheet/how_often_do_earthquakes_occur?zoombox=0.

65. Vickiie Oliphant, "Huge Tsunami Could WIPE OUT Asia: Man with 'Sixth Sense' Predicts DEVASTATING Earthquake," Express, November 6, 2017, https://www.express.co.uk/news/world/875825/Tsunami-earthquake-warning-Indian-Ocean-Asia-sixth-sense-prediction-India-Pakistan-China.

66. Scott LaFee, "Quake Myths Rely on Cloudy Facts," San Diego Union Tribune, April 9, 2010, https://www.sandiegouniontribune.com/sdut-quake-myths-rely-cloudy-facts-2010apr09-story.html; Brian Clark Howard, "Bizarre Earthquake Lights Finally Explained," National Geographic, January 7, 2014, https://www.nationalgeographic.com/science/article/140106-earthquake-lights-earthquake-prediction-geology-science; Julia Rosen, "Can Electric Signals in Earth's Atmosphere Predict Earthquakes?," Science, December 21, 2015, https://www.science.org/news/2015/12/can-electric-signals-earth-s-atmosphere-predict-earthquakes.

67. "CEPEC Keeps Eye on Earthquake Predictions," State of California Department of Conservation, October 23, 2009, https://web.archive.org/web/20100619180755/http://www.consrv.ca.gov/index/news/Pages/CEPECKeepsEyeOnEarthquakePredictions.aspx.

68. "Rumpology," Wikipedia, https://en.wikipedia.org/wiki/Rumpology.

69. Jerry Hopkins, "Earthquake! California Fears Fear Itself!," Rolling Stone, April 19, 1969, https://www.rollingstone.com/culture/culture-news/earthquake-california-fears-fear-itself-40898/.

70. "Iben Browning," Wikipedia, https://en.wikipedia.org/wiki/Iben_Browning.

71. William J. Spence, R. B. Herrmann, A. C. Johnston, and B. G. Reagor, Responses to Iben Browning's Prediction of a 1990 New Madrid, Missouri, Earthquake, USGS Survey Circular 1083 (Washington, DC: US Department of the Interior, 1993), https://pubs.usgs.gov/circ/1993/1083/report.pdf.

72. R. Snieder and T. van Eck, "Earthquake Prediction: A Political Problem?," Geologische Rundschau 86, no. 2 (August 1997): 458, doi:10.1007/s005310050153, ISSN 1432-1149.

73. Aagaard et al., "Earthquake Outlook for the San Francisco Bay."

74. "Northridge Earthquake Remembered as One of Costliest Natural Disasters in U.S. History," California Earthquake Authority, January 10, 2019, https://www.earthquakeauthority.com/Press-Room/Press-Releases/2019/Northridge-earthquake-remembered.

75. Michael Ertl, "Christchurch Earthquake: The Battle to Rebuild, Five Years On," BBC News, February 21, 2016, https://www.bbc.com/news/world-asia-35612298.

EARTHQUAKES TO TSUNAMIS

1. Jakob Eckstein, How Amtrak Makes Money, Investopedia, June 13, 2021, https://www.investopedia.comE/articles/investing/072115/how-amtrak-works-makes-money.asp.

2. "Amtrak," Wikipedia, https://en.wikipedia.org/wiki/Amtrak#Public_funding.

3. "Amtrak Host Railroad Report Card 2018," Amtrak, https://media.amtrak.com/wp-content/uploads/2019/03/Amtrak-Host-Railroad-Report-Card-and-Route-Grades-2018-with-FAQs.pdf.

4. Ibid.

5. Amtrak Office of Inspector General, *Train Operations: Better Estimates Needed of the Financial Impacts of Poor On-Time Performance*, Report OIG-A-2020-001, October 14, 2019, https://amtrakoig.gov/sites/default/files/reports/OIG-A-2020-001%20OTP%20mandate.pdf.

6. Pierre-Paul Biron, "En plus d'être en retard, ils coûtent une petite fortune," *Journal de Québec*, December 14, 2019, https://www.journaldequebec.com/2019/12/14/en-plus-detre-en-retard-ils-coutent-une-petite-fortune.

7. Pierre-Paul Biron, "Les trains de VIA Rail Canada en retard une fois sur trois," *Journal de Québec*, December 14, 2019, https://www.journaldequebec.com/2019/12/14/les-trains-de-via-rail-en-retard-une-fois-sur-trois.

8. Carlye Jones, "How Fast Is a Donkey?," https://animals.mom.com/fast-donkey-6161.html.

9. Amtrak Host Railroad Report Card 2018.

10. Biron, "Les trains de VIA Rail Canada."

11. N. Tomii, "How the Punctuality of the Shinkansen Has Been Achieved," International Conference on Computer System Design and Operation in the Railway and Other Transit Systems, in *Computers in Railways XII: Computer System Design and Operation in the Railway and Other Transit Systems*, ed. B. Ning, C. A. Brebbia, and N. Tomii (Southampton, UK: WIT, 2010), 111–20, https://www.witpress.com/Secure/elibrary/papers/CR10/CR10011FU1.pdf.

12. Danielle Demetriou, "Why Is Japan So Obsessed with Punctuality?," *Telegraph*, September 24, 2019, https://www.telegraph.co.uk/travel/destinations/asia/japan/articles/why-japan-so-obsessed-with-punctuality/.

13. "Japanese Train Departs 25 Seconds Early—Again," BBC News, May 17, 2018, https://www.bbc.com/news/world-asia-44149791.

14. Scott Neuman, "Japanese Rail Operator Says Sorry for 'Inexcusable' Departure 25 Seconds Early," NPR, May 17, 2018, https://www.npr.org/sections/thetwo-way/2018/05/17/611860169/japanese-rail-operator-says-sorry-for-inexcusable-departure-25-seconds-off-sched.

15. N. Tomii, "How the Punctuality of the Shinkansen Has Been Achieved."

16. "Hokkaido and Pacific Coast Now Understood to Face Higher Risk of Powerful Earthquakes," *Japan Times*, June 26, 2008, https://www.japantimes.co.jp/news/2018/06/26/national/hokkaido-pacific-coast-now-understood-face-higher-risk-powerful-earthquakes/#.Xfk1q_x7mhc.

17. Joshua Hammer, "The Great Japan Earthquake of 1923," *Smithsonian*, May 2011, https://www.smithsonianmag.com/history/the-great-japan-earthquake-of-1923-1764 539/.

18. Kathleen J. Tierney and James D. Goltz, "Emergency Response: Lessons Learned from the Kobe Earthquake," University of Delaware Disaster Research Center Preliminary Paper #260 1997, http://citeseerx.ist.psu.edu/viewdoc/download?doi=10.1.1.100 0.7939&rep=rep1&type=pdf.

19. Ibid.

20. "Kobe Earthquake of 1995," Facts and Details, http://factsanddetails.com/japan/cat26/sub160/item863.html.

21. Y. Shiozaki, E. Nishikawa, and T. Deguchi, *Lessons from the Great Hanshin Earthquake* (Tokyo: Creates-Kamagawa Publishers, 2005). Formerly available online at http://www.shinsaiken.jp/oldweb/hrc-e/publish/lessons_ghe/lghe17.html.

22. Seen with my own eyes.

23. Ryan Osbourne, "10 Things to Know about the Japanese Bullet Train, Coming Soon to Texas," *Dallas Business Journal*, November 30, 2018, https://www.bizjournals.com/dallas/news/2018/11/30/10-things-to-know-about-the-japanese-bullet-train.html.

24. See http://www.shinkansen.co.jp/jikoku_hyo/en/sanyou/sdh_shinkobe.html.

25. Shiozaki, et al., *Lessons from the Great Hanshin Earthquake*.

26. David Edgington, "Lessons for Japan from Kobe Quake," The Diplomat, March 23, 2011, https://thediplomat.com/2011/03/lessons-for-japan-from-kobe-quake/.

27. The event that changed the course of local history, The Great Hanshin-Awaji Earthquake Memorial Museum, Disaster Reduction and Human Renovation Institute, http://www.dri.ne.jp/en.

28. Ibid.

29. See https://www.youtube.com/watch?v=D_K8_cunPsI.

30. Masayoshi Nakashima, Takuya Nagae, Ryuta Enokida, Koichi Kajiwara, "Experiences, Accomplishments, Lessons, and Challenges of E-Defense—Tests Using World's Largest Shaking Table," *Japan Architectural Review* 1, no. 1 (January 2018): 16, https://doi.org/10.1002/2475-8876.10020.

31. Ibid.

32. "The World's Largest Earthquake Simulator: Japan's E-Defense," Real Estate Japan, August 21, 2015, https://resources.realestate.co.jp/news/the-worlds-largest-earthquake-simulator-japans-e-defense/.

33. Jake Adelstein, "Mobsters on a Mission: How Japan's Mafia Launched an Aid Effort," *Independent*, October 23, 2011 https://www.independent.co.uk/news/world/asia/mobsters-on-a-mission-how-japans-mafia-launched-an-aid-effort-2264031.html.

34. Nicholas D. Kristof, "Quake in Japan: The Scene; Kobe's Survivors Try to Adjust: Hand-Ringing, Relief, Laughter," *New York Times*, January 22, 1995, https://www.nytimes.com/1995/01/22/world/quake-japan-scene-kobe-s-survivors-try-adjust-hand-ringing-relief-laughter.html.

35. G. S. Fukushima, "The Great Hanshin Earthquake," JPRI Occasional Paper, no. 2, https://web.archive.org/web/20180723225034/http://www.jpri.org/publications/occasionalpapers/op2.html.

36. Sharon Begley, "Lessons of the Kobe Earthquake," *Newsweek*, January 29, 1995, https://www.newsweek.com/lessons-kobe-earthquake-182190.

37. "1995 Kobe Earthquake 10-Year Retrospective," Risk Management Solutions, 2005, http://forms2.rms.com/rs/729-DJX-565/images/eq_1995_kobe_eq.pdf.

38. Ibid.

39. Rachel Leng, "Japan's Civil Society from Kobe to Tohoku, Impact of Policy Changes on Government-NGO Relationship and Effectiveness of Post-Disaster Relief," *Electronic Journal of Contemporary Japanese Studies*, 15, no. 1 (2015), https://scholar.harvard.edu/files/rachel_leng/files/ejcjs_-_japans_civil_society_from_kobe_to_tohoku_rachel_leng.pdf.

40. Ibid.

41. Mitsuyoshi Akiyama, Dan M. Frangopol, and Keita Mizuno, "Performance Analysis of Tohoku-Shinkansen Viaducts Affected by the 2011 Great East Japan Earthquake," *Structure and Infrastructure Engineering*, 10, no. 9 (September 2, 2014), https://doi.org/10.1080/15732479.2013.806559.

42. "The State of Recovery in Tōhoku Eight Years after 3/11," Nippon.com, March 7, 2019, https://www.nippon.com/en/japan-data/h00404/the-state-of-recovery-in-tohoku-eight-years-after-311.html.

43. Leng, "Japan's Civil Society from Kobe to Tohoku."

44. The Splat Calculator—A Free Fall Calculator, https://www.angio.net/personal/climb/speed.html.

45. George Pararas-Carayannis, "The Mega-Tsunami of July 9, 1958 in Lituya Bay, Alaska, Analysis of Mechanism," http://www.drgeorgepc.com/Tsunami1958LituyaB.html.

46. See https://websites.pmc.ucsc.edu/~ward/lituya-es-dir.mov.

47. Robert Valdes, Nathan Halabrin, and Robert Lamb, "How Tsunamis Work: The Birth of a Tsunami," How Stuff Works, September 11, 2019, https://science.howstuffworks.com/nature/natural-disasters/tsunami2.htm.

48. Robert Valdes, Nathan Halabrin and Robert Lamb, "How Tsunamis Work: When a Tsunami Makes Landfall," How Stuff Works, September 11, 2019, https://science.howstuffworks.com/nature/natural-disasters/tsunami4.htm.

49. "International Early Warning Programme," Wikipedia, https://en.wikipedia.org/wiki/International_Early_Warning_Programme.

50. Anjali Singhwi, Bedel Saget, and Jasmine C. Lee, "What Went Wrong with Indonesia's Tsunamic Early Warning System," *New York Times*, October 2, 2018, https://www.nytimes.com/interactive/2018/10/02/world/asia/indonesia-tsunami-early-warning-system.html.

51. "28 September 2018 Tsunami—Technical," International Tsunami Information Center, http://itic.ioc-unesco.org/index.php?option=com_content&view=article&id=2034&Itemid=2840.

52. Shunichi Koshimura and Nobuo Shuto, "Response to the 2011 Great East Japan Earthquake and Tsunami Disaster," *Philosophical Transactions of the Royal Society A*, October 28, 2015, https://doi.org/10.1098/rsta.2014.0373.

53. "Tsunami in Japan Filmed by a Driver from His Car, June 20, 2011," YouTube, https://www.youtube.com/watch?v=Sp2DP1cLXKM.

54. Reconstruction Agency, http://www.reconstruction.go.jp/english/topics/GEJE/index.html.

55. "Efforts toward Reconstruction of Tohoku," Reconstruction Agency, http://www.reconstruction.go.jp/english/.

56. "The Road to Recovery: Recovery and Reconstruction from the Great East Japan Earthquake," Reconstruction Agency, October 2017, http://www.reconstruction.go.jp/english/topics/Progress_to_date/pdf/201710_The_Road_to_Recovery_E.pdf.

57. Sarah Jacobs, "'It Feels Like We're In Jail': Japan Spent $12 Billion on Seawalls after the Devastating 2011 Tsunami—and Now Locals Are Feeling Like Prisoners," *Insider*, March 12, 2018, https://www.businessinsider.com/japan-seawalls-cost-12-billion-since-fukushima-disaster-2018-3.

58. Koshimura and Shuto, "Response to the 2011."

59. "Museum Dedicated to March 2011 Tsunami Disaster Opens in Iwate Prefecture," *Japan Times*, September 23, 2019, https://www.japantimes.co.jp/news/2019/09/23/national/museum-311-tsunami-opens-iwate/#.XkYBhSN7mUk.

60. Iwate Tsunami Memorial Website, https://translate.google.com/translate?hl=ja&sl=auto&tl=en&u=https://iwate-tsunami-memorial.jp/.

61. Lukas Sundermann, Oliver Schelske, Peter Hausmann, "Mind the Risk, A Global Ranking of Cities under Threat from Natural Disasters," SwissRe, 2014, https://media.swissre.com/documents/Swiss_Re_Mind_the_risk.pdf.

62. *Typhoon Neoguri Disaster Risk Reduction Situation Report*, PreventionWeb, United Nations Office for Disaster Risk Reduction, UNDRR, DRR sitrep 2014-001, July 8, 2014, https://www.preventionweb.net/files/38384_pwsitrep2014001.pdf.

63. "1995 Kobe Earthquake 10-Year Retrospective," Risk Management Solutions, 2005, http://forms2.rms.com/rs/729-DJX-565/images/eq_1995_kobe_eq.pdf.

64. M. Maeda, Kanako Takahashi, Hamood Al-Washali, Akira Tasai, Hitoshi Shiohara, Kazuhiro Kitayama, Susumu Kono, and Tetsuya Nishida, "Damage to Reinforced Concrete School Buildings in Miyagi after the 2011 Great East Japan Earthquake," 15, World Conference in Earthquake Engineering, 2012, https://www.iitk.ac.in/nicee/wcee/article/WCEE2012_2591.pdf.

65. "House Buying Guide," Japan Property Central, https://japanpropertycentral.com/real-estate-faq/house-buying-guide/.

66. Bridget Mallon, "How Big Is the Average House Size around the World?—From U.K. Flats to American McMansions," *Elle Décor*, August 26, 2015, https://www.elledecor.com/life-culture/fun-at-home/news/a7654/house-sizes-around-the-world/ or https://www.msn.com/en-in/lifestyle/smart-living/how-big-is-the-average-house-size-around-the-world/ar-AAdKEhh.

67. Darrin Qualman, "Home Grown: 67 Years of US and Canadian House Size Data," May 8, 2018, https://www.darrinqualman.com/house-size/.

68. Mallon, "How Big."

69. "How Much Living Space Does the Average Household Have in Japan?" Real Estate Japan, August 12, 2017, https://resources.realestate.co.jp/living/how-much-living-space-does-the-average-household-have-in-japan/.

70. See https://www.mhlw.go.jp/stf/shingi/2r98520000012t0i-att/2r98520000012t75.pdf.

71. Mun Keat Looi, "Why Japan Has More Old-Fashioned Music Stores Than Anywhere Else in the World," Quartz, August 19, 2016, https://qz.com/711490/why-japan
-has-more-music-stores-than-the-rest-of-the-world/.

72. "1995 Kobe Earthquake 10-Year Retrospective."

73. R. Chung et al., "January 17, 1995 Hyogoken-Nanbu (Kobe) Earthquake: Performance of Structures, Lifelines, and Fire Protection Systems (NIST SP 901)," Special Publication (NIST SP), National Institute of Standards and Technology, Gaithersburg, MD, https://doi.org/10.6028/NIST.SP.901 and https://nvlpubs.nist.gov/nistpubs/Legacy/SP/nistspecialpublication901.pdf.

74. Jason Contant, "This Is How Much Fire Following an Earthquake in Montreal Could Cost," Canadian Underwriter, November 19, 2019, https://www.canadianunder writer.ca/claims/this-is-how-much-fire-following-an-earthquake-in-montreal-could
-cost-1004170934/.

THE WATER MAGNET

1. In some cases, a fatal attraction. Contrary to what is depicted in the 1998 computer-animated movie *A Bug's Life*, insects mesmerized by the glowing blue light of an "electrical discharge insect control system," also known as a Bug Zapper, when touching the radiating screen, do not scream Goofy-like, fall down dazzled, bounce on the porch stairs, and land into the grass. Rather, they actually explode, spreading parts and bacteria over a radius of up to six feet—which would have been educational to show, but maybe too gory for Pixar/Disney's business model.

2. "Biophilia hypothesis," Wikipedia, https://en.wikipedia.org/wiki/Biophilia_hypo thesis.

3. A few minutes on Zillow.com or Realtor.com is all that is needed to see that a small house built in the 1940s on a narrow lot wedged between two condo towers and located in a non-stellar school district, but oceanfront, can sell for ten times the cost of a brand-new and bigger one in a manicured, gated community miles away from the shore.

4. Online Etymology Dictionary, s.v. "hurricane," https://www.etymonline.com/search?q=hurricane.

5. "Hurricanes Are Named after Huracán the Mayan God of Wind, Fire, and Storms," *Yucatan Times*, August 17, 2019, https://www.theyucatantimes.com/2019/08/hurri canes-are-named-after-huracan-the-mayan-god-of-wind-fire-and-storms/.

6. Eric Anderson and Almendra Gutiérrez, "Mayan Gods: Huracan," *Yucatan Living*, June 22, 2017, https://yucatanliving.com/culture/this-is-the-story-of-the-huracan.

7. Online Etymology Dictionary, s.v. "hurricane."

8. "The Sphinx in Greek Mythology," Greek Legends and Myths, https://www.greek legendsandmyths.com/sphinx.html.

9. Sanujit, "Cultural Links between India and the Greco-Roman World," *World History Encyclopedia*, February 12, 2011, https://www.ancient.eu/article/208/cultural
-links-between-india--the-greco-roman-worl/.

10. Vandita Kapoor, "8 Incredible Similarities between Hindu and Greek Mythology," The Better India, January 4, 2016, https://www.thebetterindia.com/41620/similarities-hindu-greek-mythology/.

11. "What Is the Difference between a Hurricane and a Typhoon?," National Ocean Service, National Oceanic and Atmospheric Administration (NOAA), February 26, 2021, https://oceanservice.noaa.gov/facts/cyclone.html.

12. Ken Thomas, "Hurricanes Have Shaped Life of Scale Inventor," *Ellensburg Daily Record*, August 23, 2001, https://web.archive.org/web/20160417061701/https://news.google.com/newspapers?id=TTEfAAAAIBAJ&sjid=lccEAAAAIBAJ&pg=2625%2C7395281.

13. "Hurricanes in History," NOAA, https://www.nhc.noaa.gov/outreach/history/#katrina.

14. Sean Cornell, Duncan Fitzgerald, Nathan Frey, Ioannis Georgiou, Kevin C. Hanegan, Li-San Hung, Mark Kulp, Diane Maygarden, David Retchless, Brent Yarnal, and Tim Bralower, "Factors Determining the Size of the Storm Surge, Coastal Processes, Hazards, and Society," Penn State's College of Earth and Mineral Sciences' OER Initiative, https://www.e-education.psu.edu/earth107/node/1515.

15. "The Saffir-Simpson Hurricane Wind Scale," NOAA, May 2021, https://www.nhc.noaa.gov/pdf/sshws.pdf.

16. "Why Are Hurricanes Classified by Category?," History, September 12, 2017, https://www.history.com/news/why-are-hurricanes-classified-by-category.

17. "Minor Modification to Saffir-Simpson Hurricane Wind Scale for the 2012 Hurricane Season," NOAA, https://www.nhc.noaa.gov/pdf/sshws_2012rev.pdf.

18. "How Do Hurricanes Form?," NASA Science Space Place, December 4, 2019, https://spaceplace.nasa.gov/hurricanes/en/.

19. "Hurricane Movement," Hurricanes: Science and Society, University of Rhode Island's Graduate School of Oceanography, http://www.hurricanescience.org/science/science/hurricanemovement/.

20. *A Summary Report on Debris Impact Resistance of Building Assemblies*, Windstorm Mitigation Initiative, The Wind Science and Engineering Research Center, Texas Tech University, Florida A&M University, Florida State University, and University of Florida, August 2006, https://www.depts.ttu.edu/nwi/research/DebrisImpact/Reports/TTU_Final_NIST_Report_numbered.pdf.

21. Here is a movie file demonstrating a tornado cannon firing a two-by-four: https://www.depts.ttu.edu/nwi/research/DebrisImpact/downloads/Tornado_Cannon.mov.

22. "Impact (mechanics)," Wikipedia, https://en.wikipedia.org/wiki/Impact_(mechanics).

23. "Wind Load vs. Wind Speed," The Engineering Toolbox, https://www.engineeringtoolbox.com/wind-load-d_1775.html.

24. Luis Alvarez, "Are Impact Windows Required in Florida?," Alco, December 2020, https://info.alcoimpact.com/blog/are-impact-windows-required-in-florida.

25. FEMA, *Safe Rooms for Tornadoes and Hurricanes—Guidance for Community and Residential Safe Rooms*, FEMA P-361, fourth edition (Washington, DC: FEMA, April 2021), B8-8, https://www.fema.gov/sites/default/files/documents/fema_safe-rooms-for-tornadoes-and-hurricanes_p-361.pdf.

26. "State and Local Building Codes, Coastal Resilience," Texas A&M GriLife Extension, https://coastalresilience.tamu.edu/home/introduction-to-coastal-resilience/legal-framework-for-planning/stateandlocal/building-codes/.

27. Mark Powell, and Samuel Houston, "Hurricane Andrew's Landfall in South Florida, Part II: Surface Wind Fields and Potential Real-Time Applications," *Weather and Forecasting* 11, no. 3 (1996): 329–49, doi:10.1175/1520-0434(1996)011<0329:HAL ISF>2.0.CO;2.

28. "Hurricane Andrew (1992)," Series: Coastal Geomorphology—Storms of Record, National Park Service, https://www.nps.gov/articles/hurricane-andrew-1992.htm.

29. "Hurricane Andrew," Wikipedia, https://en.wikipedia.org/wiki/Hurricane_Andrew #Florida_2.

30. Amy O'Connor, "25 Years Later: How Florida's Insurance Industry Has Changed Since Hurricane Andrew," *Insurance Journal*, August 24, 2017, https://www.insurance journal.com/news/southeast/2017/08/24/462204.htm.

31. Elliott Mittler, "A Case Study of Florida's Homeowners' Insurance since Hurricane Andrew," Working Paper 96, Natural Hazards Center, University of Colorado, 1997, https://hazards.colorado.edu/archive/publications/wp/wp96.html.

32. Alistair Barr, "Allstate Pulls Back from Florida," MarketWatch, May 18, 2005, https://www.marketwatch.com/story/allstate-pulls-back-from-florida-insurance -market-20055181753490.

33. Greg Allen, "State Farm Abandons Florida's Homeowners Market," NPR, January 28, 2009, https://www.npr.org/templates/story/story.php?storyId=99942808.

34. "Background on: Hurricane and Windstorm Deductibles," Insurance Information Institute, June 23, 2021, https://www.iii.org/article/background-on-hurricane-and -windstorm-deductibles.

35. Elliott Mittler, "A Case Study of Florida's Homeowners' Insurance since Hurricane Andrew," Working Paper 96, Natural Hazards Center, University of Colorado, 1997, https://hazards.colorado.edu/archive/publications/wp/wp96.html.

36. Curtis Morgan, "Impact of Hurricane Andrew: Better Homes," *Miami Herald*, August 19, 2014, https://www.miamiherald.com/news/special-reports/hurricane -andrew/article1940341.html

37. "Eye (cyclone)," Wikipedia, https://en.wikipedia.org/wiki/Eye_(cyclone).

38. Powell and Houston, "Hurricane Andrew's Landfall in South Florida."

39. "Hurricane Michael," Wikipedia, https://en.wikipedia.org/wiki/Hurricane_Mich ael.

40. "Gov. Rick Scott: Hurricane Michael a 'Monstrous Storm,'" *Tallahassee Democrat*, October 9, 2018, https://www.tallahassee.com/story/news/2018/10/09/category-hurri cane-michael-monstrous-storm-gov-rick-scott-panhandle/1575200002/.

41. Doyle Rice, "Hurricane Michael Could Rival Worst Storms on Record to Hit Florida Panhandle," *USA Today*, October 9, 2018, https://www.usatoday.com/story/ news/2018/10/09/hurricane-michael-florida-preps-potentially-catastrophic-impact/1 574610002/.

42. "Hurricane Michael Intermediate Advisory Number 15A," NWS National Hurricane Center Miami FL, 700 AM CDT Wed Oct 10 2018 (before landfall), https://www .nhc.noaa.gov/archive/2018/al14/al142018.public_a.015.shtml.

43. Jeff Masters, "Hurricane Michael Brought Water Levels Over 20' High to the Coast," Wunderground, November 16, 2018, https://www.wunderground.com/cat6/ Hurricane-Michael-Brought-Water-Levels-Over-20-High-Coast.

44. "Panama City Beach Building Partially Collapses as Hurricane Michael Lashes Florida," ABC News, October 10, 2018, https://abc13.com/weather/video-building-collapses-as-hurricane-michael-lashes-florida/4455468/.

45. "Hurricane Michael Exposes Building-Code Weakness in Florida's Panhandle," *The Real Deal*, October 13, 2018, https://therealdeal.com/miami/2018/10/13/hurricane-michael-exposes-building-code-weakness-in-floridas-panhandle/.

46. Ibid.

47. Patricia Mazzei, "Among the Ruins of Mexico Beach Stands One House, Built 'for the Big One,'" *New York Times*, October 14, 2018, https://www.nytimes.com/2018/10/14/us/hurricane-michael-florida-mexico-beach-house.html.

48. Doug Allen, "Choosing Resiliency: Lessons from Hurricane Michael," *Simpson Strong-Tie Structural Engineering Blog*, February 21, 2019, https://seblog.strongtie.com/tag/hurricane-michael/.

49. Matthew Gewing and Sarah Bobby, "How the Sand Palace Survived Hurricane Michael," AIR, November 29, 2018, https://www.air-worldwide.com/blog/posts/2018/11/how-the-sand-palace-survived-hurricane-michael/.

50. John Dal Pino, "The Story of a Survivor—Sand Palace of Mexico Beach," *Structures Magazine*, March 2019, https://www.structuremag.org/wp-content/uploads/2019/02/261903-SF-Palace-1.pdf.

51. Ibid.

52. Ibid.

53. Tim Sharp, "Superstorm Sandy: Facts about the Frankenstorm," Live Science, November 27, 2012, https://www.livescience.com/24380-hurricane-sandy-status-data.html.

54. "Hurricane Sandy," Wikipedia, https://en.wikipedia.org/wiki/Hurricane_Sandy#Mid-Atlantic.

55. Kristen Meriwether, "The Storm Nobody Expected," *Epoch Times*, October 25, 2013, https://www.theepochtimes.com/hurricane-sandy-storm-nobody-expected_329831.html.

56. John Homans, "The City and the Storm," *New York Magazine*, November 3, 2012, https://nymag.com/news/features/hurricane-sandy-2012-11/.

FLOOD

1. OECD, *Financial Management of Flood Risk* (Paris: OECD Publishing, 2016), 14, https://doi.org/10.1787/9789264257689-en.

2. "Flood Myth," Myths and Folklore Wiki, https://mythology.wikia.org/wiki/Flood_myth.

3. "Noah's Ark," Wikipedia, https://simple.wikipedia.org/wiki/Noah%27s_Ark.

4. "Flood Myth."

5. "The Story of the Great Flood According to Ancient Aztec Mythology," Ancient Code, https://www.ancient-code.com/the-story-of-the-great-flood-according-to-ancient-aztec-mythology/.

6. "15 Flood Myths Similar to the Story of Noah," Mytho Religio Series, November 11, 2015, https://www.mythoreligio.com/15-flood-myths-similar-to-the-story-of-noah-2/.

7. "Floodplain," Science Clarified, http://www.scienceclarified.com/landforms/Faults-to-Mountains/Floodplain.html.

8. Ibid.

9. "Levee Systems," US Army Corps of Engineers, Mississippi Valley Division, https://www.mvd.usace.army.mil/About/Mississippi-River-Commission-MRC/Mississippi-River-Tributaries-Project-MR-T/Levee-Systems/.

10. "Great Wall of China," Wikipedia, https://en.wikipedia.org/wiki/Great_Wall_of_China.

11. Erica Ho, "Is the Great Wall of China Longer Than Previously Thought?," *Time*, July 22, 2012, http://newsfeed.time.com/2012/07/22/is-the-great-wall-of-china-longer-than-previously-thought/.

12. "Earth's circumference," Wikipedia, https://en.wikipedia.org/wiki/Earth%27s_circumference.

13. "Great Mississippi Flood of 1927," Wikipedia, https://en.wikipedia.org/wiki/Great_Mississippi_Flood_of_1927.

14. "When the Levee Breaks," Wikipedia, https://en.wikipedia.org/wiki/When_the_Levee_Breaks.

15. ASCE, *So, You Live behind a Levee* (Reston, VA: American Society of Civil Engineers, 2010), https://ascelibrary.org/doi/pdf/10.1061/9780784410837.

16. "Living with Levees," FEMA, https://www.fema.gov/flood-maps/living-levees.

17. ASCE, *So, You Live behind a Levee*.

18. Rebecca Hersher, "Levees Make Mississippi River Floods Worse, but We Keep Building Them," NPR, May 21, 2018, https://www.npr.org/2018/05/21/610945127/levees-make-mississippi-river-floods-worse-but-we-keep-building-them.

19. National Flood Insurance Program, US Government Accountability Office, https://www.gao.gov/highrisk/national_flood_insurance/why_did_study#t=1.

20. ASCE, *So, You Live behind a Levee*.

21. "The 100-Year Flood," USGS, https://www.usgs.gov/special-topic/water-science-school/science/100-year-flood?qt-science_center_objects=0#qt-science_center_objects.

22. Rebecca Hersher, "When '1-in-100-Year' Floods Happen Often, What Should You Call Them?," NPR, May 8, 2019, https://www.npr.org/2019/05/08/720737285/when-1-in-100-year-floods-happen-often-what-should-you-call-them.

23. "The 1993 Great Midwest Flood: Voices 10 Years Later," in *10th-Anniversary Anthology of Stories of Hardship and Triumph* (Washington, DC: US Dept. of Homeland Security, FEMA, 2003), https://purl.fdlp.gov/GPO/LPS45535.

24. "Mississippi River Floods," Wikipedia, https://en.wikipedia.org/wiki/Mississippi_River_floods.

25. "Floodplain," Science Clarified, http://www.scienceclarified.com/landforms/Faults-to-Mountains/Floodplain.html.

26. FEMA Media Library, https://www.fema.gov/media-library/assets/documents/3799?id=1789.

27. "Mississippi River Floods," Wikipedia, https://en.wikipedia.org/wiki/2011_Mississippi_River_floods.

28. "Floodplain."

29. Fernand Grenier, "1773 à aujourd'hui: Catastrophes hydrologiques en Beauce," *Cap-aux-Diamants*, no. 82 (2005): 14–19, https://www.erudit.org/fr/revues/cd/2005 -n82-cd1045506/7068ac.pdf.

30. Ibid.

31. Ibid.

32. Gabriel Béland, "Inondations: Plus de 3000 demandes d'indemnisation," *La Presse*, May 16, 2019, http://plus.lapresse.ca/screens/54d3ac20-9096-40a2-8582-d9de 5dbb4fd6__7C___0.html.

33. Pierre Saint-Arnaud, "Inondations: L'idée de la relocalisation soulevée par la ministre Guilbault," *La Presse*, April 16, 2019, https://www.lapresse.ca/actualites/ regional/201904/16/01-5222376-inondations-lidee-de-la-relocalisation-soulevee-par -la-ministre-guilbault.php.

34. "The 1993 Great Midwest Flood."

35. Lisa Song, Al Shaw, and Neena Satija, "Buyouts Won't Be the Answer for Many Frequent Flooding Victims," ProPublica and *Texas Tribune*, November 1, 2017, https://features.propublica.org/houston-buyouts/hurricane-harvey-home-buyouts -harris-county/.

36. Ibid.

37. "Flood Insurance," FEMA, https://www.fema.gov/national-flood-insurance-pro gram.

38. Abigail Peralta and Jonathan B. Scott, *Moving to Flood Plains: The Unintended Consequences of the National Flood Insurance Program on Population Flows* (American Economic Association Annual Conference, February 8, 2019), https://www.lsu.edu/ business/economics/files/microecon-conf-lsu-peralta.pdf.

39. Tim Henderson, "Rising Tide of State Buyouts Fights Flooding," PEW Charitable Trust, December 3, 2015, https://www.pewtrusts.org/en/research-and-analysis/blogs/ stateline/2015/12/03/rising-tide-of-state-buyouts-fights-flooding.

40. "Netherlands," Wikipedia, https://en.wikipedia.org/wiki/Netherlands.

41. Matt Rosenber, "How the Netherlands Reclaimed Land from the Sea," ThoughtCo, September 9, 2019, https://www.thoughtco.com/polders-and-dikes-of-the-netherland s-1435535.

42. "List of Floods in the Netherlands," Wikipedia, https://en.wikipedia.org/wiki/ Floods_in_the_Netherlands.

43. "Flood Control in the Netherlands," Wikipedia, https://en.wikipedia.org/wiki/ Flood_control_in_the_Netherlands.

44. "New Orleans—Extreme Daily Rainfall for Each Year," Current Results, https:// www.currentresults.com/Yearly-Weather/USA/LA/New-Orleans/extreme-annual -new-orleans-precipitation.php.

45. Richard Campanella, "How Humans Sank New Orleans," *Atlantic*, February 6, 2018, https://www.theatlantic.com/technology/archive/2018/02/how-humans-sank -new-orleans/552323/.

46. Ibid.

47. "Louisiana Base and Elevation Maps," Netstate, https://www.netstate.com/states/ geography/mapcom/la_mapscom.htm.

48. S. E. Flynn, *The Edge of Disaster: Rebuilding a Resilient Nation* (New York: Random House, 2007).

49. "Hurricane Katrina," Wikipedia, https://en.wikipedia.org/wiki/Hurricane_Katrina.

50. Thomas Frank, "After a $14-Billion Upgrade, New Orleans' Levees Are Sinking," *Scientific American*, April 11, 2019, https://www.scientificamerican.com/article/after-a-14-billion-upgrade-new-orleans-levees-are-sinking/.

51. "Notice of Intent to Prepare a Draft Environmental Impact Statement for the Lake Pontchartrain and Vicinity General Re-Evaluation Report, Louisiana," US Army Corps of Engineers, *Federal Register* 84, no.64 (April 2, 2019), https://www.eenews.net/assets/2019/04/11/document_ew_01.pdf.

52. Jonathan Rauch, "Struggling to Survive," *Atlantic*, August 2008, https://www.theatlantic.com/magazine/archive/2006/08/struggling-to-survive/305182.

53. Mark Schleifstein, "New Maps Could Be Good News for Many New Orleans–Area Flood-Insurance Customers," *Times-Picayune*, January 26, 2013, https://www.nola.com/news/environment/article_53f12df8-b57d-5dc2-9113-f4660278e7ff.html.

54. Richard Thompson, "St. Bernard Parish Rebuilds Itself from Utter Devastation after Hurricane Katrina," *Advocate*, August 23, 2015, https://www.theadvocate.com/baton_rouge/news/article_8b548335-05ab-51c0-aa88-4ba5400a1c58.html.

55. "New Orleans September 2005," *New York Times*, https://www.nytimes.com/interactive/2015/08/26/us/ten-years-after-katrina.html.

56. "To California by Sea," Smithsonian National Museum of American History, https://americanhistory.si.edu/on-the-water/maritime-nation/california-sea.

57. "Treaty of Guadalupe Hidalgo," History, November 9, 2009, https://www.history.com/topics/mexican-american-war/treaty-of-guadalupe-hidalgo.

58. "Census Research Guide: 1850, Historical Census Populations of Counties and Cities in California," https://ucsd.libguides.com/census/1850 and data: https://dof.ca.gov/Reports/Demographic_Reports/documents/2010-1850_STCO_IncCities-FINAL.xls.

59. "California Becomes the 31st State in Record Time," History, September, 9, 2009, https://www.history.com/this-day-in-history/california-becomes-the-31st-state-in-record-time.

60. "California Gold Rush." *Encyclopedia Britannica*, https://www.britannica.com/topic/California-Gold-Rush.

61. "Sports Money: 2021 NFL Valuations," *Forbes*, https://www.forbes.com/nfl-valuations/list/.

62. Matt Rosenberg, "California Population," ThoughtCo, August 27, 2020, https://www.thoughtco.com/california-population-overview-1435260.

63. "San Joaquin River Delta," Wikipedia, https://en.wikipedia.org/wiki/Sacramento%E2%80%93San_Joaquin_River_Delta.

64. "Northern California," Wikipedia, https://en.wikipedia.org/wiki/Great_Flood_of_1862#Northern_California.

65. Walter W. Weir, "Subsidence of Peat Lands of the Sacramento-San Joaquin Delta, California," *Hilgardia Journal* 20, no. 3 (June 1950): 37–56, DOI#10.3733/hilg.v20n03p037.

66. Matthew Bates and Jay Lund, "Delta Subsidence Reversal, Levee Failure, and Aquatic Habitat—A Cautionary Tale," *San Francisco Estuary and Watershed Science* 11, no. 1 (March 2013): figure 2. https://doi.org/10.15447/sfews.2013v11iss1art1.

67. Flynn, *The Edge of Disaster.*

68. Stephen Stock, Julie Putnam, Jeremy Carroll, and Scott Pham, "Will California's Levee System Hold Up in an Earthquake?," NBC Bay Area Sacramento, February 2, 2015, https://www.nbcbayarea.com/news/local/will-californias-levee-system-hold-up-in-an-earthquake/115301/.

69. Mac Taylor, *Achieving State Goals for the Sacramento–San Joaquin Delta, Legislative Analyst's Office*, The California Legislature's Nonpartisan Fiscal and Policy Advisor, January 15, 2015, https://lao.ca.gov/reports/2015/res/Delta/sac-sj-delta-011515.pdf.

70. "When the Levee Breaks: Cascading Failures in the Sacramento-San Joaquin River Delta, California," https://www.youtube.com/watch?time_continue=12&v=BtlfbhMp02Q&feature=emb_logo.

71. Flynn, *The Edge of Disaster.*

72. Stock et al., "Will California's Levee System Hold Up in an Earthquake?"

73. Ibid.

74. "California Water Fix and Eco Restore: Legislation," Wikipedia, https://en.wikipedia.org/wiki/California_Water_Fix_and_Eco_Restore#Legislation.

75. M. Reisner, *Cadillac Desert: The American West and Its Disappearing Water*, revised edition (New York: Penguin Books, 2014).

76. "Champlain Sea," Wikipedia, https://en.wikipedia.org/wiki/Champlain_Sea.

77. "Elevation of Fallingbrook, Ottawa," ON K4A, Canada, Worldwide Elevation Map Finder, https://elevation.maplogs.com/poi/fallingbrook_ottawa_on_k4a_canada.256909.html.

78. "Apollo Crater Park," BringFido, https://www.bringfido.com/attraction/9953.

79. See https://www.google.com/maps/place/Fallingbrook,+Ottawa,+ON,+Canada/@45.4760533,-75.4886318,18.69z/data=!4m5!3m4!1s0x4cce12f58bbabfbd:0x66b8f5ba9c23ec7e!8m2!3d45.4737643!4d-75.4780294.

80. EPA, *Protecting Water Quality from Urban Runoff*, Document EPA 841-F-03-003, February 2003, https://www3.epa.gov/npdes/pubs/nps_urban-facts_final.pdf.

TORNADO ALLEY

1. "Camp Fire (2018)," Wikipedia, https://en.wikipedia.org/wiki/Camp_Fire_(2018).

2. Lisa M. Krieger and David Debolt, "Camp Fire: Paradise Residents Say They Received No Mass Cellphone Alerts to Evacuate, or to Warn of Fires," *Mercury News*, November 13, 2018, https://www.mercurynews.com/2018/11/13/camp-fire-paradise-residents-say-they-received-no-mass-cellphone-alerts-to-evacuate-or-to-warn-of-fires/.

3. "Home Prices in the 100 Largest Metro Areas," Kiplinger, January 23, 2021, https://www.kiplinger.com/tool/real-estate/T010-S003-home-prices-in-100-top-u-s-metro-areas/index.php.

4. "How to Design a Tornado-Safe Room," *Reader's Digest*, March 29, 2021, https://www.rd.com/home/improvement/how-to-design-a-tornado-safe-room/.

5. "Above Ground vs. In-Ground Storm Shelters," Ground Zero Storm Shelters, https://www.groundzeroshelters.com/above-ground-vs-in-ground.

6. "Safe Rooms," FEMA, https://www.fema.gov/media-library/assets/documents/ 2009.

7. Maggie Koerth, "Tornado Town, USA—Four Devastating Tornadoes Hit Moore, Oklahoma, in 16 Years. Was It Geography or Just Bad Luck?," Five ThirtyEight, May 26, 2016, https://fivethirtyeight.com/features/tornadoes/.

8. John Schwartz, "Why No Safe Room to Run To? Cost and Plains Culture," *New York Times*, May 21, 2013, https://www.nytimes.com/2013/05/22/us/shelter-requirements -resisted-in-tornado-alley.html.

9. Bob Tschudi, "How Much Does a Storm Shelter Cost?," HomeAdvisor, December 23, 2020, https://www.homeadvisor.com/cost/safety-and-security/build-a-storm -shelter/.

10. "Tornado Climatology," Wikipedia, https://en.wikipedia.org/wiki/Tornado_cli-matology#/media/File:Improved_Average_Annual_Tornado_Reports.svg.

11. "Historical Records and Trends," National Centers for Environmental Information, NOAA, https://www.ncdc.noaa.gov/climate-information/extreme-events/us -tornado-climatology/trends.

12. "Tornado Climatology."

13. "Tornado Records," Wikipedia, https://en.wikipedia.org/wiki/Tornado_records #Earliest-known_tornado_in_the_Americas.

14. Jake Carpenter and Catherine E. Shoichet, "'Unpredictable' Storm in Oklahoma Turned on Three Chasers," CNN, June 3, 2013, https://www.cnn.com/2013/06/02/us/ midwest-weather/index.html.

15. Ibid.

16. "Historical Records and Trends."

17. "*Sharknado*," Wikipedia, https://en.wikipedia.org/wiki/Sharknado.

18. "F5 and EF5 Tornadoes of the United States, 1950–Present," Storm Prediction Center, NOAA, https://www.spc.noaa.gov/faq/tornado/f5torns.html.

19. Michelle Star, "New Evidence Shows Tornadoes Don't Form the Way Scientists Have Always Thought," Science Alert, https://www.sciencealert.com/new-evidence -has-turned-our-knowledge-about-tornado-formation-upside-down.

20. "All about Tornadoes," Earth Networks, https://www.earthnetworks.com/tor nado/.

21. "Tornadoes FAQ," National Weather Service, NOAA, https://www.weather.gov/ lmk/tornadoesfaq.

22. "Tornado Records: Longest Damage Path and Duration," Wikipedia, https://en .wikipedia.org/wiki/Tornado_records#Longest_damage_path_and_duration.

23. Jessica Dillinger, "The Most Devastating Tornados by Fatalities, Environment," World Atlas, April 25, 2017, https://www.worldatlas.com/articles/a-history-of-destruc tion-the-10-deadliest-tornadoes-in-the-us.html.

24. Harold Brooks, "Mobile Home Tornado Fatalities: Some Observations," NOAA/ ERL/National Severe Storms Laboratory, https://www.nssl.noaa.gov/users/brooks/ public_html/essays/mobilehome.html.

25. "Mobile Homes 2500% More Deadly in Tornadoes," Modular Today, https:// www.modulartoday.com/mobilehomes.html.

26. "Are Manufactured Homes Safe in Tornadoes and Destructive Force Winds?," Wholesale Manufactured Homes, https://www.wholesalemobilehomes.net/are-manufactured-homes-safe-in-destructive-force-winds-and-tornadoes/.

27. Stephen Strader and Walker Ashley, "Fine-Scale Assessment of Mobile-Home Tornado Vulnerability in the Central and Southeast United States," *Weather, Climate, and Society* 10, no. 4 (2018): 797–812, http://dx.doi.org/10.1175/WCAS-D-18-0060.1.

28. "Tornado Myths," Wikipedia, https://en.wikipedia.org/wiki/Tornado_myths.

29. Steve Tracton, "Killing Killer Tornadoes before They Strike: Is It Possible?," *Washington Post*, May 6, 2011, https://www.washingtonpost.com/blogs/capital-weather-gang/post/killing-killer-tornadoes-before-they-strike-is-it-possible/2011/05/06/AFePEn8F_blog.html.

30. "Tornado Tour Frequent Questions," Jason Weingart Photography, http://www.jasonrweingart.com/tornado-tour-faqs.

31. Jeff Lieberman, "Bloodied and Bruised, I Survived the Inside of a Kansas Tornado. But There 'Ain't Gonna Be No Rematch,'" *Washington Post*, June 19, 2019, https://www.washingtonpost.com/weather/2019/06/19/bloodied-bruised-i-survived-inside-kansas-tornado-there-aint-gonna-be-no-rematch/.

32. *"Twister,"* IMDb, https://www.imdb.com/title/tt0117998/?ref_=fn_tt_tt_7.

SITTING ON A VOLCANO

1. "Eruptions," Kagoshima City, http://www.city.kagoshima.lg.jp/soumu/shichoshitu/kokusai/en/emergency/sakurajima.html.

2. Arimura Lava Observatory, Kagoshima—The Official Tourism Website, http://www.kagoshima-kankou.com/for/attractions/10509.

3. "Kurokami Buried Shrine Gate: The Tale of Sakurajima's Fury Laid Bare for Future Generations," http://www.kagoshima-kankou.com/for/attractions/10531.

4. Sakurajime Volcano Hazard Map, http://www.city.kagoshima.lg.jp/soumu/shichoshitu/kokusai/en/emergency/documents/sakurazimahm_eng.pdf.

5. K. Ishihara, "Dynamical Analysis of Volcanic Explosion," *Journal of Geodynamics* 3, nos. 3–4 (1985); 327–49, https://doi.org/10.1016/0264-3707(85)90041-9. And https://www.sciencedirect.com/science/article/pii/0264370785900419.

6. "Volcanic bomb," Wikipedia, https://en.wikipedia.org/wiki/Volcanic_bomb.

7. See Google Maps: https://www.google.com/maps/place/Mount+Vesuvius.

8. Giuseppe Vilardo and Guido Ventura, "Geomorphological Map of the 1944 Vesuvius Lava Flow (Italy)," *Journal of Maps* 4, no. 1 (2008): 225–34, DOI: 10.4113/jom.2008.

9. See Google Maps: https://www.google.com/maps/place/Mount+Vesuvius.

10. Fraser Cain, "What Is the Temperature of Lava?" Universe Today, March 25, 2009, https://www.universetoday.com/27891/temperature-of-lava/.

11. "Concrete under Fire," *Concrete Construction Magazine*, September 1, 2004, https://www.concreteconstruction.net/concrete-production-precast/concrete-under-fire_o.

12. "Spalling of Concrete at High Temperatures," *Concrete Construction Magazine*, March 1, 1980, https://www.concreteconstruction.net/how-to/spalling-of-concrete-at-high-temperatures_o.

13. "More about Etna," Volcano World, Oregon State University, https://volcano.oregonstate.edu/more-about-etna.

14. Taylor Kate Brown, "How Do You Stop the Flow of Lava?," BBC News, 11 September 2014, https://www.bbc.com/news/magazine-29136747.

15. Robin Andrews, "How to Stop a Lava Flow," *Forbes*, March 29, 2017, https://www.forbes.com/sites/robinandrews/2017/03/29/how-to-stop-a-lava-flow/#25b3b8cc29ec.

16. David Bressan, "When the U.S. Army Bombed Hawaii's Volcanoes," *Forbes*, May 27, 2018, https://www.forbes.com/sites/davidbressan/2018/05/27/when-the-u-s-army-bombed-hawaiis-volcanoes/#18e346633077.

17. "How Fast Does Lava Flow?," Volcano World, Oregon State University, https://volcano.oregonstate.edu/faq/how-fast-does-lava-flow.

18. "Pompei," Wikipedia, https://en.wikipedia.org/wiki/Pompeii.

19. "A Day in Pompeii—Full Length Animation" (video), YouTube, December 9, 2013, https://www.youtube.com/watch?v=dY_3ggKg0Bc.

20. "Atlas Obscura, Imprisoned in Ash: The Plaster Citizens of Pompeii," *Slate*, March 12, 2015, http://www.slate.com/blogs/atlas_obscura/2015/03/12/plaster_casts_of_victims_bodies_at_pompeii_in_italy.html.

21. W. Feemster Jashemski, "Pompeii," *Encyclopedia Britannica*, https://www.britannica.com/place/Pompeii.

22. "Eruption of Mount Vesuvius during 1944 World War II Naples Italy" (video), YouTube, https://www.youtube.com/watch?v=iZc7Mtid9a4.

23. Andrew Sagerson, "Mount Vesuvius, One of the Most Dangerous Volcanoes in the World," Wanted in Rome, May 5, 2021, https://www.wantedinrome.com/news/mount-vesuvius-one-of-the-most-dangerous-volcanoes-in-the-world.html.

24. "Mount Vesuvius," Wikipedia, https://en.wikipedia.org/wiki/Mount_Vesuvius.

25. Lucy Pasha-Robinson, "Mount Vesuvius Eruption Risk: Emergency Plans to Evacuate 700,000 Finalised," *Independent*, October 14, 2016, https://www.independent.co.uk/news/world/europe/mount-vesuvius-emergency-evacuation-eruption-plans-finalised-a7360686.html.

26. Peter Tyson, "Can We Predict Eruptions?," NOVA Online, PBS, November 2000, https://www.pbs.org/wgbh/nova/vesuvius/predict.html.

27. S. Carlino, R. Somma, and G. C. Mayberry, "Volcanic Risk Perception of Young People in The Urban Areas of Vesuvius: Comparisons with Other Volcanic Areas and Implications for Emergency Management," *Journal of Volcanology and Geothermal Research* 172, nos. 3–4 (2008): 229–43, https://doi.org/10.1016/j.jvolgeores.2007.12.010. and http://pages.mtu.edu/~raman/papers2/CarlinoVesuvPercep.pdf.

28. Ladan Cher, "Meet the Squatters of Mount Vesuvius," *Foreign Policy*, August 22, 2017, https://foreignpolicy.com/2017/08/22/meet-the-squatters-of-mt-vesuvius/.

29. Ellen Hale, "Italians Trying to Prevent a Modern Pompeii," *USA Today*, October 20, 2003, https://usatoday30.usatoday.com/news/world/2003-10-20-vesuvius-usat_x.htm.

30. "How Many Active Volcanoes Are There on Earth?," USGS, https://www.usgs.gov/faqs/how-many-active-volcanoes-are-there-earth?qt-news_science_products=0#qt-news_science_products.

31. "Large Amount of Trash and Waste Left on Mt Fuji," *Japan Today*, September 17, 2014, https://japantoday.com/category/national/large-amount-of-trash-and-waste -left-on-mt-fuji.

32. Julian Ryall, "Japan's Sacred Mount Fuji Risks Turning into Trash Mountain," *South China Morning Post*, September 17, 2014, https://www.scmp.com/news/asia/ article/1594823/japans-sacred-mount-fuji-risks-turning-trash-mountain.

33. "Why You Should Never Climb Mount Fuji," Halfway Anywhere, https://www .halfwayanywhere.com/asia/japan/why-you-should-never-climb-mount-fuji/.

34. "Fuji Five Lakes," Wikipedia, https://en.wikipedia.org/wiki/Fuji_Five_Lakes.

35. Volcanoes of Japan, https://gbank.gsj.jp/volcano/Quat_Vol/volcano_data/F01 .html.

36. Uli Schmetzer, "Mt. Fuji's Rumblings Have Japan Nervous," *Chicago Tribune*, February 7, 2001, https://www.chicagotribune.com/news/ct-xpm-2001-02-07-010 2070098-story.html.

37. "Volcanic Activity and Eruption Prediction in Japan," Mt. Fuji Research Institute, https://jp.ambafrance.org/IMG/pdf/prof._fujii_cs_20-01-2017.pdf?22540/.

38. Sara Gates, "Mount Fuji Is in a 'Critical State' and Could Be Ready to Blow, Researchers Say," *Huffington Post*, July 17, 2014, https://www.huffingtonpost.com/ 2014/07/17/mount-fuji-critical-state-ready-erupt_n_5596656.html.

39. "Mount Fuji 'Under More Pressure Than Last Eruption,'" Phys Org, September 6, 2012, https://phys.org/news/2012-09-mount-fuji-pressure-eruption.html

40. Toyohiro Ichioka, "Could a Mount Fuji Eruption Paralyze Tokyo?," Japan Forward, June 21, 2018, https://japan-forward.com/could-a-mount-fuji-eruption -paralyze-tokyo/.

41. "Mt Fuji Eruption Expected to Cause 750,000 Refugees," *Japan Today*, February 9, 2014, https://japantoday.com/category/national/mt-fuji-eruption-expected-to-cause -750000-refugees.

42. "Demographic History of Japan before the Meiji Restoration," Wikipedia, https:// en.wikipedia.org/wiki/Demographic_history_of_Japan_before_the_Meiji_Restoration #Total_population_2.

43. "Greater Tokyo Area," Wikipedia, https://en.wikipedia.org/wiki/Greater_Tokyo _Area#Various_definitions_of_Tokyo,_Greater_Tokyo_&_Kant%C5%8D.

44. "Greater Tokyo Area: National Capital Region," Wikipedia, https://en.wikipedia .org/wiki/Greater_Tokyo_Area#National_Capital_Region.

45. Julian Ryall, "Mount Fuji Eruption Fears Prompt Japan to Draw Up Disas- ter Plans," *Telegraph*, July 31, 2018, https://www.telegraph.co.uk/news/2018/07/31/ mount-fuji-eruption-fears-prompt-japan-draw-disaster-plans/.

46. David Cyranoski, "Why Japan Missed Volcano's Warning Signs," *Nature*, Sep- tember 29, 2014, https://doi.org/10.1038/nature.2014.16022 or https://www.nature .com/news/why-japan-missed-volcano-s-warning-signs-1.16022.

47. "Pele (deity)," Wikipedia, https://en.wikipedia.org/wiki/Pele_(deity).

48. Fraser Cain, "Vulcan and Volcanoes," *Universe Today*, May 25, 2009, https://www .universetoday.com/31365/vulcan-and-volcanoes/.

49. *"Journey to the Center of the Earth,"* Wikipedia, https://en.wikipedia.org/wiki/ Journey_to_the_Center_of_the_Earth.

50. Alicia Ault, "What's the Deepest Hole Ever Dug?," *Smithsonian Magazine*, February 19, 2015, https://www.smithsonianmag.com/smithsonian-institution/ask -smithsonian-whats-deepest-hole-ever-dug-180954349/.

51. "What Is the Longest, Deepest and Largest Hole Ever Drilled on Earth?," Drilling Formulas, December 21, 2014, http://www.drillingformulas.com/what-is-the-longest -deepest-and-largest-hole-ever-drilled-on-earth/.

52. "How Does Directional Drilling Work?," Rig Zone, https://www.rigzone.com/ training/insight.asp?insight_id=295&c_id=.

53. Tom Levitt, "The $1 Billion Mission to Reach the Earth's Mantle," CNN, October 2, 2012, https://www.cnn.com/2012/10/01/tech/mantle-earth-drill-mission/index .html.

54. "Subduction," Wikipedia, https://en.wikipedia.org/wiki/Subduction.

55. Dominique Mosbergen, "Man Dives into an Active Volcano with a GoPro, Calls It a 'Window into Hell.'" *Huffington Post*, September 5, 2014, https://www.huffpost.com/ entry/volcano-gopro-george-kourounis-sam-cossman_n_5774706.

56. "Paricutin," Wikipedia, https://en.wikipedia.org/wiki/Par%C3%ADcutin.

57. "Paricutín," *Encyclopedia Britannica*, https://www.britannica.com/place/Paricu tin.

58. Amy Woodyatt, "New Zealand Family Forced to Flee as Bubbling Mud Pool Appears in Garden," CNN, June 28, 2019, https://www.cnn.com/travel/article/new -zealand-mud-pool-intl-scli/index.html.

59. Andrew Evans, "Is Iceland Really Green and Greenland Really Icy? A Long-standing Rumor Claims the Names Are a Bait and Switch." *National Geographic*, June 30, 2016, https://www.nationalgeographic.com/news/2016/06/iceland-greenland -name-swap/; "Greenland Ice Sheet," Wikipedia, https://en.wikipedia.org/wiki/ Greenland_ice_sheet.

60. Nanna Gunnarsdóttir, "The Ultimate Guide to Volcanoes in Iceland + The Impact of COVID-19, Guide to Iceland," https://guidetoiceland.is/nature-info/the -deadliest-volcanoes-in-iceland.

61. *Guide to Iceland*, Katla Travel Guide, https://guidetoiceland.is/travel-iceland/ drive/katla.

62. Valur Gunnarsson, "Iceland Volcano: Hundreds Evacuated after Eruption," *Guardian*, April 15, 2010, https://www.theguardian.com/world/2010/apr/15/iceland -volcano-eruption-evacuated.

63. "Vatnajökull Eruption—Iceland, 1996," Earth Watching, https://earth.esa.int/ web/earth-watching/natural-disasters/volcanoes/sub/-/asset_publisher/Uz2lmG15Iz WD/content/vatnajokull-eruption-iceland-1996.

64. "The 1996 Iceland Jökulhlaup," International Council for Science, https://www .wdcgc.spri.cam.ac.uk/news/jokulhlaup/.

65. "Niagara Falls Characteristics," Wikipedia, https://en.wikipedia.org/wiki/Niagara _Falls#Characteristics.

66. Dave Petley, "Extraordinary Video of a Jokulhlaup in Iceland," Blogosphere, American Geophysical Union, April 17, 2010, https://blogs.agu.org/landslideblog/ 2010/04/17/extraordinary-video-of-a-jokulhlaup-in-iceland/.

67. Hobart M. King, "Volcanic Explosivity Index (VEI)," Geology.com, https://geology .com/stories/13/volcanic-explosivity-index/.

68. R. S. Bradley, "The Explosive Volcanic Eruption Signal in Northern Hemisphere Continental Temperature Records," *Climatic Change* 12 (1988): 221–43 (1988), https://doi.org/10.1007/BF00139431.

69. "Mount Tambora," Wikipedia, https://en.wikipedia.org/wiki/Mount_Tambora.

70. Victoria Jaggard, "Why Did the Dinosaurs Go Extinct?," *National Geographic*, July 31, 2019, https://www.nationalgeographic.com/science/prehistoric-world/dino saur-extinction/.

71. "Supervolcano," Wikipedia, https://en.wikipedia.org/wiki/Supervolcano.

72. Jana Howden, "Seven of the Biggest Volcanic Explosions to Rock the Earth," *Cosmos*, December 13, 2016, https://cosmosmagazine.com/earth/earth-sciences/ seven-of-the-biggest-volcanic-explosions/.

73. Remy Melina, "What's the Biggest Volcanic Eruption Ever?," Live Science, November 10, 2010, https://www.livescience.com/11113-biggest-volcanic-eruption.html.

74. P. M. Ng'walali et al., "Fatalities by Inhalation of Volcanic Gas at Mt. Aso Crater in Kumamoto, Japan," *Legal Medicine* (Tokyo, Japan) 1, no. 3 (1999): 180–84, doi:10.1016/s1344-6223(99)80034-0.

75. "Whakaari / White Island," Wikipedia, https://en.wikipedia.org/wiki/Whakaari _/_White_Island.

TECHNOLOGICAL DISASTERS

1. "Springfield Nuclear Power Plant," Simpsons Wiki, Fandom, https://simpsons .fandom.com/wiki/Springfield_Nuclear_Power_Plant.

2. "Chernobyl Accident 1986," World Nuclear Association, May 2021, https:// www.world-nuclear.org/information-library/safety-and-security/safety-of-plants/ chernobyl-accident.aspx.

3. Ibid.

4. "Chernobyl disaster," Wikipedia, https://en.wikipedia.org/wiki/Chernobyl_disas ter.

5. "Chernobyl disaster: Delayed announcement," Wikipedia, https://en.wikipedia .org/wiki/Chernobyl_disaster#Delayed_announcement.

6. "25 Years after Chernobyl, How Sweden Found Out," Sveriges Radio, April 22, 2011, https://sverigesradio.se/artikel/25-years-after-chernobyl-how-sweden-found-out; Richard D. North, "Living with Catastrophe," *Independent*, Sunday October 23, 2011, https://www.independent.co.uk/arts-entertainment/living-with-catastrophe-1524915 .html.

7. "Fukushima Daiichi Nuclear Disaster," Wikipedia, https://en.wikipedia.org/wiki/ Fukushima_Daiichi_nuclear_disaster.

8. Bodhana Kurylo, "The Role of Chernobyl in the Breakdown of the USSR," *Armstrong Undergraduate Journal of History* 6, no. 1, article 5 (2016), doi: 10.20429/aujh .2016.060105.

9. Mark Joseph Stern, "Did Chernobyl Cause the Soviet Union to Explode?," *Slate*, January 25, 2013, https://slate.com/technology/2013/01/chernobyl-and-the-fall-of-the

-soviet-union-gorbachevs-glasnost-allowed-the-nuclear-catastrophe-to-undermine
-the-ussr.html.

10. Stern, "Did Chernobyl Cause the Soviet Union to Explode?"

11. CHNP, "Visits to SSE CHNPP," https://visit.chnpp.gov.ua/; Chernobyl Tour, https://www.chernobyl-tour.com/english/48-one-day-trip-to-the-chernobyl-zone-and -prypyat-town.html.

12. CHNP, "Visits to SSE CHNPP," Conditions for Route "Basic Plus" and Terms of Service, https://visit.chnpp.gov.ua/page/terms_basic_plus.

13. Frank von Hippel, "Chernobyl Didn't Kill Nuclear Power," *Scientific American*, April 1, 2016, https://www.scientificamerican.com/article/chernobyl-didn-apos-t-kill -nuclear-power/.

14. Saburoh Midorikawa, Hiroyuki Miura, and Tomohiro Atsumi, "Characteristics of Strong Ground Motion from the 2011 Gigantic Tohoku, Japan Earthquake," http:// ares.tu.chiba-u.jp/peru/pdf/meeting/120817/M4_Midorikawa.pdf.

15. "Fukushima Daiichi Accident," World Nuclear Association, April 2021, https:// www.world-nuclear.org/information-library/safety-and-security/safety-of-plants/ fukushima-accident.aspx.

16. Ibid.

17. "Fukushima Daiichi nuclear disaster."

18. Ibid., see https://en.wikipedia.org/wiki/Fukushima_Daiichi_nuclear_disaster# 1991:_Backup_generator_of_Reactor_1_flooded.

19. "Putting Tsunami Countermeasures on Hold at Fukushima Nuke Plant 'Natural': ex-TEPCO VP," *Mainichi*, October 20, 2018, https://mainichi.jp/english/articles/ 20181020/p2a/00m/0na/018000c.

20. Martin Fackler, "Nuclear Disaster in Japan Was Avoidable, Critics Contend," *New York Times*, March 9, 2012, https://www.nytimes.com/2012/03/10/world/asia/critics -say-japan-ignored-warnings-of-nuclear-disaster.html.

21. "Nuclear power phase-out," Wikipedia, https://en.wikipedia.org/wiki/Nuclear _power_phase-out.

22. Ibid.

23. NEA, *Impacts of the Fukushima Daiichi Accident on Nuclear Development Policies, Nuclear Development* (Paris: OECD Publishing, 2017), https://doi .org/10.1787/9789264276192-en. or https://www.oecd-nea.org/ndd/pubs/2017/7212 -impacts-fukushima-policies.pdf.

24. Michael Corradini, "The Future of Nuclear Power after Fukushima" (PowerPoint presentation), American Nuclear Society, http://www.ans.org/about/officers/docs/ FutureOfNuclearPowerAfterFukushima.pdf.

25. Satoshi Sugiyama, "In Aftermath of Fukushima Triple Meltdown, Japan's Nuclear Industry Faces Fierce Headwind," *Japan Times*, March 11, 2019, https://www.world nuclearreport.org/In-aftermath-of-Fukushima-triple-meltdown-Japan-s-nuclear -industry-faces-fierce.html.

26. Tom Hals and Emily Flitter, "How Two Cutting Edge U.S. Nuclear Projects Bankrupted Westinghouse," Reuters, May 2, 2017, https://www.reuters.com/article/us -toshiba-accounting-westinghouse-nucle/how-two-cutting-edge-u-s-nuclear-projects -bankrupted-westinghouse-idUSKBN17Y0CQ.

27. Sugiyama, "In Aftermath of Fukushima Triple Meltdown."

28. NEA/IEA *Nuclear Energy 2015, IEA Technology Roadmaps* (Paris: OECD Publishing, 2015), https://doi.org/10.1787/9789264229938-en or https://www.oecd-nea.org/ndd/reports/2010/nea6962-nuclear-roadmap.pdf.

29. He Zuoxiu, "Chinese Nuclear Disaster 'Highly Probable' by 2030," China Dialogue, March 19, 2013, https://www.chinadialogue.net/article/show/single/en/5808-Chinese-nuclear-disaster-highly-probable-by-2-3-.

30. "Elevator: Pre-industrial Era," Wikipedia, https://en.wikipedia.org/wiki/Elevator?utm_medium=website&utm_source=archdaily.com#Pre-industrial_era.

31. "Elevator," *Encyclopedia Britannica*, https://www.britannica.com/technology/elevator-vertical-transport#ref90007.

32. "Fission vs. Fusion—What's the Difference?" Duke Energy Nuclear Education, January 30, 2013, https://nuclear.duke-energy.com/2013/01/30/fission-vs-fusion-whats-the-difference.

33. "Bubble Bobble Revolution," Wikipedia, https://en.wikipedia.org/wiki/Bubble_Bobble_Revolution.

34. "Map 256 Glitch," Pac-Man wiki, Fandom, https://pacman.fandom.com/wiki/Map_256_Glitch.

35. Ian Bogost, "Programmers: Stop Calling Yourselves Engineers," *Atlantic*, November 5, 2015, https://www.theatlantic.com/technology/archive/2015/11/programmers-should-not-call-themselves-engineers/414271/; Michael Davis, "Will Software Engineering Ever Be Engineering?," *Communications of the ACM*, 54, no. 11 (November 2011): 32–34, doi:10.1145/2018396.2018407 or https://cacm.acm.org/magazines/2011/11/138199-will-software-engineering-ever-be-engineering/fulltext; Kathleen Vignos, "Programmers, Let's Earn the Right to Be Called Engineers," *Wired*, November 15, 2015, https://www.wired.com/2015/11/programmers-lets-earn-the-right-to-be-called-engineers/.

36. Mariella Moon, "Software Bug Forced Nest Thermostats Offline," Engadget, January 14, 2016, https://www.engadget.com/2016/01/14/nest-software-bug/.

37. Devin Coldewey, "Empty Nest: Glitch Disables Smart Thermostats Nationwide," NBC News, January 14, 2016, https://www.nbcnews.com/tech/gadgets/empty-nest-glitch-disables-smart-thermostats-nationwide-n496641.

38. James Titcomb, "Nest Thermostat Owners Left without Heating after Software Glitch," *Telegraph*, January 14, 2016, https://www.telegraph.co.uk/technology/news/12099033/Nest-thermostat-owners-left-without-heating-after-software-glitch.html.

39. Bart Jansen, "Samsung Galaxy Note 7 Banned on all U.S. Flights Due to Fire Hazard," *USA Today*, October. 14, 2016, https://www.usatoday.com/story/news/2016/10/14/dot-bans-samsung-galaxy-note-7-flights/92066322/; Alyssa Newcomb, "Samsung Finally Explains the Galaxy Note 7 Exploding Battery Mess," NBC News, January 22, 2017, https://www.nbcnews.com/tech/tech-news/samsung-finally-explains-galaxy-note-7-exploding-battery-mess-n710581.

40. "Samsung Galaxy Note 7," Wikipedia, https://en.wikipedia.org/wiki/Samsung_Galaxy_Note_7.

41. "How Samsung Tripped on Quality Control in Its Rush to Beat Apple," Reuters, September 6, 2016, https://fortune.com/2016/09/06/samsung-apple-galaxy-note -quality/.

42. Natalie Kitroeff and David Gelles, "Claims of Shoddy Production Draw Scrutiny to a Second Boeing Jet," *New York Times*, April 20, 2019, https://www.nytimes .com/2019/04/20/business/boeing-dreamliner-production-problems.html.

43. Chris Isidore, "A New Software Glitch Was Discovered on Boeing's 737 Max," CNN, February 6, 2020, https://www.cnn.com/2020/02/06/business/boeing-737-max -software/index.html; "Boeing 737 MAX certification," Wikipedia, https://en.wiki pedia.org/wiki/Boeing_737_MAX_certification#US_Congress_inquiries; Gregory Wallace and Rene Marsh, "Boeing Pilots Discussed 'Fundamental Issues' with 737 MAX in Internal Messages," CNN, October 18, 2019, https://www.cnn.com/2019/10/18/ politics/boeing-737-max-faa-documents/index.html.

44. Simson Garfinkel, "History's Worst Software Bugs," *Wired*, November 8, 2005, https://www.wired.com/2005/11/historys-worst-software-bugs/.

45. Sara Baase, *A Gift of Fire* (New York: Pearson Prentice Hall, 2008).

46. Nancy G. Leveson and Clark S Turner, "An Investigation of the Therac-25 Accidents," *IEEE Computer* 26, no. 7: 18–41, doi:10.1109/MC.1993.274940.

47. "Therac 25," Wikipedia, https://en.wikipedia.org/wiki/Therac-25.

48. "Theodore Cooper," Wikipedia, https://en.wikipedia.org/wiki/Theodore_Cooper.

49. "Norman Medal Past Award Winners," American Society of Civil Engineers, https://www.asce.org/career-growth/awards-and-honors/norman-medal/norman -medal-past-award-winners.

50. Cynthia Pearson and Norbert Delatte "Collapse of the Quebec Bridge, 1907," *Journal of Performance of Constructed Facilities* 20, no. 1 (2006).

51. Garson O'Toole, "With Great Power Comes Great Responsibility," Quote Investigator, July 23, 2015, https://quoteinvestigator.com/2015/07/23/great-power/.

52. "The Day the Sun Brought Darkness," NASA, March 13, 2009, https://www.nasa .gov/topics/earth/features/sun_darkness.html.

53. "Hydro-Québec Successfully Manages the Effects of Solar Storms," Hydro Quebec Newsroom, September 12, 2014, http://news.hydroquebec.com/en/news/153/ hydro-quebec-successfully-manages-the-effects-of-solar-storms/.

54. William D. Middleton, *The Bridge at Québec* (Bloomington: Indiana University Press, 2001).

55. "The Judgement," https://web.archive.org/web/20060206204605/; http://www.civ eng.carleton.ca/ECL/reports/ECL270/Judgement.html.

56. "List of Bridge Failures," Wikipedia, https://en.wikipedia.org/wiki/List_of_bridge _failures.

57. "Yarmouth Suspension Bridge," Wikipedia, https://en.wikipedia.org/wiki/Yar mouth_suspension_bridge.

58. "Joseph John Scoles," Wikipedia, https://en.wikipedia.org/wiki/Joseph_John _Scoles.

59. Tabea Tietz, "The Broughton Suspension Bridge and the Resonance Disaster," *SciHi Blog*, April 12, 2020, http://scihi.org/broughton-suspension-bridge-resonance -disaster/.

60. "Gasconade Bridge Train Disaster," Wikipedia, https://en.wikipedia.org/wiki/Gasconade_Bridge_train_disaster.

61. Desjardins Canal Disaster, Local History & Archives, "Reproduction of 'The Calamitous Railroad Accident at Burlington Bridge! Over the Des Jardines Canal, Canada,'" *Frank Leslie's Illustrated Newspaper*, April 4, 1857: 277–78, https://lha.hpl.ca/articles/desjardins-canal-disaster.

62. Mike Hoey, "Hidden History: Train Wrecks," CNY News, Jun 6, 2017, https://www.cnyhomepage.com/news/hidden-history-train-wrecks/.

63. "Ashtabula River railroad disaster," Wikipedia, https://en.wikipedia.org/wiki/Ashtabula_River_railroad_disaster.

64. Wikipedia, https://en.wikipedia.org/wiki/King_Street_Bridge_(Melbourne).

65. "Mianus River Bridge," Wikipedia, https://en.wikipedia.org/wiki/Mianus_River_Bridge.

66. Jack Taylor, "Colorado History: The Eden Train Wreck," 99.9 KEKB, October 16, 2018, https://kekbfm.com/colorado-history-the-eden-train-wreck/?utm_source=tsmclip&utm_medium=referral, https://kekbfm.com/colorado-history-the-eden-train-wreck/.

67. "Collapse of the Honeymoon Bridge," Niagara Falls Museums, https://niagarafallsmuseums.ca/discover-our-history/history-notes/honeymoonbridge.aspx.

68. "The Collapsing Bridge," https://www.youtube.com/watch?v=kT1yNERACtQ.

69. Ethan Siegel, "Science Busts the Biggest Myth Ever about Why Bridges Collapse," *Forbes*, May 24, 2017, https://www.forbes.com/sites/startswithabang/2017/05/24/science-busts-the-biggest-myth-ever-about-why-bridges-collapse/#aef0c2a1f4c0.

70. Nanette South Clark, "The Duplessis Bridge Collapse—January 31, 1951," An Engineer's Aspect, http://anengineersaspect.blogspot.com/2013/04/the-duplessis-bridge-collapse-january.html.

71. "List of Bridge Failures."

72. "Sunshine Skyway Bridge: 1980 Collapse," Wikipedia, https://en.wikipedia.org/wiki/Sunshine_Skyway_Bridge#1980_collapse.

73. "Cypress Street Viaducts," Engineering.com, October 13, 2006, https://www.engineering.com/Library/ArticlesPage/tabid/85/ArticleID/73/Cypress-Street-Viaducts.aspx.

74. Bernard J. Feldman, "The Nimitz Freeway Collapse," *The Physics Teacher* 42 (October 2004), http://www.umsl.edu/~feldmanb/pdfs/Nimitz_Freeway.pdf.

75. Brett Jackson, "Replacing Oakland's Cypress Freeway," *Public Roads* 61 no. 5 (March/April 1998), https://www.fhwa.dot.gov/publications/publicroads/98marapr/cypress.cfm.

76. "Cypress Street Viaducts."

77. Jackson, "Replacing Oakland's Cypress Freeway."

78. "China Bridge Collapse Caught on Camera," YouTube, https://www.youtube.com/watch?v=knaNXcKS-qM.

79. "Shocking Video Shows Greece Bridge Collapse on Live TV," *Today*, NBC News, February 27, 2019, https://www.today.com/video/shocking-video-shows-greece-bridge-collapse-on-live-tv-1449085507653.

80. "The Moment a Bridge Collapsed into a Harbour in Taiwan," YouTube, https://www.youtube.com/watch?v=ZC9_X2mEhqg.

81. *I Am Legend*, Warner Bros. https://www.warnerbros.com/movies/i-am-legend/.

82. Erin McCarthy, "*I Am Legend*'s Junk Science: Hollywood Sci-Fi vs. Reality," *Popular Mechanics*, December 14, 2007, https://www.popularmechanics.com/culture/movies/a2399/4236920/.

83. "Brooklyn Bridge—*I Am Legend*—New York City, New York, USA," Waymarking.com, https://www.waymarking.com/waymarks/WMWAKD_Brooklyn_Bridge_I_am_legend_New_York_City_New_York_USA.

84. "The Brooklyn Bridge," Seven Wonders of the Age of Steam, http://www.unmuseum.org/7wonders/brooklyn_bridge.htm.

TERRORIST ATTACKS

1. Grace Chen, "Longer Lunches, Smarter Students? The Controversy of 10 Minute or 1 Hour Lunch Periods," *Public School Review*, December 4, 2020, https://www.publicschoolreview.com/blog/longer-lunches-smarter-students-the-controversy-of-10-minute-or-1-hour-lunch-periods.

2. Meghan Keneally, "The 11 Mass Deadly School Shootings That Happened since Columbine," ABC News, April 19, 2019, https://abcnews.go.com/US/11-mass-deadly-school-shootings-happened-columbine/story?id=62494128.

3. Nicholas D. Kristof, "Martyrs, Virgins and Grapes," *New York Times*, August 4, 2004, https://www.nytimes.com/2004/08/04/opinion/martyrs-virgins-and-grapes.html.

4. Veronica Ward, "What Do We Know about Suicide Bombing?," *Politics and the Life Sciences* 37, no. 1 (May 2, 2018): 88–112, https://www.cambridge.org/core/journals/politics-and-the-life-sciences/article/what-do-we-know-about-suicide-bombing/9643234008F6BA05DA337C91696B5C75.

5. "Use of Children as 'Human Bombs' Rising in North East Nigeria," Unicef, August 22, 2017, https://www.unicef.org/press-releases/use-of-children-human-bombs-rising-north-east-nigeria.

6. "Amerithrax or Anthrax Investigation," FBI, https://www.fbi.gov/history/famous-cases/amerithrax-or-anthrax-investigation.

7. S. E. Flynn, *America the Vulnerable: How Our Government Is Failing to Protect Us from Terrorism* (New York: Harper Perennial, 2005).

8. Flynn, *America the Vulnerable*, 126; A. K. Erenler, M. Güzel, and A. Baydin, "How Prepared Are We for Possible Bioterrorist Attacks: An Approach from Emergency Medicine Perspective," *Scientific World Journal*, 7849863 (2018), https://doi.org/10.1155/2018/7849863; D. Shapiro, "Surge Capacity for Response to Bioterrorism in Hospital Clinical Microbiology Laboratories," *Journal of Clinical Microbiology* 41, no. 12 (2020) https://journals.asm.org/doi/full/10.1128/jcm.41.12.5372-5376.2003.

9. Flynn, *America the Vulnerable*.

10. Stephen Flynn, *The Edge of Disaster* (New York: Random House 2007).

11. "Port of Long Beach," Wikipedia, https://en.wikipedia.org/wiki/Port_of_Long_Beach.

12. Flynn, *America the Vulnerable*, 54.

13. "Chemical weapons in World War I," Wikipedia, https://en.wikipedia.org/wiki/Chemical_weapons_in_World_War_I.

14. "Regulatory Information by Topic: Toxic Substances," EPA, https://www.epa.gov/regulatory-information-topic/regulatory-information-topic-toxic-substances; "DOT Announces Final Rule to Strengthen Safe Transportation of Flammable Liquids by Rail," US Department of Transportation, May 1, 2015, https://www.transportation.gov/briefing-room/dot-announces-final-rule-strengthen-safe-transportation-flamm able-liquids-rail-1.

15. "Chemical Sector Regulatory Authorities and Executive Orders," Cybersecurity and Infrastructure Security Agency, https://www.cisa.gov/chemical-sector-regulator y-authorities-and-eos.

16. "Summary of the Toxic Substances Control Act," EPA, https://www.epa.gov/laws-regulations/summary-toxic-substances-control-act.

17. J. Tollefson, "Why the Historic Deal to Expand US Chemical Regulation Matters," *Nature* 534 (2016), 18–19, https://doi.org/10.1038/nature.2016.19973.

18. Eben Kaplan, "Targets for Terrorists: Chemical Facilities," Council on Foreign Relations, December 11, 2006, https://www.cfr.org/backgrounder/targets-terrorists -chemical-facilities; Flynn, *America the Vulnerable*, 119.

19. Flynn, *America the Vulnerable*, 120.

20. Bill Gasperetti, "Security Since 9/11: Creating the Maritime Transportation Security Act and the ISPS Code," February 9, 2018, Homeland Security Today, https://www .hstoday.us/uncategorized/security-since-9-11-creating-maritime-transportation -security-act-isps-code/.

21. "Chemical Facility Anti-Terrorism Standards (CFATS) Program," Cybersecurity and Infrastructure Agency, https://www.cisa.gov/chemical-facility-anti-terrorism -standards and https://www.cisa.gov/sites/default/files/publications/fs_cfats-overview -508.pdf.

22. Protecting and Securing Chemical Facilities from Terrorist Attacks Act of 2014, Public Law 113–254—Dec. 18, 2014, US Congress, https://www.govinfo.gov/content/pkg/PLAW-113publ254/pdf/PLAW-113publ254.pdf; "CFATS Statutes," Cybersecurity and Infrastructure Security Agency, https://www.cisa.gov/cfats-statutes.

23. Alex Scott, "Terrorist Attack Hits U.S.-Owned Chemical Plant in France," *Chemical & Engineering News*, June 26, 2015, https://cen.acs.org/articles/93/web/2015/06/Terrorist-Attack-Hits-US-Owned.html.

24. Mark Griffiths and Jill Kickul, "The Tragedy of the Commons," *Ivey Business Journal*, March/April 2013, https://iveybusinessjournal.com/publication/the-tragedy -of-the-commons/.

25. Garrett Hardin, "The Tragedy of the Commons: The Population Problem Has No Technical Solution; It Requires a Fundamental Extension in Morality," *Science* 162, no. 3859 (December 13, 1968), 1243–48, doi: 10.1126/science.162.3859.1243; "Tragedy of the Commons: Modern Commons," Wikipedia, https://en.wikipedia.org/wiki/Tragedy_of_the_commons#Modern_commons.

26. Flynn, *America the Vulnerable*, 55.

27. "Quick Facts 2016," National Center for Statistics and Analysis, National Highway Traffic Safety Administration, https://crashstats.nhtsa.dot.gov/Api/Public/View Publication/812451.

28. "Gun Violence in the United States: Mass Shootings," Wikipedia, https://en.wiki pedia.org/wiki/Gun_violence_in_the_United_States#Mass_shootings.

29. "Post Tropical Storm Sandy Event Overview," Office of the New Jersey State Climatologist, Rutgers University, https://climate.rutgers.edu/stateclim/?target=sandy; "Super Storm Sandy Report," National Weather Service, NOAA, https://www.weather .gov/phi/sandyreport.

30. Doyle Rice and Alia E. Dastagir, "One Year After Sandy, 9 Devastating Facts," *USA Today*, October 29, 2013, https://www.usatoday.com/story/news/nation/2013/10/29/ sandy-anniversary-facts-devastation/3305985/.

31. "Hurricane Sandy: United States 2, Wikipedia, https://en.wikipedia.org/wiki/ Hurricane_Sandy#United_States_2.

32. Sabina Zawadzki and Anna Louie Sussman, "Analysis—Six Months after Sandy, NY Fuel Supply Chain Still Vulnerable," Reuters, April 30, 2013, https://www.reuters .com/article/uk-usa-sandy-fuel/analysis-six-months-after-sandy-ny-fuel-supply-chain -still-vulnerable-idUKBRE93T0DI20130430.

33. A. Kenward, D. Yawitz, and U. Raja, *Sewage Overflows from Hurricane Sandy* (Princeton, NJ: Climate Central, 2013), 10. https://www.climatecentral.org/pdfs/Sewage.pdf.

34. David Sandalow, "Hurricane Sandy and Our Energy Infrastructure," Department of Energy, November 30, 2012, https://www.energy.gov/articles/hurricane-sandy-and -our-energy-infrastructure.

35. "Twenty Years Ago, Québec Was Battered by an Ice Storm," *Hydro Québec News*, January 5, 2018, http://news.hydroquebec.com/en/press-releases/1313/twenty-years -ago-quebec-was-battered-by-an-ice-storm/; see https://www.google.com/search? client=firefox-b-1-d&q=1998+population+of++quebec.

36. "January 1998 North American ice storm," Wikipedia, https://en.wikipedia.org/ wiki/January_1998_North_American_ice_storm.

37. Laurie Fagan, "Farmer Reflects on Devastating Loss in 1998 Ice Storm," CBC News, January 5, 2018, https://www.cbc.ca/news/canada/ottawa/farmer-dairy-eastern -ontario-ice-storm-cows-1.4473431.

38. "Ice Storm Devastates Farm Livestock," CBC News, April 1, 1999, https://www .cbc.ca/news/canada/ice-storm-devastates-farm-livestock-1.165771.

39. Lesley-Ann Dupigny-Giroux, "Impacts and Consequences of the Ice Storm of 1998 for the North American North-East," Royal Meteorological Society, April 30, 2012, https://doi.org/10.1002/j.1477-8696.2000.tb04012.x.

40. Eugene L. Lecomte, James W. Russell, Alan W. Pang, *Ice Storm '98* (Toronto, ON, and Boston: Institute for Catastrophic Loss Reduction and Institute for Business and House Security, 1998), https://www.iclr.org/wp-content/uploads/PDFS/1998_ice _storm_english.pdf.

41. Dupigny-Giroux, "Impacts and Consequences of the Ice Storm of 1998."

42. Darian Lorat, "What Happens If You Don't Milk Cows?," Farm It Yourself, https://farmityourself.com/what-happens-if-you-dont-milk-cows/.

43. "Mastitis in dairy cattle," Wikipedia, https://en.wikipedia.org/wiki/Mastitis_in_dairy_cattle.

44. Fagan, "Farmer Reflects."

45. Laura Neilson Bonikowsky and Niko Block, "Tempête de verglas de 1998," *L'Encyclopédie canadienne*, November 8, 2012, https://www.thecanadianencyclopedia.ca/fr/article/la-crise-du-verglas-1998.

46. R. Dagger and Terence Ball, "Communism." *Encyclopedia Britannica*, https://www.britannica.com/topic/communism#ref276323.

47. Rokas Laurinavičius, "The First McDonald's in Moscow Opened in 1990, and These 27 Pics Show How Insane It All Was," BoredPanda, https://www.boredpanda.com/first-mcdonald-restaurant-opens-soviet-union-moscow-russia-1990/?utm_source=google&utm_medium=organic&utm_campaign=organic.

ANNOYING DOOMSDAY SCENARIOS

1. Damien Fletcher, "After the Asparagus Fortune Teller, Here Are 13 Other Wacky Ways of Predicting Future," *Mirror*, January 20, 2012, https://www.mirror.co.uk/news/weird-news/after-the-asparagus-fortune-teller-here-159749.

2. "COVID-19—China," World Health Organization, January 5, 2020, https://www.who.int/emergencies/disease-outbreak-news/item/2020-DON229.

3. *"A World at Risk": Annual Report on Global Preparedness for Health Emergencies*, Global Preparedness Monitoring Board, 2019, https://www.gpmb.org/annual-reports/overview/item/2019-a-world-at-risk.

4. "Apocalypticists," Wikipedia, https://en.wikipedia.org/wiki/Category:Apocalypticists.

5. Andrew Clayman, "The Cracker Jack Co., est. 1871," Made in Chicago Museum, https://www.madeinchicagomuseum.com/single-post/cracker-jack-co/.

INTERLUDE

1. "Disaster film: 1970s," Wikipedia, https://en.wikipedia.org/wiki/Disaster_film#1970s.

2. Mr. Cranky, *Shadowculture's Mr. Cranky Presents: The 100 Crankiest Movie Reviews Ever* (Authorhouse, 2004); Also, https://web.archive.org/web/*/http://www.mrcranky.com.

3. "20 Largest Earthquakes in the World," USGS, https://www.usgs.gov/natural-hazards/earthquake-hazards/science/20-largest-earthquakes-world?qt-science_center_objects=0#qt-science_center_objects.

4. "Can 'MegaQuakes' Really Happen Like a Magnitude 10 or Larger?" USGS, https://www.usgs.gov/faqs/can-megaquakes-really-happen-a-magnitude-10-or-larger?qt-news_science_products=0#qt-news_science_products.

5. "Disaster Wars: Earthquake vs Tsunami, User Reviews," IMDb, https://www.imdb.com/title/tt2571362/reviews?ref_=tt_ov_rt.

6. "Attack on the Golden Gate Bridge—Supercut," YouTube, https://www.youtube.com/watch?v=jw7K6zg3fhI.

7. "San Andreas," http://www.sanandreasmovie.com/.

8. "California Dreamin'—Sia," YouTube, https://www.youtube.com/watch?v=q-rx 9RuGlDM.

9. "The Asylum," Wikipedia, https://en.wikipedia.org/wiki/The_Asylum.

10. "The ONLY Reason to Watch Megafault," YouTube, https://www.youtube.com/ watch?v=R14sMMgUDKs.

11. "Golden Raspberry Awards," Wikipedia, https://en.wikipedia.org/wiki/Golden _Raspberry_Awards.

12. "*Skjelvet*," The Numbers, https://www.the-numbers.com/movie/Skjelvet-(Norway) -(2018)#tab=summary.

13. "Saved by the Belle," Wikipedia, https://en.wikipedia.org/wiki/Saved_by_the _Belle.

14. Steven Short, "San Francisco's Two Official Songs or, The Day Tony Bennett Hid in His Hotel," KALW Public Media, February 2011, https://www.kalw.org/show/ crosscurrents/2012-02-14/san-franciscos-two-official-songs-or-the-day-tony-bennett -hid-in-his-hotel.

THE WONDERFUL ABILITY TO FORGET

1. "Alquist-Priolo Earthquake Fault Zones," California Department of Conservation, https://www.conservation.ca.gov/CGS/rghm/ap/.

2. EQ Zapp: "California Earthquake Hazards Zone Application," California Department of Conservation, https://www.conservation.ca.gov/cgs/geohazards/eq-zapp.

3. "Alquist-Priolo Earthquake Fault Zones."

4. Robert Sanders, "Memorial Stadium Gets Renovated with Help from Berkeley's Own," *Berkeley News*, September 1, 2011, https://news.berkeley.edu/2011/09/01/ memorial-stadium-renovated-with-help-of-berkeleys-own/.

5. Horst Rademacher, "The Hayward Fault at the Campus of the University of California, Berkeley: A Guide to a Brief Walking Tour," Seismological Laboratory, University of California, Berkeley, http://earthquakes.berkeley.edu/docs/HF_Tour _Stadium-1.1-Protected.pdf.

6. https://www.zillow.com/homes/for_sale/Daly-City-CA/pmf,pf_pt/31163_rid/ globalrelevanceex_sort/37.665634,-122.487461,37.663931,-122.490345_rect/18_zm/.

7. Katherine Wiles, "The Housing Market Is So Hot, a Burnt-Out Bay Area Home Just Sold for $1 Million in Cash," Market Watch, August 12, 2021, https://www.mar-ketwatch.com/story/the-housing-market-is-so-hot-a-burnt-out-bay-area-home-is-dr awing-cash-bids-above-850-000-11628791388.

8. Alix Martichoux, "691,000 People Moved Out of California Last Year. Here's Where They Went," SFGATE, November 5, 2019, https://www.sfgate.com/expensive -san-francisco/article/move-california-where-to-go-cheap-states-best-14811246.php.

9. Mark DiCamillo, "Release #2019-08: Leaving California: Half of State's Voters Have Been Considering This; Republicans and Conservatives Three Times as likely as Democrats and Liberals to Be Giving Serious Consideration to Leaving the State," Institute of Governmental Studies, UC Berkeley, 2019, https://escholarship.org/uc/ item/96j2704t.

10. "Why Are So Many People Leaving California?," Quora, https://www.quora.com/ Why-are-so-many-people-leaving-California.

11. Stacy R. Stewart, *Tropical Cyclone Report, Hurricane Matthew (AL142016), 28 September—9 October 2016*, National Hurricane Center (April 7, 2017), 4, https://www.nhc.noaa.gov/data/tcr/AL142016_Matthew.pdf.

12. Jeffrey Kluger, "Why We Keep Ignoring Even the Most Dire Climate Change Warnings," *Time*, October 8, 2018, https://time.com/5418690/why-ignore-climate-change-warnings-un-report/.

13. David Heath, "Contesting the Science of Smoking," *Atlantic*, May 4, 2016, https://www.theatlantic.com/politics/archive/2016/05/low-tar-cigarettes/481116/; Ryan Jaslow, "Big Tobacco Kept Cancer Risk in Cigarettes Secret Study," CBS News, September 30, 2011, https://www.cbsnews.com/news/big-tobacco-keptcancer-risk-in-cigarettes-secret-study/.

14. "Smoking and Tobacco Use, Data and Statistics," Centers for Disease Control and Prevention, https://www.cdc.gov/tobacco/data_statistics/index.htm.

15. C.-H. Geschwind, *California Earthquakes: Science, Risk, and the Politics of Hazard Mitigation* (Baltimore: Johns Hopkins University Press, 2008), 20–21.

16. Memorial Web Page for the 1926 Great Miami Hurricane, National Weather Service, Weather Forecast Office Miami, Florida, https://www.weather.gov/mfl/miami_hurricane.

17. Elliott Mittler, "A Case Study of Florida's Homeowners' Insurance Since Hurricane Andrew," Working Paper 96, Natural Hazards Center, University of Colorado, 1997, https://hazards.colorado.edu/archive/publications/wp/wp96.html.

18. "Miami Hurricane Damage," Wikipedia, https://en.wikipedia.org/wiki/1926_Miami_hurricane#/media/File:1926_Miami_Hurricane_damage.jpg; Memorial Web Page for the 1926 Great Miami Hurricane.

19. Geschwind, *California Earthquakes*, 105–6.

20. Ibid., 115.

21. Ibis., 189.

22. "Funding Federal-Aid Highways, Publication No. FHWA-PL-17-011," Office of Policy and Governmental Affairs, Federal Highway Administration, U.S. Department of Transportation, January 2017, https://www.fhwa.dot.gov/policy/olsp/fundingfederalaid/07.cfm.

23. "Gasoline Prices Adjusted for Inflation," U.S. Inflation Calculator, https://www.usinflationcalculator.com/gasoline-prices-adjusted-for-inflation/.

24. "The Highway Trust Fund Explained," Peter G. Peterson Foundation, August 14, 2020, https://www.pgpf.org/budget-basics/budget-explainer-highway-trust-fund.

25. Samuel Barradas, "Foreign Company Now Owns Six Major US Tolls Roads, Truckers Report," https://www.thetruckersreport.com/foreign-company-now-owns-six-major-us-tolls-roads/.

26. Michael A. Cohen, "Bloated Defense Budget Passes Easily but Congress Fights over Safety Net Programs," MSNBC, October 1, 2021.

27. H.R.3684—Infrastructure Investment and Jobs Act 117th Congress (2021–2022), Congress.gov, https://www.congress.gov/bill/117th-congress/house-bill/3684.

28. *2017 Infrastructure Report Card, A Comprehensive Assessment of America's Infrastructure* (Reston, VA: American Society of Civil Engineers, 2017), 7–8, https://2017.

infrastructurereportcard.org/wp-content/uploads/2019/02/Full-2017-Report-Card
-FINAL.pdf.

29. "United States Housing Starts," Trading Economics, https://tradingeconomics
.com/united-states/housing-starts; "United States New Home Sales," Trading Economics, https://tradingeconomics.com/united-states/new-home-sales.

30. Christine Bartsch, "From the Global Stage to Your Backyard: 10 Factors That
Influence Real Estate Property Values," HomeLight, July 31, 2019, https://www.home
light.com/blog/real-estate-property-value/.

31. Geschwind, *California Earthquakes*, 17.

32. "Study: 1868 Earthquake Larger Than Once Thought," *Eastbay Times*, February 7,
2007, https://www.eastbaytimes.com/2007/02/07/study-1868-earthquake-larger-than
-once-thought/.

33. Geschwind, *California Earthquakes*, 17.

34. John H. Bennett, "Vacaville-Winters Earthquakes . . . 1892," *California Geology*,
40, no. 4 (April 1987), http://www.johnmartin.com/earthquakes/eqpapers/00000068
.htm.

35. Geschwind, *California Earthquakes*, 106.

36. Ibid., 17–18.

37. "Oversize and Overweight Fines and Penalties by State," Oversize.io, https://over
size.io/regulations/oversize-overweight-fines-by-state.

38. Tripti Lahiri, "Mexico Held an Earthquake Drill Two Hours before Its Latest
Deadly Quake Hit," Quartz, September 20, 2017, https://qz.com/1082239/mexico
-held-an-earthquake-drill-two-hours-before-its-latest-deadly-quake-hit/.

39. "1985 Mexico City Earthquake," History.com, November 9, 2009, https://www
.history.com/topics/natural-disasters-and-environment/1985-mexico-city-earthquake.

40. "Carrying Out an Effective Fire Drill," Euro Fire Protection and Maintenance Service, May 9, 2013, http://www.eurofireprotection.com/blog/carrying-out-an-effective
-fire-drill/.

41. David Adler, "The Mexico City Earthquake, 30 Years On: Have the Lessons Been
Forgotten?," *Guardian*, September 18, 2015, https://www.theguardian.com/cities/2015/
sep/18/mexico-city-earthquake-30-years-lessons.

42. "Puebla earthquake," Wikipedia, https://en.wikipedia.org/wiki/2017_Puebla_earth
quake.

43. Moisés Juárez-Camarena, Gabriel Auvinet-Guichard, and Edgar Méndez-
Sánchez, "Geotechnical Zoning of Mexico Valley Subsoil," *Ingeniería, Investigación y
Tecnología* 17, no. 3 (2016), 297–308, https://doi.org/10.1016/j.riit.2016.07.001.

44. "How Earthquakes Break the Speed Limit," Seismology Lab, UC Berkeley, March
8, 2019, https://seismo.berkeley.edu/blog/2019/03/08/how-earthquakes-break-the
-speed-limit.html.

45. Gerardo Suárez, J. M. Espinosa-Aranda, Armando Cuéllar, Gerardo Ibarrola,
Armando García, Martín Zavala, Samuel Maldonado, and Roberto Islas, "A Dedicated Seismic Early Warning Network: The Mexican Seismic Alert System (SAS-
MEX)," *Seismological Research Letters* 89, no. 2A (2018): 382–91. doi: https://doi
.org/10.1785/0220170184, https://pubs.geoscienceworld.org/ssa/srl/article-abstract/
89/2A/382/527311/A-Dedicated-Seismic-Early-Warning-Network-The.

46. Kathryn Miles, "Mexico City Earthquake Is a Warning for Americans," *Time*, September 21, 2017, https://time.com/4952173/mexico-city-earthquake-warning -americans/.

47. *Smyth-Fernwald Property, Historic Structures Report* (Siegel & Strain Architects, March 2011), https://capitalstrategies.berkeley.edu/sites/default/files/hsr_smythfern wald_march2011.pdf.

48. Doris Sloan, D. Wells, Glenn Borchardt, John Caulfield, D. M. Doolin, J. Eidinger, Lind Gee, Russell W. Graymer, Peggy Hellweg, Alan L. Kropp, James Lienkaemper, Charles Rabamad, N. Sitar, Heidi D. Stenner, Stephen Tobriner, David Tsztoo, and M. L. Zoback, "The Hayward Fault," *GSA Field Guides*, https://doi .org/10.1130/2006.1906SF(17) and https://earthquake.usgs.gov/earthquakes/events/ 1868calif/pdf/fld007_17e.pdf.

49. Tracey Taylor, "'Temporary' 1945 UC Berkeley Housing Demolished" (video), Berkeley Side, March 7, 2013, https://www.berkeleyside.com/2013/03/07/video -temporary-1945-cal-housing-units-demolished.

AIRPORT PROCTOLOGISTS

1. My eight- and ten-year-old sons enjoyed driving that 747 on a trip to Japan.

2. Happened to me in the 1990s.

3. Elliott Neal Hester, "Cockpit Assault," *Salon*, April 8, 2000, https://www.salon .com/2000/04/08/cockpits/.

4. "First Hijack of an Aircraft," Guinness World Records, https://www.guinness worldrecords.com/world-records/first-hijack-of-an-aircraft/.

5. "List of aircraft hijackings," Wikipedia, https://en.wikipedia.org/wiki/List_of_air craft_hijackings.

6. "Aircraft hijacking," Wikipedia, https://en.wikipedia.org/wiki/Aircraft_hijacking.

7. Ibid.

8. Ibid.

9. "All Nippon Airways Flight 61," Wikipedia, https://en.wikipedia.org/wiki/All_Nip pon_Airways_Flight_61.

10. "TWA Flight 847," Wikipedia, https://en.wikipedia.org/wiki/TWA_Flight_847.

11. Jack Nelson, "Reagan Rules Out negotiations with Muslim Hijackers," *Los Angeles Times*, June 19, 1985, https://www.latimes.com/archives/la-xpm-1985-06-19-mn -9185-story.html.

12. "Lebanese Civil War," Wikipedia, https://en.wikipedia.org/wiki/Lebanese_Civil _War.

13. "Air France Flight 8969," Wikipedia, https://en.wikipedia.org/wiki/Air_France _Flight_8969.

14. Christophe Cornevin, "54 heures d'angoisse, 16 minutes de guerre," *Le Figaro*, http://www.lefigaro.fr/assets/marignane/.

15. Carole Guéchi, "Air France reprend ses vols vers Alger," *Le Parisien*, June 28, 2003, https://www.leparisien.fr/economie/air-france-reprend-ses-vols-vers-alger-28 -06-2003-2004207031.php.

16. Jonathan V. Last, "The Foreseeable Past," *Washington Examiner*, May 3, 2004, https://www.washingtonexaminer.com/weekly-standard/the-foreseeable-past.

17. Hester, "Cockpit assault."

18. Ibid.

19. "September 11 Attacks," Wikipedia, https://en.wikipedia.org/wiki/September_11
_attacks.

20. Last, "The Foreseeable Past."

21. "Southern Airways Flight 49," Wikipedia, https://en.wikipedia.org/wiki/Southern
_Airways_Flight_49.

22. Last, "The Foreseeable Past."

23. "Richard Reid," Wikipedia, https://en.wikipedia.org/wiki/Richard_Reid#Bombing
_attempt_on_American_Airlines_Flight_63.

24. "Umar Farouk Abdulmutallab," Wikipedia, https://en.wikipedia.org/wiki/Umar
_Farouk_Abdulmutallab#Attack.

25. "Full body scanner," Wikipedia, https://en.wikipedia.org/wiki/Full_body_scanner
#History_(US).

26. Abraham Tekippe, "How to Hijack an Airplane in 3 Seconds," *Atlantic*, August 16,
2011, https://www.theatlantic.com/national/archive/2011/08/how-to-hijack-an-air
plane-in-3-seconds/243631/.

27. David Kerley and Jeffrey Cook, "TSA Fails Most Tests in Latest Undercover
Operation at US Airports," ABC News, November 9, 2017, https://abcnews.go.com/
US/tsa-fails-tests-latest-undercover-operation-us-airports/story?id=51022188.

28. Eric Katz, "One in Five TSA Screeners Quits within Six Months," Government
Executive, April 3, 2019, https://www.govexec.com/pay-benefits/2019/04/
one-four-tsa-screeners-quits-within-six-months/156045/.

29. Brian Naylor, "TSA Officers among Lowest Paid of Federal Workers," NPR, January
26, 2010, https://www.npr.org/templates/story/story.php?storyId=122948752.

30. "2004 Madrid train bombings," Wikipedia, https://en.wikipedia.org/wiki/2004
_Madrid_train_bombings.

31. Ron Nixon, "With Its Focus on Air Travel, U.S. Leaves Trains Vulnerable to
Attack, Experts Say," *New York Times*, December 11, 2015, https://www.nytimes.com/
2015/12/12/us/us-trains-vulnerable-to-attack-security-experts-say.html.

32. Ibid.

33. The Literature Page, Homer: *Iliad*, Book XVII, http://www.literaturepage.com/
read/theiliad-249.html.

34. "1989 World Series," Wikipedia, https://en.wikipedia.org/wiki/1989_World
_Series#The_Loma_Prieta_earthquake.

35. Thomas M. Brocher, Jack Boatwright, James J. Lienkaemper, Carol S. Prentice,
David P. Schwartz, and Howard Bundock, "The Hayward Fault—Is It Due for a Repeat
of the Powerful 1868 Earthquake?," USGS, October 17, 2018, https://www.usgs.gov/
news/hayward-fault-it-due-a-repeat-powerful-1868-earthquake?qt-news_science
_products=3#qt-news_science_products.

36. Shane T. Detweiler and Anne M. Wein, *The HayWired Earthquake Scenario*,
Scientific Investigations Report 2017-5013, https://pubs.er.usgs.gov/publication/sir
20175013.

37. East Bay Municipal Utility District, *Annex to 2010 Association of Bay Area Governments, Local Hazard Mitigation Plan, Taming Natural Disasters* (Oakland, CA: EBMUD,

October 25, 2011), 10–11, https://abag.ca.gov/sites/default/files/ebmud-annex-2011 .pdf; *Seismic Improvement Program*, Case Studies, East Bay Municipal District (Oakland, CA: East Bay Municipal Utility District, Earthquake Engineering Research Institute), https://mitigation.eeri.org/lifelines/east-bay-municipal-utility-district-ca -case-study and https://mitigation.eeri.org/files/resources-for-success/00023.pdf.

38. George W. Housner, and Charles C. Thiel Jr., "Competing against Time: Report of the Governor's Board of Inquiry on the 1989 Loma Prieta Earthquake," *Earthquake Spectra*, 6, no. 4 (1990): 681–711, https://authors.library.caltech.edu/5013/.

39. C. C. Thiel, (ed.), *Competing against Time: The Governor's Board of Inquiry on the 1989 Loma Prieta Earthquake* (Sacramento: State of California Office of Planning and Research, 1990), 4.

40. Adam B. Summers, "California Spending by the Numbers: A Historic Look at State Spending from Gov. Pete Wilson to Gov. Arnold Schwarzenegger," Reason Foundation, 2009, https://reason.org/wp-content/uploads/files/a2ec7caccc5d660e 870c4a21526ef5f8.pdf.

41. Malena Carollo, "Florida Utilities Will Start Burying More Power Lines. You'll Pay the Bill," *Tampa Bay Times*, July 31, 2019, https://www.tampabay.com/business/ florida-utilities-will-start-burying-more-power-lines-youll-pay-the-bill-20190801/.

42. Marcia Heroux Pounds, "Plan to Bury Power Lines Advances—at a Cost to Residents," *South Florida Sun Sentinel*, April 26, 2019, https://www.sun-sentinel.com/ business/fl-bz-fpl-undergrounding-bill-20190422-story.html.

STATISTICAL HOCUS-POCUS

1. Ulrike Hahn and Paul A. Warren, "Perceptions of Randomness: Why Three Heads Are Better Than Four," *Psychological Review* 116, no. 2 (2009): 454–61 https://pdfs .semanticscholar.org/c08a/b58e5cdeaa040ac209ac6d66cd802d9c7492.pdf.

2. "Gambler's Fallacy," Wikipedia, https://en.wikipedia.org/wiki/Gambler%27s _fallacy#Coin_toss.

3. Daniel B. Murray and Scott W. Teare, "Probability of a Tossed Coin Landing on Edge," *Physical Review E*. 48, no. 4 (1993): 2547–52, doi:10.1103/PhysRevE.48.2547.

4. "'Carolina Pick 4' Drawing Results In $7.8 Million Win for Lottery Players," WFMY News, June 22, 2019, https://www.wfmynews2.com/article/news/lottery/ carolina-pick-4-drawing-results-in-78-million-win-for-lottery-players/83-a8d2d49f -bee3-4946-9b69-af4928bb31f5.

5. "Jackpot: More Than 2,000 Winners in NC Lottery after Picking Numbers 0-0-0- 0," NBC News, June 24, 2019, https://www.nbc4i.com/news/local-news/jackpot-more -than-2000-winners-in-nc-lottery-after-picking-numbers-0-0-0-0/.

6. Alan Branch, "Alfred Kinsey: A Brief Summary and Critique," ERLC, May 21, 2014, https://erlc.com/resource-library/articles/alfred-kinsey-a-brief-summary-and -critique.

7. S. Singh, *Critical Reasons for Crashes Investigated in the National Motor Vehicle Crash Causation Survey*, Traffic Safety Facts Crash Stats. Report No. DOT HS 812 115 (Washington, DC: National Highway Traffic Safety Administration, 2015), https:// crashstats.nhtsa.dot.gov/Api/Public/ViewPublication/812115.

8. Clifford Law, "The Dangers of Driverless Cars," *The National Law Review*, Wednesday, May 5, 2021, https://www.natlawreview.com/article/dangers-driverless-cars.

9. National Highway Traffic Safety Administration, *National Motor Vehicle Crash Causation Survey*, Report to Congress, US Department of Transportation, Report HS 811 059 July 2008, https://crashstats.nhtsa.dot.gov/Api/Public/ViewPublication/811059.

10. Stephen Ornes, "How to Hack a Self-Driving Car," *Physics World*, August 18, 2020, https://physicsworld.com/a/how-to-hack-a-self-driving-car/; "Top 20 Pros and Cons Associated with Self-Driving Cars," Auto Insurance Center, https://www.auto insurancecenter.com/top-20-pros-and-cons-associated-with-self-driving-cars.htm; Jason Kornwitz, "Hackers vs Cars, The Cybersecurity Risk of Self-Driving Cars," News@Northeastern, February 15, 2017, https://news.northeastern.edu/2017/02/15/the-cybersecurity-risk-of-self-driving-cars/.

11. Ellen Edmonds, "AAA: Today's Vehicle Technology Must Walk So Self-Driving Cars Can Run," AAA Newsroom, February 25, 2021, https://newsroom.aaa.com/2021/02/aaa-todays-vehicle-technology-must-walk-so-self-driving-cars-can-run/.

12. Alissa Walker, "How Likely You Are to Get Killed by a Car, Depending on Its Speed," Gizmodo, May 26, 2016, https://gizmodo.com/how-likely-you-are-to-get-killed-by-a-car-depending-on-1778993900.

13. D. C. Richards, *Relationship between Speed and Risk of Fatal Injury: Pedestrians and Car Occupants*, Road Safety Web Publication No. 16 (London: Transport Research Laboratory, Department for Transport, 2010), https://nacto.org/docs/usdg/relationship_between_speed_risk_fatal_injury_pedestrians_and_car_occupants_richards.pdf.

14. Daniel A. Gross, "How Elite Us Schools Give Preference to Wealthy and White 'Legacy' Applicants," *The Guardian*, January 23, 2019, https://www.theguardian.com/us-news/2019/jan/23/elite-schools-ivy-league-legacy-admissions-harvard-wealthier-whiter.

15. "Misuse of Statistics," Wikipedia, https://en.wikipedia.org/wiki/Misuse_of_statistics#Discarding_unfavorable_observations.

16. Hayley C. Cuccinello, "World's Highest-Paid Authors 2018: Michael Wolff Joins List Thanks to 'Fire and Fury,'" *Forbes*, Dec 11, 2018, https://www.forbes.com/sites/hayleycuccinello/2018/12/11/worlds-highest-paid-authors-2018-michael-wolff/#3446a9a12517.

17. Alison Flood, "Median Earnings of Professional Authors Fall Below the Minimum Wage," *Guardian*, April 20, 2015, https://www.theguardian.com/books/2015/apr/20/earnings-authors-below-minimum-wage.

18. "Books published per country per year," Wikipedia, https://en.wikipedia.org/wiki/Books_published_per_country_per_year.

19. "Number of New Books and Editions Published in the United States in the Category 'Fiction' from 2002 to 2013," Statista, https://www.statista.com/statistics/194678/us-book-production-by-subject-since-2002-fiction/.

20. B. J. Gallagher, "The Ten Awful Truths—and the Ten Wonderful Truths—about Book Publishing," *Huffington Post*, 04/05/2012, https://www.huffpost.com/entry/book-publishing_b_1394159.

21. Jordan Weissmann, "The Decline of the American Book Lover," *Atlantic*, January 21, 2014, https://www.theatlantic.com/business/archive/2014/01/the-decline-of-the -american-book-lover/283222/.

22. "Governor Ron DeSantis Declares State of Emergency, Urges Floridians to Prepare for Hurricane Dorian," August 28, 2019, https://www.flgov.com/2019/08/28/ governor-ron-desantis-declares-state-of-emergency-urges-floridians-to-prepare-for -hurricane-dorian/.

23. Amber Roberson, "Could Hurricane Dorian Become Category 6? No. Here's Why." *USA Today*, September 1, 2019, https://www.usatoday.com/story/news/nation/ 2019/09/01/hurricane-dorian-category-5-right-now-could-category-6-no/2186661001/.

24. "Palm Beach County Orders Mandatory Evacuations for Zones A and B," CBS12 news, September 1, 2019, https://cbs12.com/news/local/palm-beach-county-ordering -evacuations-and-opening-shelters.

25. Brian Resnick, "Hurricane Dorian Is a Category 3 'Major Hurricane' Just off the Coast of the Carolinas," Vox, Sept. 5, 2019, https://www.vox.com/energy-and -environment/2019/9/2/20844602/hurricane-dorian-florida-forecast-track-rainfall -windspeed-georgia-carolinas-bahamas.

26. Patrick Oppmann, "Hurricane Dorian Is Lashing the East Coast of Central Florida," CNN, September 3, 2019, https://www.cnn.com/2019/09/03/us/hurricane -dorian-tuesday-wxc/index.html.

27. Jeffrey Collins and Ben Finley, "Stranded North Carolina Residents Take Stock of Hurricane Dorian's Damage," *Huffington Post*, September 7, 2019, https://www .huffpost.com/entry/hurricane-dorian-north-carolina-damage_n_5d73caa5e4b0645 1356ed96c.

28. Aly Thomson, "Crane Toppled, Trees Uprooted as Dorian Hammers Nova Scotia," CBC News, September 7, 2019, https://www.cbc.ca/news/canada/nova-scotia/hurri cane-dorian-nova-scotia-destruction-1.5274887.

29. "Definition of the NHC Track Forecast Cone," National Hurricane Center, NOAA, https://www.nhc.noaa.gov/aboutcone.shtml.

30. Hurricanes Frequently Asked Questions, https://www.aoml.noaa.gov/hrd-faq/ #forecast-accuracy (formerly https://www.aoml.noaa.gov/hrd/tcfaq/F6.html).

31. "Dorian Graphics Archive: 5-day Forecast Track, Initial Wind Field and Watch/ Warning Graphic," National Hurricane Center, NOAA, https://www.nhc.noaa.gov/ archive/2019/DORIAN_graphics.php?product=5day_cone_with_line_and_wind.

32. "Cassandra (metaphor)," Wikipedia, https://en.wikipedia.org/wiki/Cassandra _(metaphor).

33. Wikipedia, https://en.wikipedia.org/wiki/Geology_of_the_Rocky_Mountains.

34. Angus M. Thuermer Jr., "It's the Same Old Grand, Only Now It's 5 Feet Grander," *Jackson Hole News & Guide*, December 11, 2013, https://www.jhnewsandguide.com/ news/environmental/article_318d8cda-f3aa-5f6c-bde0-7f97a67881b9.html.

35. J. Tobiason, "Geologic Activity," National Park Service, May 6, 2020, https://www .nps.gov/grte/learn/nature/geology.htm.

36. Andrew Alden, "Earthquake Magnitudes," Thoughtco., October 2, 2019, https:// www.thoughtco.com/what-are-earthquake-magnitudes-1439115.

37. "Earthquake: Could it Happen Here?," Rocky Mountain National Park, National Park Service, March 31, 2012, https://www.nps.gov/romo/earthquake.htm.

38. Lisa Wald and Kate Scharer, "Introduction to Paleoseismology," USGS, https://www.usgs.gov/natural-hazards/earthquake-hazards/science/introduction-paleo seismology?qt-science_center_objects=0#qt-science_center_objects.

39. Thomas M. Brocher, Jack Boatwright, James J. Lienkaemper, Carol S. Prentice, David P. Schwartz, and Howard Bundock, "The Hayward Fault—Is It Due for a Repeat of the Powerful 1868 Earthquake?," U.S. Geological Survey Fact Sheet 2008-3019, USGS, 2008, https://pubs.usgs.gov/fs/2008/3019/.

40. "What to Expect from an Earthquake along the Hayward Fault," California Earthquake Authority, May 31, 2019, https://www.earthquakeauthority.com/Blog/2019/hayward-fault-earthquake-prediction.

41. "PubTalk 9/2018—Hayward Earthquake" (video), USGS, September 27, 2018, https://www.usgs.gov/media/videos/pubtalk-92018-hayward-earthquake.

42. "History of the California Earthquake Authority (CEA)—It All Started with an Earthquake," California Earthquake Authority, https://www.earthquakeauthority.com/About-CEA/CEA-History.

43. "Get Your Free California Earthquake Insurance Estimate—How Much Does Earthquake Insurance Cost?," California Earthquake Authority, https://www.earthquakeauthority.com/California-Earthquake-Insurance-Policies/Earthquake-Insurance-Premium-Calculator.

44. Eric Westervelt, "Quake Insurance? California Wants People to Say Yes to Coverage," NPR, October 18, 2018, https://www.npr.org/2018/10/18/658570642/quake-insurance-california-pushing-people-to-say-yes-to-coverage.

45. "Notice of Maximum Amount of Assistance Under the Individuals and Households Program: A Notice by the Federal Emergency Management Agency," Federal Register, October 16, 2019, https://www.federalregister.gov/documents/2019/10/16/2019-22471/notice-of-maximum-amount-of-assistance-under-the-individuals-and-house holds-program.

46. Chad Calder, "How Much Money Can You Expect from FEMA? Disaster Grants Sure to Disappoint, Analysis Finds," *Advocate*, August 17, 2016, https://www.theadvocate.com/louisiana_flood_2016/article_22c86fe0-64cd-11e6-9bb2-07f95d36ee28.html.

47. "Home Values in Berkeley, CA," Realtor.com, https://www.realtor.com/realestate andhomes-search/Berkeley_CA/overview.

48. "National Census of Fatal Occupational Injuries in 2019," USDL-20-2265, US Department of Labor, Bureau of Labor Statistics, https://www.bls.gov/news.release/pdf/cfoi.pdf.

49. Jeffrey S. Rosenthal, "Margins of Error in Opinion Polls," Probability, 2003, http://probability.ca/jeff/writing/pollerror.html.

50. Fabio Turone, "The Trouble with Health Statistics," *Cancer World*, 6 (September 2018), https://archive.cancerworld.net/cover-story/the-trouble-with-health-statistics/.

51. William Kremer, "Do Doctors Understand Test Results?," BBC World Service, July 7, 2014, https://www.bbc.com/news/magazine-28166019.

52. "Types of Sharks," Shark Sider, https://www.sharksider.com/types-of-sharks/.

53. Mark Mancini, "13 Facts about Nurse Sharks," Mental Floss, October 23, 2018, http://mentalfloss.com/article/559319/nurse-shark-facts.

54. Brian Handwerk, "Shark Facts: Attack Stats, Record Swims, More," *National Geographic*, June 13, 2005, https://www.nationalgeographic.com/animals/2005/06/shark-facts/.

55. "David Copperfield (illusionist)," Wikipedia, https://en.wikipedia.org/wiki/David _Copperfield_(illusionist).

56. "Laurentia," Wikipedia, https://en.wikipedia.org/wiki/Laurentia.

57. Russell L. Wheeler, "Sizes of the Largest Possible Earthquakes in the Central and Eastern United States—Summary of a Workshop," September 8–9, 2008, Golden, Colorado, Open-File Report 2009-1263, U.S. Geological Survey and U.S. Nuclear Regulatory Commission, https://pubs.usgs.gov/of/2009/1263/downloads/OF09-1263.pdf.

58. C. Fenton, J. Adams, and S. Halchuk, "Seismic Hazards Assessment for Radioactive Waste Disposal Sites In Regions of Low Seismic Activity," *Geotechnical and Geological Engineering* 24 (2006):585–86 and figure 4, https://link.springer.com/content/pdf/10.1007%2Fs10706-005-1148-4.pdf.

59. Anthony Crone, Michael Machette, and JR Bowman, "Geologic Investigations of the 1988 Tennant Creek, Australia, Earthquakes-Implications for Paleoseismicity in Stable Continental Regions," USGS Bulletin, 2032-A (1992), https://pubs.usgs.gov/bul/2032a/report.pdf.

60. "Uncertainty Quantification," Wikipedia, https://en.wikipedia.org/wiki/Uncertainty_quantification#Aleatoric_and_epistemic_uncertainty.

61. "Log-log Plot," Wikipedia, https://en.wikipedia.org/wiki/Log%E2%80%93log _plot#/media/File:LogLog_exponentials.svg.

62. See "World War One Hidden Stories: Canada's Soldier (Full Network Special)," https://www.youtube.com/watch?v=McsEN1_HfQM (from 4:30 to 5:00 min).

63. See "*Hockey: A People's History*," https://www.youtube.com/watch?v=1ZucC2DY pf4 (from 3:45 min).

64. Ken Carriere, "*Hockey: A People's History*," *The Globe and Mail*, September 16, 2006, https://www.theglobeandmail.com/arts/hockey-a-peoples-history/article966968/.

65. See "*Wide World of Sports* Intro (1978)," https://www.youtube.com/watch?v =6k1gajZeegg.

LIFE'S CASINO

1. "How Much Does It Cost to Remove Asbestos?," Home Advisor, April 14, 2021, https://www.homeadvisor.com/cost/environmental-safety/remove-asbestos/.

2. "Working with Asbestos Containing Roofs," Asbestos Information Center, https://www.aic.org.uk/asbestos-roofs/.

3. Joe Lahav, "No Asbestos Ban in the US," The Mesothelioma Center, Asbestos.com, September 21, 2021 https://www.asbestos.com/mesothelioma-lawyer/legislation/ban/.

4. "Asbestos Cement FAQs," Asbestos Information Center, https://www.aic.org.uk/asbestos-cement-faqs/.

5. "Asbestos and Health: Frequently Asked Questions," Agency for Toxic Substances and Disease Registry, CDC, https://www.atsdr.cdc.gov/asbestos/docs/asbestos_fact sheet_508.pdf.

6. L. S. Siegal, "As the Asbestos Crumbles: A Look at New Evidentiary Issues in Asbestos-Related Property Damage Litigations," *Hofstra Law Review* 20, no. 1132 (1992): 1146, https://pdfs.semanticscholar.org/7cd0/cf7102e4ddc2cf8ecdb43d21c329 dd9755f9.pdf.

7. "OSH Answers Fact Sheets, Hazard and Risk, Canadian Centre for Occupational Health and Safety," https://www.ccohs.ca/oshanswers/hsprograms/hazard_risk.html.

8. "Death by coconut," Wikipedia, https://en.wikipedia.org/wiki/Death_by_coconut.

9. R. E. Kasperson and S. C. Moser, "Risk Conundrums in a Fast and Complex World," in *Risk Conundrums: Solving Unsolvable Problems* ed. R. E. Kasperson (New York: Routledge, 2017), 1–10.

10. "Glossary," Society for Risk Analysis, https://www.sra.org/wp-content/uploads/ 2020/04/SRA-Glossary-FINAL.pdf.

11. "Let's Talk about BASE Jumping," Digg, March 26, 2014, https://digg.com/2014/ what-is-base-jumping.

12. A. Westman, M. Rosén, P. Berggren et al., "Parachuting from Fixed Objects: Descriptive Study of 106 Fatal Events in BASE Jumping 1981–2006," *British Journal of Sports Medicine* 42 (2008): 431-436, https://bjsm.bmj.com/content/42/6/431.

13. Barney Ronay, "Base Jumping Is Absurdly Dangerous—So Why Do It?," *Guardian*, August 6, 2010, https://www.theguardian.com/sport/2010/aug/06/base -jumping-why-do-it.

14. Jenna Stadsvold, "Effects of Adrenaline and Why We Love the Thrill," Head Rush Technologies, https://headrushtech.com/blogs/effects-of-adrenaline-why-we-love-the -thrill/.

15. "Alpine Skiing," Wikipedia, https://en.wikipedia.org/wiki/Alpine_skiing.

16. Mona Chalabi, "Is Skiing the World's Most Dangerous Sport?," *Guardian*, December 30, 2013, https://www.theguardian.com/news/datablog/2013/dec/30/is-skiing-the -worlds-most-dangerous-sport.

17. Gavin Naylor and Tyler Bowling, "Yearly Worldwide Shark Attack Summary," International Shark Attack File, The ISAF 2020 shark attack report, https://www .floridamuseum.ufl.edu/shark-attacks/yearly-worldwide-summary/.

18. Marco Santana, "Visit Florida Reports Record 126.1 Million Visitors to State in 2018," *Orlando Sentinel*, February 20, 2019, https://www.orlandosentinel.com/ business/tourism/os-bz-visit-florida-tourism-record-20190220-story.html.

19. James Houston, "The Economic Value of Beaches: A 2013 Update," *Shore Beach* 81 (2013): 3–11. Also summarized in this PowerPoint: https://www.fsbpa.com/13 AnnualConfPresentations/HoustonValueofFloridaBeaches.pdf.

20. "Your Chances of Dying," Best Health Degrees, http://www.besthealthdegrees .com/health-risks/.

21. Joseph Hooper, "The Asbestos Mess," *New York Times*, November 25, 1990, https://www.nytimes.com/1990/11/25/magazine/the-asbestos-mess.html.

22. "TC Energy Confirms Termination of Keystone XL Pipeline Project," TC Energy, June 9, 2021, https://www.tcenergy.com/announcements/2021-06-09-tc-energy-con firms-termination-of-keystone-xl-pipeline-project/.

23. W. K. Viscusi, "The Value of Risks to Life and Health," *Journal of Economic Literature* 31, no. 4 (1993): 1912–46.

24. Y. Y. Haimes, *Risk Modeling, Assessment, and Management* (New York: Wiley Interscience, 2004), 23.

25. "Drunk Driving," National Highway Traffic Safety Administration, https://www .nhtsa.gov/risky-driving/drunk-driving.

26. Paul Weissler, "Can Tech Prevent Drunk-Driving Fatalities?," SAE International, May 30, 2018, https://www.sae.org/news/2018/05/tech-for-drunk-driving-prevention.

27. See for example, "Ice Management," Hibernia, https://www.hibernia.ca/ice.html.

28. "Hurricanes and the Offshore Oil and Natural Gas industry," National Ocean Industries Association, http://www.noia.org/wp-content/uploads/2013/03/326.pdf.

29. "California Earthquake Early Warning System—Frequently Asked Questions," California Governor's Office of Emergency Services, https://www.caloes.ca.gov/Earth quakeTsunamiVolcanoProgramsSite/Documents/California%20Earthquake%20 Early%20Warning%20System%20Fact%20Sheet.pdf.

30. "America's Infrastructure Scores a D+," American Society of Civil Engineers, 2017, https://2017.infrastructurereportcard.org/.

31. "America's Infrastructure Scores—Roads," American Society of Civil Engineers, 2017, https://2017.infrastructurereportcard.org/cat-item/roads/.

32. "America's Infrastructure Scores—Dams," American Society of Civil Engineers, 2017, https://2017.infrastructurereportcard.org/cat-item/dams/.

33. "America's Infrastructure Scores—Bridges," American Society of Civil Engineers, 2017, https://2017.infrastructurereportcard.org/cat-item/bridges/.

34. "America's Infrastructure Scores—Drinking Water," American Society of Civil Engineers, 2017, https://2017.infrastructurereportcard.org/cat-item/drinking_water/.

35. Yao Gang, Li Hang, Yang Yang, Pu Wei, "Seismic Responses and Dynamic Characteristics of Boom Tower Crane Basing on Measured Strong Earthquake Excitation," *Journal of Vibroengineering* 21, no. 1 (2019): 154–69, https://doi.org/10.21595/ jve.2018.19626 https://www.jvejournals.com/article/19626.

36. "Taipei 101," Wikipedia, https://en.wikipedia.org/wiki/Taipei_101.

37. Y. Ushio, M. Okano, and Y. Nagano, "The Earthquake Responses of Climbing-Type Tower Cranes Installed in High-Rise Buildings in Consideration of Various Situations under Construction," paper no. 3403, Proceedings of the 16th World Conference on Earthquake, Santiago Chile, 2016. https://www.wcee.nicee.org/wcee/ article/16WCEE/WCEE2017-3403.pdf.

38. Seiji Takanashi, Hiromi Adachi, and Mitsukazu Nakanishi, "Study on the Seismic Performance of the Tower Crane for Construction," *AIJ Journal of Technology and Design* 13, no. 26 (2007): 415–20, https://doi.org/10.3130/aijt.13.415, https://www .jstage.jst.go.jp/article/aijt/13/26/13_26_415/_article.

39. Maria Lazarte, "Earthquake-Resistant Cranes with New ISO Standard," International Organization for Standardization, August 25, 2016, https://www.iso.org/ news/2016/08/Ref2108.html.

BLACK SWANS

1. "Bin Laden Preparing to Hijack US Aircraft and Other Attacks," Declassified and Approved for Release, July 12 2004, DCI Counterterrorist Center, 4 December 1998,

https://web.archive.org/web/20121015182609/, http://www.foia.cia.gov/docs/DOC
_0001110635/0001110635_0001.gif.

2. "Black Swan," Investopia, https://www.investopedia.com/terms/b/blackswan.asp.

3. Brendan Shea, "Karl Popper: Philosophy of Science," *The Internet Encyclopedia of Philosophy*, https://www.iep.utm.edu/pop-sci/.

4. Malcolm Gladwell, "Blowing Up," *New Yorker*, April 14, 2002, https://www.new yorker.com/magazine/2002/04/22/blowing-up.

5. Nassim Nicholas Taleb, *The Black Swan: The Impact of the Highly Improbable* (New York: Random House, 2010).

6. Stephanie Baker, "Taleb Says Crisis Makes Nobel Panel Liable for Legitimizing Economists," *Bloomberg*, October 8, 2010, https://www.bloomberg.com/news/articles/ 2010-10-08/taleb-says-crisis-makes-nobel-panel-liable-for-legitimizing-economists.

7. Will Hutton, "Now We Know the Truth. The Financial Meltdown Wasn't a Mistake—It Was a Con," *Guardian*, April 18. 2010, https://www.theguardian.com/ business/2010/apr/18/goldman-sachs-regulators-civil-charges.

8. Jesse Eisinger, "Why Only One Top Banker Went to Jail for the Financial Crisis," Money Issue, *New York Times Magazine*, April 30, 2014, https://www.nytimes .com/2014/05/04/magazine/only-one-top-banker-jail-financial-crisis.html.

9. "9 Wall Street Execs Who Cashed In on the Crisis," *Mother Jones*, January/February 2010, https://www.motherjones.com/politics/2010/01/wall-street-bailout-executive -compensation/.

10. "Black Swan," Investopia.

11. J. B. Aron, *Licensed to Kill? The Nuclear Regulatory Commission and the Shoreham Power Plant* (Pittsburgh, PA: University of Pittsburgh Press, 1997), 85.

12. Dan Fagin, "Lights Out at Shoreham—Anti-nuclear Activism Spurs the Closing of a New $6 Billion Plant," *Newsday*, November 30, 2007, https://web.archive .org/web/20071201005429/http://www.newsday.com/community/guide/lihistory/ny -history-hs9shore,0,563942.story.

13. "Shoreham Nuclear Power Plant," Wikipedia, https://en.wikipedia.org/wiki/ Shoreham_Nuclear_Power_Plant#cite_note-nora2011-10.

14. Denis Omelchenko, "Nassim Taleb Says Coronavirus Pandemic Is Not a 'Black Swan,'" IHODI.com, April 1, 2020, https://ihodl.com/opinion/2020-04-01/nassim -taleb-says-coronavirus-pandemic-not-black-swan/https://fpov.com/2020/03/27/ black-swan/.

15. "Bill Gates: The Next Outbreak? We're Not Ready," https://www.youtube.com/ watch?time_continue=1&v=6Af6b_wyiwI&feature=emb_logo.

16. "Ebola Virus Disease," World Health Organization, February 23, 2021, https:// www.who.int/news-room/fact-sheets/detail/ebola-virus-disease.

17. Nicolas Berrod, "Bill Gates, la CIA, Jacques Attali . . . ils avaient alerté sur une épidémie mondiale," *Le Parisien*, March 23, 2020, http://www.leparisien.fr/societe/ coronavirus-bill-gates-la-cia-jacques-attali-ils-avaient-alerte-sur-une-epidemie -mondiale-23-03-2020-8285973.php.

18. Dan Diamond, "Inside America's 2-Decade Failure to Prepare for Coronavirus," *Politico Magazine*, April 11, 2020, https://www.politico.com/news/magazine/2020/ 04/11/america-two-decade-failure-prepare-coronavirus-179574.

19. Ibid.

20. Samuel Brannen and Kathleen Hicks, "We Predicted a Coronavirus Pandemic: Here's What Policymakers Could Have Seen Coming," *Politico Magazine*, March 7, 2020, https://www.politico.com/news/magazine/2020/03/07/coronavirus-epidemic -prediction-policy-advice-121172.

21. "Public Trust in Government: 1958–2021," Pew Research Center, May 17, 2021, https://www.people-press.org/2019/04/11/public-trust-in-government-1958-2019/.

22. Brannen and Hicks, "We Predicted a Coronavirus Pandemic."

23. "Dark Winter," Johns Hopkins Center for Health Security, https://www.centerfor healthsecurity.org/our-work/events-archive/2001_dark-winter/index.html.

24. Ibid.

25. "Clade X Livestream (Archived)," Johns Hopkins Center for Health Security, https://www.centerforhealthsecurity.org/our-work/events/2018_clade_x_exercise/ livestream.

26. "Clade X Players," Johns Hopkins Center for Health Security, https://www.center forhealthsecurity.org/our-work/events/2018_clade_x_exercise/players/.

27. "Clade X Livestream (Archived)."

28. *The Global Risks Report 2019*, 14th Edition, World Economic Forum, 2019, http:// www3.weforum.org/docs/WEF_Global_Risks_Report_2019.pdf.

29. Faraz Toor, "A Look at New York City's Earthquake Risks," Spectrum News NY1, April 2, 2018, https://www.ny1.com/nyc/all-boroughs/news/2018/04/01/ new-york-city-is-due-for-magnitude-5-or-stronger-earthquake.

30. "Expectations for Charleston's Next Great Unexpected Disaster," *Charleston City Paper*, March 28, 2018, https://charlestoncitypaper.com/expectations-for -charlestons-next-great-unexpected-disaster/.

31. Kristine Pankow, Walter J. Arabasz, Robert Carey, Gary Christenson, Josh Groeneveld, Brent Maxfield, Peter W. McDonough, Barry Welliver, and T. Leslie Youd, *Scenario for a Magnitude 7.0 Earthquake on the Wasatch Fault–Salt Lake City Segment Hazards and Loss Estimates*, Earthquake Engineering Research Institute, Utah Chapter, June 4, 2015, https://dem.utah.gov/wp-content/uploads/sites/18/2015/03/ RS1058_EERI_SLC_EQ_Scenario.pdf.

32. Glynn Cosker, "With Potential of 'Big One' Earthquake Hitting Memphis and St. Louis Area, Preparedness Is Vital," *Tennessean*, August 21, 2019, https://www .tennessean.com/story/opinion/2019/08/21/earthquake-memphis-be-ready/2059573 001/.

33. Laura Geggel, "What If a 9.0-Magnitude Earthquake Hit Seattle?," Live Science, November 16, 2017, https://www.livescience.com/60959-magnitude-9-earthquake -pacific-northwest-study.html.

34. Robert B. Horner, Maurice Lamontagne, and Robert J. Wetmiller, "Rock and Roll in the N.W.T.: The 1985 Nahanni Earthquakes," Earthquakes Canada, Natural Resources Canada, http://www.earthquakescanada.nrcan.gc.ca/historic-historique/ events/19851223-en.php.

35. "Region 4, Northwest Territories," Wikipedia, https://en.wikipedia.org/wiki/ Region_4,_Northwest_Territories.

36. Horner, Lamontagne, and Wetmiller, "Rock and Roll in the N.W.T."

37. J. Adams, "Seismic Hazard Estimation in Canada and Its Contribution to the Canadian Building Code Implications for Code Development in Countries such as Australia," *Australian Journal of Structural Engineering* 11 (2011): 267–82, doi: 10.1080/13287982.2010.11465072.

38. Geoff Manaugh and Nicola Twilley, "The Philosophy of SimCity: An Interview with the Game's Lead Designer," *Atlantic*, May 9, 2013, https://www.theatlantic.com/technology/archive/2013/05/the-philosophy-of-simcity-an-interview-with-the-games-lead-designer/275724/.

39. "Cheats and Secrets, SimCity Wiki Guide," IGN, August 28, 2017, https://www.ign.com/wikis/simcity/Cheats_and_Secrets.

40. U. Wilensky, and W. Rand, *An Introduction to Agent-Based Modeling: Modeling Natural, Social, and Engineered Complex Systems with NetLogo* (Cambridge, MA: MIT Press, 2015).

41. P. Grim, R. Rosenberger, A. Rosenfeld, et al., "How Simulations Fail," *Synthese* 190, (2013): 2367–90, https://doi.org/10.1007/s11229-011-9976-7.

42. "How do I build large, healthy cities in Simcity 4?," Arquade, https://gaming.stackexchange.com/questions/1526/how-do-i-build-large-healthy-cities-in-simcity-4.

43. CynicalDriver, "Traffic 'AI' . . . This Is Why Services and Traffic Are Broken!," Answers HQ, March 2013, https://answers.ea.com/t5/SimCity-2013/Traffic-quot-AI-quot-This-is-why-services-and-traffic-are-broken/m-p/737060.

44. "SimCity GlassBox at Work," YouTube, https://www.youtube.com/watch?v=fDeSRdHvefw.

45. "Sim City Additional Pathfinding Experiment," YouTube, https://www.youtube.com/watch?v=zcEaHT9mt-Y.

46. "SimCity: The Intersection Trap," YouTube, https://www.youtube.com/watch?v=-d0b41H-Lnk.

47. At least, that is the response I received from a prominent economist.

48. "Salem Witch Trials," Wikipedia, https://en.wikipedia.org/wiki/Salem_witch_trials#Formal_prosecution:_The_Court_of_Oyer_and_Terminer.

49. Alexandra Sifferlin, "4 Diseases Making a Comeback Thanks to Anti-Vaxxers," *Time*, March 17, 2014, https://time.com/27308/4-diseases-making-a-comeback-thanks-to-anti-vaxxers/.

THE BRAIN

1. Susan Millstein and Bonnie Halpern-Felsher, "Perceptions of Risk and Vulnerability," *Journal of Adolescent Health* 31 (2002): 10–27, doi:10.1016/S1054-139X(02)00412-3.

2. Jenna Mullins, "28 Stupid Things Teenagers Have Already Done This Year," Eonline, May 19, 2014, https://www.eonline.com/news/543126/28-stupid-things-teenagers-have-already-done-this-year.

3. Jessica Firger, "Why Teenage Boys Do Stupid Things," CBS News, June 12, 2014, https://www.cbsnews.com/news/whats-wrong-with-the-teen-brain/.

4. J. M. S. Pearce, "The 'Split Brain' and Roger Wolcott Sperry (1913–1994)," *Revue Neurologique* 175, no. 4 (2019): 217–20, https://doi.org/10.1016/j.neurol.2018.07.007.

5. Garson O'Toole, "The Intuitive Mind Is a Sacred Gift and the Rational Mind Is a Faithful Servant," Quote Investigator, September 18, 2013, https://quoteinvestigator .com/2013/09/18/intuitive-mind/.

6. Stephen M. Kosslyn and G. Wayne Miller, "Left Brain, Right Brain: Two Sides, Always Working Together," *Psychology Today*, May 7, 2014, https://www.psychology today.com/us/blog/the-theory-cognitive-modes/201405/left-brain-right-brain-two -sides-always-working-together.

7. Stacey Vanek Smith, "Why We Sign Up for Gym Memberships but Never Go to the Gym," NPR, December 30, 2014 https://www.npr.org/sections/money/2014/ 12/30/373996649/why-we-sign-up-for-gym-memberships-but-don-t-go-to-the-gym.

8. "Grid Illusion," Wikipedia, https://en.wikipedia.org/wiki/Grid_illusion.

9. Michael Bach, "144 Optical Illusions and Visual Phenomena," https://michael bach.de/ot/.

10. "Optical Illusions," Optics 4 Kids, https://www.optics4kids.org/illusions.

11. Richard L Gregory, "Knowledge in Perception and Illusion," *Philosophical Transactions of the Royal Society of London*, Series B, Biological sciences, 352, no. 1358 (1997): 1121–27, http://www.richardgregory.org/papers/knowl_illusion/knowledge -in-perception.pdf.

12. Romie Stott, "When Tomatoes Were Blamed for Witchcraft and Werewolves," Gastro Obscura, October 24, 2016, https://www.atlasobscura.com/articles/when -tomatoes-were-blamed-for-witchcraft-and-werewolves.

13. "Black Cats and Werewolves Once Thought to Be Witches in Disguise," *Johns Hopkins University Newswire*, October 8, 2003, https://www.newswise.com/articles/ black-cats-and-werewolves-once-thought-to-be-witches-in-disguise.

14. Becky Little, "The Wildest Moon Landing Conspiracy Theories, Debunked," History.com, https://www.history.com/news/moon-landing-fake-conspiracy-theories.

15. "Moon Base," Clavius, http://www.clavius.org/index.html.

16. Raymond Nickerson, "Confirmation Bias: A Ubiquitous Phenomenon in Many Guises," *Review of General Psychology* 2 (1998): 175–220. 10.1037/1089-2680.2.2.175.

17. Steven Smith, Leandre Fabrigar, and Meghan Norris, "Reflecting on Six Decades of Selective Exposure Research: Progress, Challenges, and Opportunities," *Social and Personality Psychology Compass* 2 (2008): 464–93, 10.1111/j.1751-9004.2007.00060 .x; "Selective exposure theory," Wikipedia, https://en.wikipedia.org/wiki/Selective _exposure_theory.

18. "Monty Python's *Life of Brian*," Wikipedia, https://en.wikipedia.org/wiki/ Monty_Python%27s_Life_of_Brian#Reception.

19. Wendell Rawls Jr., "'*Life of Brian*' Stirs Carolina Controversy," *New York Times*, October 24, 1979, https://www.nytimes.com/1979/10/24/archives/life-of-brian-stirs -carolina-controversy-minister-is-outraged.html.

20. Robert Sellers, "Welease Bwian," *Guardian*, March 27, 2003, https://www.the guardian.com/culture/2003/mar/28/artsfeatures1.

21. Christopher Peterson, "When Prophecy Fails: The Sequel," *Psychology Today*, May 22, 2011, https://www.psychologytoday.com/us/blog/the-good-life/201105/when -prophecy-fails-the-sequel.

22. "List of UFO religions," Wikipedia, https://en.wikipedia.org/wiki/List_of_UFO _religions#List.

23. "Heaven's Gate Cult Members Found Dead," History.com, March 26, 1997, https:// www.history.com/this-day-in-history/heavens-gate-cult-members-found-dead.

24. M. Rousselet, O. Duretete, J. B. Hardouin, and M. Grall-Bronnec, "Cult Membership: What Factors Contribute to Joining or Leaving?," *Psychiatry Research* 257 (2017): 27–33, https://doi.org/10.1016/j.psychres.2017.07.018.

25. Diane Benscoter, "Watch: I Used to Be in a Cult and Here's What It Did to My Brain," *Huffington Post*, September 26, 2013, https://www.huffpost.com/entry/ extremist-brain_b_3998314.

26. Diane Benscoter, "How Cults Rewire the Brain," TED, February 2009, https:// www.ted.com/talks/ex_moonie_diane_benscoter_how_cults_think/discussion.

27. Clyde H. Farnsworth, "Canada Seeks Money Trail of Secret Cult," *New York Times*, October 16, 1994, https://www.nytimes.com/1994/10/16/world/canada-seeks -money-trail-of-secret-cult.html; J. R. Lewis, *The Order of the Solar Temple: The Temple of Death* (Aldershot, UK: Ashgate, 2006), 29; J. R. Hall, P. D. Schuyler, and S. Trinh, *Apocalypse Observed: Religious Movements and Violence in North America, Europe and Japan* (New York: Taylor and Francis, 2005), 135.

28. Ross Laver, "Apocalysse Now—Mystery Surrounds the Violent Deaths of 53 People Linked to a Bizarre Cult," *Macleans Magazine*, October 17 1994, https://archive .macleans.ca/article/1994/10/17/apocalysse-now.

29. Clyde H. Farnsworth, "Quebec Police Say Baby Was Target of Cult," *New York Times*, November 20, 1994, https://www.nytimes.com/1994/11/20/world/quebec -police-say-baby-was-target-of-cult.html; "Temple Solaire," *Encyclopedia of Religion*, Encyclopedia.com, https://www.encyclopedia.com/environment/encycloped ias-almanacs-transcripts-and-maps/temple-solaire.

30. Brad Hunter, "Solar Temple Massacre: Mystery Endures 25 Years Later," *Toronto Sun*, October 5, 2019, https://torontosun.com/news/world/solar-temple-massacre -mystery-endures-25-years-later.

31. Glenn Collins, "The Psychology of the Cult Experience," *New York Times*, March 15, 1982, https://www.nytimes.com/1982/03/15/style/the-psychology-of-the -cult-experience.html.

32. Steve Hendrix, "Mail That Baby: A Brief History of Kids Sent through the U.S. Postal Service," *Washington Post*, May 24, 2017, https://www.washingtonpost.com/ news/retropolis/wp/2017/05/24/mail-that-baby-a-brief-history-of-kids-sent-through -the-u-s-postal-service/.

33. Ralph Lewis, "What Actually Is a Belief? And Why Is It So Hard to Change?," *Psychology Today*, October 7, 2018, https://www.psychologytoday.com/us/blog/finding -purpose/201810/what-actually-is-belief-and-why-is-it-so-hard-change.

34. "Why People Procrastinate: The Psychology and Causes of Procrastination," Solv ingProcrastination.com, https://solvingprocrastination.com/why-people-procrastinate/.

35. D. D. Reed and J. K. Luiselli, "Temporal Discounting," in *Encyclopedia of Child Behavior and Development*, ed. S. Goldstein and J. A. Naglieri (Boston: Springer, 2011), https://doi.org/10.1007/978-0-387-79061-9_3162; "Time preference," Wiki pedia, https://en.wikipedia.org/wiki/Time_preference.

36. Joseph P. Redden, "Hyperbolic Discounting," in *Encyclopedia of Social Psychology*, ed. Roy F. Baumeister and Kathleen D. Vohs (Thousand Oaks, CA: Sage, 2007), http://www.behaviorlab.org/Papers/Hyperbolic.pdf.

37. "Why People Procrastinate."

38. J. Portmann, *A History of Sin: Its Evolution to Today and Beyond* (Lanham, MD: Rowman & Littlefield, 2007).

39. Luciana Gravotta, "History of Sin: How It All Began," *Scientific American*, November 2013, https://www.scientificamerican.com/article/history-of-sin-how-it-all-began/.

40. Nelson H. Minnich, "Leo X (Pope) (1475–1521; Reigned 1513–1521), Europe, 1450 to 1789," in *Encyclopedia of the Early Modern World*, Encyclopedia.com, September 23, 2021, https://www.encyclopedia.com/people/philosophy-and-religion/roman-catholic-popes-and-antipopes/leo-x.

41. Phil Harrell and Scott Saloway, "How the Concert for Bangladesh Changed the Celebrity Fundraiser," NPR, July 30, 2021, https://www.npr.org/2021/07/30/10223 52422/concert-for-bangladesh.

42. "The Concert for Bangladesh," Wikipedia, https://en.wikipedia.org/wiki/The_Concert_for_Bangladesh; Lorraine Boissoneault, "The Genocide the U.S. Can't Remember, but Bangladesh Can't Forget," *Smithsonian Magazine*, December 16, 2016, https://www.smithsonianmag.com/history/genocide-us-cant-remember-bangladesh-cant-forget-180961490/; "Bangladesh Liberation War," Wikipedia, https://en.wiki pedia.org/wiki/Bangladesh_Liberation_War.

43. "1970—The Great Bhola Cyclone," Hurricanes: Science and Society, http://www.hurricanescience.org/history/storms/1970s/greatbhola/.

44. "Bangladesh–Pakistan relations," Wikipedia, https://en.wikipedia.org/wiki/Bangladesh%E2%80%93Pakistan_relations.

45. Ibid.

46. "The Concert for New York City," Wikipedia, https://en.wikipedia.org/wiki/The_Concert_for_New_York_City.

47. "Tsunami Relief Cardiff, 2005," *Wales Music*, BBC, August 18, 2009, http://www.bbc.co.uk/wales/music/sites/tsunami-relief-cardiff/.

48. "Shelter from the Storm: A Concert for the Gulf Coast," Wikipedia, https://en.wikipedia.org/wiki/Shelter_from_the_Storm:_A_Concert_for_the_Gulf_Coast.

49. "12-12-12: The Concert for Sandy Relief," Wikipedia, https://en.wikipedia.org/wiki/12-12-12:_The_Concert_for_Sandy_Relief.

50. "Hope for Haiti Now," Wikipedia, https://en.wikipedia.org/wiki/Hope_for_Haiti_Now.

51. "Hand in Hand: A Benefit for Hurricane Relief," Wikipedia, https://en.wikipedia.org/wiki/Hand_in_Hand:_A_Benefit_for_Hurricane_Relief.

52. "Smoke on the Water with Queen, Pink Floyd, Rush, Black Sabbath, Deep Purple, etc.," https://www.youtube.com/watch?v=1tsw3nKDlBE.

53. "*The Wall*: Live in Berlin," Wikipedia, https://en.wikipedia.org/wiki/The_Wall_%E2%80%93_Live_in_Berlin.

54. UNICEF USA Initiatives, The George Harrison Fund for UNICEF, https://www.unicefusa.org/mission/usa/george-harrison-fund.

55. "Scaling Up Social Protection," UNICEF Bangladesh, https://www.unicef.org/bangladesh/en/scaling-social-protection.

56. "Cyclone Season and Threat of Violence Loom Over 720,000 Rohingya Children in Myanmar and Bangladesh—UNICEF," UNICEF Bangladesh, https://www.unicef.org/bangladesh/en/press-releases/cyclone-season-and-threat-violence-loom-over-720000-rohingya-children-myanmar-and.

57. "Aquila," Wikipedia, https://en.wikipedia.org/wiki/L%27Aquila#Earthquakes.

58. "List of Earthquakes in Italy," Wikipedia, https://en.wikipedia.org/wiki/List_of_earthquakes_in_Italy.

59. David Bressan, "April 6, 2009: The L'Aquila Earthquake," *Scientific American*, April 6, 2012, https://blogs.scientificamerican.com/history-of-geology/april-6-2009-the-laquila-earthquake/.

60. Lisa Zyga, "Italian Scientists Who Failed to Predict L'Aquila Earthquake May Face Manslaughter Charges," Phys Org, June 24th, 2010, http://www.physorg.com/news196622867.html.

61. David Ropeik, "The L'Aquila Verdict: A Judgment Not against Science, but against a Failure of Science Communication," *Scientific American*, October 22, 2012, https://blogs.scientificamerican.com/guest-blog/the-laquila-verdict-a-judgment-not-against-science-but-against-a-failure-of-science-communication/.

62. Lauren Kurtz, "Italian Scientists' Convictions for Not Predicting Earthquake Reversed," *Climate Law Blog*, Columbia Law School, December 8, 2014, http://blogs.law.columbia.edu/climatechange/2014/12/08/italian-scientists-convictions-reversed/; Edwin Cartlidge, "Updated: Appeals Court Overturns Manslaughter Convictions of Six Earthquake Scientists," *Science*, November 10, 2014, https://www.science.org/news/2014/11/updated-appeals-court-overturns-manslaughter-convictions-six-earthquake-scientists.

63. Edwin Cartlidge, "Seven-Year Legal Saga Ends as Italian Official Is Cleared of Manslaughter in Earthquake Trial," *Science*, October 3, 2016, https://www.sciencemag.org/news/2016/10/seven-year-legal-saga-ends-italian-official-cleared-manslaughter-earthquake-trial; Edwin Cartlidge, "Italy's Supreme Court Clears L'Aquila Earthquake Scientists for Good," *Science*, November 20, 2015, https://www.science.org/news/2015/11/italy-s-supreme-court-clears-l-aquila-earthquake-scientists-good.

64. Gabriel Andrade, "René Girard (1923–2015): 3. The Scape Goat Mechanism," *The Internet Encyclopedia of Philosophy*, https://www.iep.utm.edu/girard/#H3.

65. "Scapegoating," Wikipedia, https://en.wikipedia.org/wiki/Scapegoating.

66. Shawn Ee, "'Scapegoating'—What It Is and How to Understand It," The Psychology Practice, November 15, 2017, Also, formerly: http://www.the-confidant.info/2011/feel-better-create-a-scapegoat/.

67. "Bell System," Wikipedia, https://en.wikipedia.org/wiki/Bell_System.

THE MIRRORS WE ELECT

1. Ed Rappaport, *Preliminary Report, Hurricane Andrew*, August 16–28, 1992, National Hurricane Center, December 10, 1993, https://www.nhc.noaa.gov/1992andrew.html.

2. Rick Dixon, "The Florida Building Code, Florida's Response to Hurricane Risk," Florida Building Commission, PowerPoint, https://www.sbafla.com/method/portals/methodology/WindstormMitigationCommittee/2009/20090917_DixonFLBldgCode.pdf.

3. "Development of Loss Relativities for Wind Resistive Features of Residential Structures," Applied Research Associates, Inc., IntraRisk Division, for Florida Department of Community Affairs, March 28, 2002, https://www.floir.com/siteDocuments/ARAwindmitigation.pdf.

4. "Legislator Profile, Former Florida Senator Charlie Clary (R)," Lobbytools.com, https://public.lobbytools.com/legislators/7.

5. "Charlie Clary," DAG Architects, http://www.dagarchitects.com/charlie-clary.

6. "Waterside Village North," Emporis, https://www.emporis.com/buildings/1337615/waterside-village-north-mexico-beach-fl-usa.

7. Senate Bill 0004c2, Florida Senate, Committees on Banking and Insurance; Comprehensive Planning, Local and Military Affairs; and Senators Clary, Diaz-Balart, Campbell, Lee, McKay, Casas and Sullivan, Public Records, The Florida Senate, 2000, https://www.flsenate.gov/Session/Bill/2000/4/BillText/c2/PDF.

8. "Development of Loss Relativities for Wind Resistive Features of Residential Structures."

9. *Codes Quarterly: The News Journal of the Florida Building Commission* 1, no. 1 (September 2006); Gary Fineout, "Hurricane Michael Lays Bare Florida Panhandle's Weaker Building Codes," NU Property Casualty 360, October 18, 2018, https://www.propertycasualty360.com/2018/10/18/hurricane-michael-lays-bare-panhandles-weaker-buil/?slreturn=20180927071104; Andres Viglucci, David Ovalle, Caitlin Ostroff and Nicholas Nehamas, "Florida's Building Code Is Tough, but Michael Was Tougher: Is It Time for a Rewrite?," *Miami Herald*, October 16, 2018, https://www.miamiherald.com/news/state/florida/article219862625.html; Linda Kleindienst and Mark Hollis, "Building Code in Panhandle Less Strict," *Los Angeles Times*, September 19, 2004, https://www.latimes.com/archives/la-xpm-2004-sep-19-na-code19-story.html.

10. "Parable of the Isms (1935–)," Humanitysdarkside.com, May 5, 2015, https://humanitysdarkerside.com/2014/05/15/parable-of-the-isms-1936/.

11. "You Have Two Cows: You Have n Cows," Uncyclopedia.com, https://en.uncyclopedia.co/wiki/You_have_two_cows/26.

12. "Fun: You Have Two Cows," Rational Wiki, https://rationalwiki.org/wiki/Fun:You_have_two_cows.

13. "Ilona Saller," Wikipedia, https://en.wikipedia.org/wiki/Ilona_Staller.

14. Mona Charen, "Don't Forget How Butcher of Baghdad Earned the Name," *Baltimore Sun*, March 10, 2003, https://www.baltimoresun.com/news/bs-xpm-2003-03-10-0303100303-story.html.

15. Lenny Flank, "Cicciolina: The Porn Star in Parliament," *Daily Kos*, February 12, 2015, https://www.dailykos.com/stories/2015/2/12/1328892/-Cicciolina-The-Porn-Star-in-Parliament.

16. "Who Says Washington Is 'Hollywood for ugly people'? We Trace a Cliche Back to Its Origins," *Washington Post*, December 6, 2010, http://voices.washingtonpost.com/reliable-source/2010/12/who_says_washington_is_hollywo.html.

17. As reported by Mike Harcourt, former premier of British Columbia, Practical Wisdom, November 6, 2011, https://blogs.ubc.ca/cameron/2011/11/06/why-dont-more-good-people-go-into-politics-and-what-can-be-done-about-it/.

18. Jesse Singal, "Politicians Have Different Personalities Than the Rest of Us, Personality Psychology," *New York Magazine*, February 1, 2017, https://www.thecut.com/2017/02/politicians-have-different-personalities-than-the-rest-of-us.html.

19. Christian Jarrett, "US politicians Differed from the Public on Each of the Five Main Personality Traits," *Research Digest*, January 31, 2017, https://digest.bps.org.uk/2017/01/31/us-politicians-differed-from-the-public-on-each-of-the-five-main-personality-traits/.

20. Eric W. Dolan, "Psychology Study Indicates That Narcissists Are More Involved in Politics Than the Rest of Us," *Political Psychology*, September 30, 2020, https://www.psypost.org/2020/09/psychology-study-indicates-that-narcissists-are-more-involved-in-politics-than-the-rest-of-us-58112.

21. Leon F. Seltzer, "Narcissism: Why It's So Rampant in Politics," *Psychology Today*, December 21, 2011, https://www.psychologytoday.com/us/blog/evolution-the-self/201112/narcissism-why-its-so-rampant-in-politics.

22. David Rosen, "The 6 Political Personality Types," Campaigns & Elections, October 7, 2013, https://www.campaignsandelections.com/campaign-insider/the-6-political-personality-types.

23. "California Floods," Wikipedia, https://en.wikipedia.org/wiki/2017_California_floods; Tom Di Liberto, "Soaking Rains and Massive Snows Pile Up in California in January 2017," Climate.gov, NOAA, January 30, 2017, https://www.climate.gov/news-features/event-tracker/soaking-rains-and-massive-snows-pile-california-january-2017.

24. "Lake Oroville Spillway Incident: Timeline of Major Events February 4–25," California Department of Water Resources, http://www.science.earthjay.com/instruction/HSU/2017_spring/GEOL_106/lectures/lecture_13/Lake_Oroville_events_timeline.pdf.

25. J. France, I. Alvi, P. Dickson, H. Falvey, S. Rigbey, and J. Trojanowski, *Independent Forensic Team Report for Oroville Dam Spillway Incident* (Dam Safety, May 18, 2001), https://damsafety.org/sites/default/files/files/Independent%20Forensic%20Team%20Report%20Final%2001-05-18.pdf.

26. Ibid., S01.

27. Ibid., S03–S04.

28. "Oroville Dam Crisis," Wikipedia, https://en.wikipedia.org/wiki/Oroville_Dam_crisis#cite_note-42.

29. See https://www.google.com/search?client=firefox-b-1-d&q=3.5+million+acre-feet.

30. "America's Infrastructure Scores a D+," *2017 Infrastructure Report Card*, American Society of Civil Engineers, https://2017.infrastructurereportcard.org/.

31. *2021 Report Card for America's Infrastructure: A Comprehensive Assessment of America's Infrastructure*, American Society of Civil Engineers, https://

infrastructurereportcard.org/wp-content/uploads/2020/12/National_IRC_2021-report.pdf.

32. "Dams," in *2021 Report Card for America's Infrastructure*, American Society of Civil Engineers, 2021, https://infrastructurereportcard.org/wp-content/uploads/2020/12/Dams-2021.pdf.

33. "Dams," in *2017 Report Card for America's Infrastructure*, American Society of Civil Engineers, 2017, https://2017.infrastructurereportcard.org/wp-content/uploads/2017/01/Dams-Final.pdf.

34. "Dams," *2013 Report Card for America's Infrastructure*, American Society of Civil Engineers, 2013, http://2013.infrastructurereportcard.org/dams/.

35. "Dams," *2009 Report Card for America's Infrastructure*, American Society of Civil Engineers, 2009, https://www.infrastructurereportcard.org/wp-content/uploads/2018/01/Report-Card-for-Americas-Infrastructure-Full-Book.pdf.

36. "Case Studies, Lessons Learned from Dam Incidents and Failures," Association of State Dam Safety Officials, https://damfailures.org/case-study/.

37. "South Fork Dam," Wikipedia, https://en.wikipedia.org/wiki/South_Fork_Dam; "Case Study: South Fork Dam (Pennsylvania, 1889)," Association of State Dam Safety Officials, https://damfailures.org/case-study/south-fork-dam-pennsylvania-1889/.

38. "St. Francis Dam," Wikipedia, https://en.wikipedia.org/wiki/St._Francis_Dam; "Case Study: St. Francis Dam (California, 1928)," Association of State Dam Safety Officials, https://damfailures.org/case-study/st-francis-dam-california-1928/.

39. Scott Harrison, "California Retrospective: St. Francis Dam Collapse Left a Trail of Death and Destruction," *Los Angeles Times*, March 19, 2016, https://www.latimes.com/local/california/la-me-stfrancis-dam-retrospective-20160319-story.html.

40. "Programs," Division of Safety of Dams, California Department of Water Resources, https://water.ca.gov/Programs/All-Programs/Division-of-Safety-of-Dams.

41. California Department of Water Resources, https://water.ca.gov/About.

42. California Department of Water Resources, https://water.ca.gov/About/History.

43. "State Water Project," California Department of Water Resources, https://water.ca.gov/Programs/State-Water-Project; Nick DeBar, Lisa Maddaus, and Dave Stoughton, "Oroville Dam," https://watershed.ucdavis.edu/shed/lund/dams/Oroville/OrovilleDam.html.

44. France et al. *Independent Forensic Team Report for Oroville Dam Spillway Incident*, 59.

45. Ibid.

46. Ibid., 64.

47. Ibid., 66–67.

48. Ibid., K2-5–K2-6.

49. Ibid., 61.

50. Ibid., 61.

51. Ibid., 62.

52. Ellen Edmonds, "Roadside Breakdowns Preventable with Proper Maintenance, Finds AAA," AAA Newsroom, October 8, 2015, https://newsroom.aaa.com/2015/10/roadside-breakdowns-preventable-with-proper-maintenance-finds-aaa/.

53. "What Actually Happens When There Is Not Enough Oil in a Car Engine?," Stack Exchange, https://mechanics.stackexchange.com/questions/24555/what-actually-happens-when-there-is-not-enough-oil-in-a-car-engine.

54. Camille Sheehan, "Americans Are Delaying Car Maintenance to the Tune of Nearly $25 Billion," Autocare Association, August 1, 2019, https://www.autocare.org/americans_delaying_maintenance_nearly_25_billion/.

55. Gemma Francis, "Nearly 75% of Millennials Unable to Change a Car Tyre, Finds Study," *Independent*, April 4, 2018, https://www.independent.co.uk/life-style/millennials-change-tyre-practical-car-maintenance-diy-rewire-plug-a8288211.html.

56. "Idiot light," Wikipedia, https://en.wikipedia.org/wiki/Idiot_light.

57. "Half Life: The Decay of Knowledge and What to Do about It," fs.com, https://fs.blog/2018/03/half-life/.

58. Robert N. Charette, "An Engineering Career: Only a Young Person's Game?," *IEEE Spectrum*, September 4, 2013, https://spectrum.ieee.org/riskfactor/computing/it/an-engineering-career-only-a-young-persons-game.

59. "GSA Chief Resigns over Extravagant Spending," NPR, April 3, 2012, https://www.npr.org/2012/04/03/149898854/gsa-chief-resigns-over-agencys-extravagant-spending.

60. Alexander Abad-Santos, "GSA Threw an $800,000 Party and All You Got Was the Bill," *Atlantic*, April 3, 2012, https://www.theatlantic.com/politics/archive/2012/04/gsa-threw-800000-party-and-all-you-got-was-bill/329797/.

61. "Pohénégamook, Plus qu'une légende," Ville de Pohénégamook, 2021, http://pohenegamook.net/.

62. "FY 2021 Budget in Brief," US Department of Homeland Security, 2021, https://www.dhs.gov/sites/default/files/publications/fy_2021_dhs_bib_0.pdf.

63. "About DHS," US Department of Homeland Security, July 13, 2021 https://www.dhs.gov/about-dhs.

64. "Creation of the Department of Homeland Security," US Department of Homeland Security, September 24, 2015, https://www.dhs.gov/creation-department-homeland-security.

65. Homeland Security Presidential Directive-3, The White House, President George W. Bush, March 12, 2002, https://georgewbush-whitehouse.archives.gov/news/releases/2002/03/20020312-5.html.

66. *Homeland Security Advisory Council, HSAS Task Force, Stakeholder Feedback*, U.S. Department of Homeland Security, 2009, https://www.dhs.gov/xlibrary/assets/hsas_task_force_stakeholder_feedback.pdf; "Homeland Security Advisory System," Wikipedia, https://en.wikipedia.org/wiki/Homeland_Security_Advisory_System.

67. "Public Trust in Government: 1958–2021," Pew Research Center, May 17, 2021, https://www.people-press.org/2019/04/11/public-trust-in-government-1958-2019/.

68. Mary Gormandy White, "Examples of Public Policy," YourDictionary.com, https://examples.yourdictionary.com/examples-of-public-policy.html.

69. S. E. Flynn, *America the Vulnerable: How Our Government Is Failing to Protect Us from Terrorism* (New York: HarperCollins, 2007), 158.

70. "Seismic Policy: What Drives the Agenda in Earthquake Policy?," YouTube, https://www.youtube.com/watch?v=GI45V7vAr1s.

71. Flynn, *America the Vulnerable*, 95.

72. Andreas Rutkauskas, "The Invisible Security of Canada's Seemingly Chill Border," *Wired*, April 1, 2016, https://www.wired.com/2016/04/invisible-security-canadas-seemingly-chill-border/.

73. Ken Klippenstein, "Exclusive: Inside Trump's Failed Plan to Surveil the Canadian Border," *The Nation*, April 10, 2020, https://www.thenation.com/article/politics/canada-border-covid-security/.

74. Wilson Ring, "Human Smuggling Getting Sophisticated on U.S.-Canada Border," *Chicago Tribune*, July 23, 2018, https://www.chicagotribune.com/nation-world/ct-human-smuggling-us-canada-border-20180723-story.html.

75. Anne Marie Knott, "The Real Reasons Companies Are So Focused on the Short Term," *Harvard Business Review*, December 13, 2017, https://hbr.org/2017/12/the-real-reasons-companies-are-so-focused-on-the-short-term.

76. Michael Brown, "Stop Blaming Me for Hurricane Katrina—Ten Years Later, the Name 'Brownie' Is Still Identified with the Government's Failures: Here's What Really Happened," *Politico*, August 27, 2015, https://www.politico.com/magazine/story/2015/08/katrina-ten-years-later-michael-brown-121782/.

77. Ibid.

78. *A Failure of Initiative, Final Report of the Select Bipartisan Committee to Investigate the Preparation for and Response to Hurricane Katrina by the Select Bipartisan Committee to Investigate the Preparation for and Response to Hurricane Katrina*, Congress.gov. H. Rept. 109-377, US House of Representatives (Washington, DC: US Government Printing Office, 2006), https://www.congress.gov/congressional-report/109th-congress/house-report/377/1 or https://www.nrc.gov/docs/ML1209/ML12093A081.pdf.

79. Frank Davies, "Ex-FEMA Chief Was Advised to Look Good," *Chicago Tribune*, November 4, 2005, https://www.chicagotribune.com/news/ct-xpm-2005-11-04-0511040114-story.html.

80. Brown, "Stop Blaming Me for Hurricane Katrina."

81. "2008 United States Presidential Election," Wikipedia, https://en.wikipedia.org/wiki/2008_United_States_presidential_election#Issues.

82. Flynn, *America the Vulnerable*, 175.

EARTHQUAKE DAMAGE HAPPENS BECAUSE . . .

1. "Iran Declares Ban on Western Music," *Guardian*, December 20, 2005, https://www.theguardian.com/world/2005/dec/20/iran.

2. The Tenth Amendment states: "The powers not delegated to the United States by the Constitution, nor prohibited by it to the States, are reserved to the States respectively, or to the people." "Tenth Amendment to the United States Constitution," Wikipedia, https://en.wikipedia.org/wiki/Tenth_Amendment_to_the_United_States_Constitution#Text.

3. Roberto Leon and James Rossberg, "Evolution and Future of Building Codes in the USA," *Structural Engineering International* 22, no. 2 (2012): 265–69, doi: 10.2749/101686612X13291382991047.

4. "Code of Hammurabi," History.com, November 9, 2009, https://www.history.com/topics/ancient-history/hammurabi.

5. Heather McCune, "The Birth of Building Codes," ProBuilder.com, October 31, 2001, https://www.probuilder.com/birth-building-codes.

6. "Did Nero Really Fiddle While Rome Burned?," History.com, November 20, 2012, https://www.history.com/news/did-nero-really-fiddle-while-rome-burned.

7. Emily Torem, "The Great Chicago Fire Changed Building Code Forever," Moss Design, October 12, 2017, http://moss-design.com/the-great-chicago-fire-changed-building-code-forever/.

8. "Great Fire of London," Wikipedia, https://en.wikipedia.org/wiki/Great_Fire_of_London; "This Day in History, September 2," History.com, February 9, 2010, https://www.history.com/this-day-in-history/great-fire-of-london-begins.

9. "Chicago Fire of 1871," History.com, March 4, 2010, https://www.history.com/topics/19th-century/great-chicago-fire.

10. Henry C. Binford, "Tenements," in *Encyclopedia of Chicago*, http://www.encyclopedia.chicagohistory.org/pages/1240.html.

11. Joseph C. Bigott, "Building Codes and Standards," in *Encyclopedia of Chicago*, http://www.encyclopedia.chicagohistory.org/pages/179.html.

12. "Beauvais Cathedral: The Gravity-Defying Church," French Moments, December 29, 2012, https://frenchmoments.eu/beauvais-cathedral/.

13. Emma J. Wells, "Bravado or Blunder? Architectural Fails of Medieval Church Builders," History Extra, July 1, 2019, https://www.historyextra.com/period/medieval/church-architecture-fails-middle-ages-cathedrals-design-how-built/.

14. Sophie Roberts, "The Big Grapple: Historic Snaps Show the Terrifying Conditions for Empire State Building Construction Workers in the 1930s," *The Sun*, May 1, 2017, https://www.thesun.co.uk/living/3456414/empire-state-building-construction-workers-photography/.

15. Chris Wild, "Empire State Building Daredevil Photos Will Give You Vertigo," Mashable, March 7, 2015, https://mashable.com/2015/03/07/empire-state-building-vertigo/.

16. "How the Golden Gate Bridge Changed Safety Standards," FixFast USA, February 12, 2016, https://www.fixfastusa.com/news-blog/golden-gate-bridge-changed-safety-standards/.

17. Colin Nabity, "How Much Does Medical Malpractice Insurance Cost in 2021?," LeverageRx, May 14, 2018, https://www.leveragerx.com/blog/medical-malpractice-insurance-cost/.

18. "Professional Liability vs. Errors and Omissions Insurance," Insureon, https://www.insureon.com/blog/professional-liability-vs-errors-omissions-insurance-whats-the-difference.

19. "History of the AISC Specification 1923–2010," AISC Live Webinar, January 22, 2016, https://www.aisc.org/globalassets/continuing-education/quiz-handouts/history-of-the-aisc-specification.pdf.

20. Roch Steinbach, "Hoover's Building Code Committee Report—1925," June 7, 2017, https://www.slideshare.net/RochSteinbach1/hoovers-building-code-committee-report-1925.

21. "Building Officials and Code Administrators International," *The BOCA Basic Building Code* (Chicago, 1950).

22. Leon and Rossberg, "Evolution and Future of Building Codes."

23. *Southern Standard Building Code*, adopted November 16, 1945, at the annual convention of the Southern Building Code Congress (Birmingham, AL: Southern Building Code Congress); "The Abridged History of Buildings Codes," Whirlwind Steel, September 8, 2014, https://www.whirlwindsteel.com/2021/10/04/the-abridged -history-of-building-codes/.

24. Pacific Coast Building Officials Conference, and International Conference of Building Officials, *Building Standards, Uniform Building Code* (Long Beach, CA: Inter-national Conference of Building Officials).

25. Leon and Rossberg, "Evolution and Future of Building Codes."

26. Ibid.

27. W. Bronson, *The Earth Shook, the Sky Burned: A Moving Record of America's Great Earthquake and Fire* (San Francisco: Chronicle Books, 1986).

28. M. Bruneau, C. M. Uang., and R. Sabelli, *Ductile Design of Steel Structures*, second edition (New York: McGraw-Hill, 2011).

29. "Historical UBC," Digital Assets Library, UC Berkeley, http://digitalassets.lib .berkeley.edu/ubc/UBC_1927.pdf.

30. I. Towhata, "History of Geotechnical Earthquake Engineering in Japan," Proceed-ings of 14th World Conference on Earthquake Engineering, October 12–17, Beijing, China, IAEE.

31. K. Suyehiro, "Engineering Seismology Notes on American Lectures," *Proceedings of ASCE* 58, no. 4 (1932): 9–110.

32. R. Iacopi, *Earthquake Country: How, Why, and Where Earthquakes Strike in Cali-fornia* (Menlo Park, CA: Sunset Books, 1981).

33. D. J. Alesch and W. J. Petak, *The Politics and Economics of Earthquake Hazard Mitigation* (Boulder: Institute of Behavioral Science, University of Colorado, 1986).

34. Leon and Rossberg, "Evolution and Future of Building Codes."

35. See Plato, Liberty Tree, http://libertytree.ca/quotes/Plato.Quote.3067.

36. "Building Codes in the New Madrid Seismic Zone (NMSZ)," FEMA, Homeland Security Digital Library, 2010, https://www.hsdl.org/?view&did=7708.

37. Max Brantley, "Legislation Waives Earthquake Design Standards for Select Proj-ects," *Arkansas Times*, May 19, 2016, https://arktimes.com/arkansas-blog/2016/05/19/ legislation-waives-earthquake-design-standards-for-select-projects.

38. "1811–1812 New Madrid Earthquakes," Wikipedia, https://en.wikipedia .org/wiki/1811%E2%80%9312_New_Madrid_earthquakes; "Science of the New Madrid Seismic Zone," USGS, https://www.usgs.gov/natural-hazards/earthquake -hazards/science/new-madrid-seismic-zone?qt-science_center_objects=0#qt-science _center_objects.

39. "Building Codes in the New Madrid Seismic Zone (NMSZ)."

40. "Building Codes by State," Insurance Institute for Business & Home Safety, https://ibhs.org/public-policy/building-codes-by-state/.

41. "Building Codes by State—Georgia," Insurance Institute for Business & Home Safety, https://ibhs.org/public-policy/building-codes-by-state/#ga.

42. "Building Codes by State—Hawaii," Insurance Institute for Business & Home Safety, https://ibhs.org/public-policy/building-codes-by-state/#hi.

43. Christopher Flavelle, "Why Storm-Prone States Continue to Balk at Tough Building Codes," *Insurance Journal*, March 19, 2018, https://www.insurancejournal.com/news/national/2018/03/19/483773.htm.

44. Ibid.

45. Ibid.

46. *2019 Chicago Building Code*, City of Chicago Department of Buildings, October 10, 2019, https://www.chicago.gov/city/en/depts/bldgs/provdrs/bldg_code/alerts/2019/october/2019-chicago-building-code-books-now-available-for-pre-order.html.

47. David Wagman, "Chicago Looks to Revise Its 1940s-Era Building Code," Engineering360, March 29, 2019, https://insights.globalspec.com/article/11529/chicago-looks-to-revise-its-1940s-era-building-code.

48. Greg Cunningham, "Mayor Emanuel Introduces Major Building Code Modernization to Enable Safer, More Cost Effective and Innovative Construction," City of Chicago Department of Buildings, March 13, 2019, https://www.chicago.gov/city/en/depts/bldgs/provdrs/bldg_code/news/2019/march/mayor-emanuel-introduces-major-building-code-modernization-to-en.html.

49. "NSPE Code of Ethics for Engineers," National Society of Professional Engineers, https://www.nspe.org/resources/ethics/code-ethics.

50. Thomas Fuller, Anjali Singhvi, Mika Gröndahl and Derek Watkins, "Buildings Can Be Designed to Withstand Earthquakes—Why Doesn't the U.S. Build More of Them?," *New York Times*, June 7, 2019, https://www.nytimes.com/interactive/2019/06/03/us/earthquake-preparedness-usa-japan.html.

51. "Mitigation Saves: Mitigation Saves up to $13 per $1 Invested," The National Institute of Building Sciences, 2020, http://2021.nibs.org/files/pdfs/mitigationsaves2019_complete.pdf.

52. Fuller et al., "Buildings Can Be Designed to Withstand Earthquakes."

53. Assembly Bill AB-1857 Building Codes: Earthquake Safety: Immediate Occupancy Standard (2017–2018), California Legislative Information, California Legislature 2017–2018 Regular Session, https://leginfo.legislature.ca.gov/faces/billTextClient.xhtml?bill_id=201720180AB1857.

54. Assembly Bill AB-393 Building Codes: Earthquake Safety: Functional Recovery Standard (2019–2020), California Legislative Information, California Legislature 2019–2020 Regular Session, https://leginfo.legislature.ca.gov/faces/billStatusClient.xhtml?bill_id=201920200AB393.

55. Assembly Bill AB-1329 Building Codes: Earthquakes: Functional Recovery Standard (2021–2022), California Legislative Information, California Legislature 2020–2021 Regular Session, https://leginfo.legislature.ca.gov/faces/billNavClient.xhtml?bill_id=202120220AB1329.

FATALISM: A REALLY SHORT CHAPTER

1. "Fatalism," Wikipedia, https://en.wikipedia.org/wiki/Fatalism.

2. "What God Wants, Part I," Wikipedia, https://en.wikipedia.org/wiki/What_God_Wants,_Part_I.

3. Jacqueline Charles and Curtis Morgan, "Lack of Construction Codes Sealed Haitian Capital's Fate," McClatchy Washington Bureau APP, June 16, 2015, https://www.mcclatchydc.com/news/nation-world/world/article24571369.html.

4. Ker Than, "Bangladesh Building Collapse Due to Shoddy Construction," *National Geographic News*, April 26, 2013, https://www.nationalgeographic.com/history/article/130425-bangladesh-dhaka-building-collapse-world.

5. "Latest Earthquakes: M 7.3—29 km S of Halabja, Iraq," USGS, November 12, 2017, https://earthquake.usgs.gov/earthquakes/eventpage/us2000bmcg/executive#executive.

6. Thomas Erdbrink, "Iranians Are Outraged over Shoddy Construction in Earthquake Zone," *New York Times*, November 15, 2017, https://www.nytimes.com/2017/11/15/world/middleeast/iran-earthquake-construction-corruption.html.

7. "James Bay Project," Wikipedia, https://en.wikipedia.org/wiki/James_Bay_Project.

GLOOMY PREDICTIONS

1. Kennette Benedict, Angela Kane, Joana Castro Pereira, Philip Osano, David Heymann, Romana Kofler, Lindley Johnson, Gerhard Drolshagen, Stephen Sparks, Ariel Conn, *Global Catastrophic Risks 2020*, Global Challenges Foundation, 2020, https://globalchallenges.org/wp-content/uploads/Global-Catastrophic-Risks-2020-Annual-Report.pdf; "Global Catastrophic Risk," Wikipedia, https://en.wikipedia.org/wiki/Global_catastrophic_risk#Non-anthropogenic.

2. H. G. Wells, *The War of the Worlds*, Project Gutenberg, https://www.gutenberg.org/files/36/36-h/36-h.htm.

3. Seth Shostak, "We Just Beamed a Signal at Space Aliens. Was That a Bad Idea?," NBC Mach, November 20, 2017, https://www.nbcnews.com/mach/science/we-just-beamed-signal-space-aliens-was-bad-idea-ncna822446.

4. Matt Williams, "What Is the Life Cycle of the Sun?," Universe Today, December 22, 2015, https://www.universetoday.com/18847/life-of-the-sun/.

5. C.-H. Geschwind, *California Earthquakes: Science, Risk, and the Politics of Hazard Mitigation* (Baltimore: Johns Hopkins University Press, 2008), 85; Susan Hough, *Predicting the Unpredictable: The Tumultuous Science of Earthquake Prediction* (Princeton, NJ: Princeton University Press, 2010), 22–24.

6. Geschwind, *California Earthquakes*, 88.

7. Ibid., 88–96.

8. Ibid.

9. Steven Erlanger, "The War to End All Wars? Hardly. But It Did Change Them Forever," *New York Times*, June 26, 2014, https://www.nytimes.com/2014/06/27/world/europe/world-war-i-brought-fundamental-changes-to-the-world.html.

10. "The War to End All Wars," BBC News, 10 November, 1998, http://news.bbc.co.uk/2/hi/special_report/1998/10/98/world_war_i/198172.stm.

11. Vaclav Smil, "Too Cheap to Meter," Nuclear Power Revisited, IEEE Spectrum, September 26, 2016, https://spectrum.ieee.org/too-cheap-to-meter-nuclear-power-revisited; Lewis L. Strauss, Remarks Prepared for Delivery at the Founder's Day Dinner of National Association of Science Writers, US Atomic Energy Commission (September 16, 1954), 9. http://www.nrc.gov/docs/ML1613/ML16131A120.pdf.

12. "Too Cheap to Meter," Wikipedia, https://en.wikipedia.org/wiki/Too_cheap_to_meter.

13. "Nuclear Power in the World Today," World Nuclear Association (updated February 2022), https://www.world-nuclear.org/information-library/current-and-future-generation/nuclear-power-in-the-world-today.aspx.

14. "Economics of Nuclear Power," World Nuclear Association (updated September 2021), https://www.world-nuclear.org/information-library/economic-aspects/economics-of-nuclear-power.aspx.

15. "Y2K Bug," *National Geographic* Resource Library, January 21, 2011, https://www.nationalgeographic.org/encyclopedia/Y2K-bug/.

16. Josh K. Elliott, "Y2K 20-20: How a New Year's Mass Scare Became an Embarrassing Joke," *Global News*, December 31, 2019, pmhttps://globalnews.ca/news/6351344/y2k-scare-new-year-2000/.

17. "Y2K Doomsayers Say They Were Wrong," AP News, January 5, 2000, https://apnews.com/919d23611bfd5ebbe8b5d2e75dee5cd9.

18. Elliott, "Y2K 20-20."

19. Joanna Glasner, "We're Ready for 2000, Really," *Wired*, December 23, 1999, https://www.wired.com/1999/12/were-ready-for-2000-really/.

20. "The Glitch That Didn't Steal New Year's," *Scientific American*, January 17, 2000, https://www.scientificamerican.com/article/the-glitch-that-didnt-ste/.

THE SILENT HEROES

1. "The Story of One of the Most Memorable Marketing Blunders Ever," The Coca-Cola Company, https://www.coca-colacompany.com/news/the-story-of-one-of-the-most-memorable-marketing-blunders-ever.

2. "New Coke," Wikipedia, https://en.wikipedia.org/wiki/New_Coke#Backlash.

3. Tim Murphy, "New Coke Didn't Fail. It Was Murdered," *Mother Jones*, July 9, 2019, https://www.motherjones.com/food/2019/07/what-if-weve-all-been-wrong-about-what-killed-new-coke/.

4. C.-H. Geschwind, *California Earthquakes: Science, Risk, and the Politics of Hazard Mitigation* (Baltimore: Johns Hopkins University Press, 2008), 155–64.

5. Ibid., 65.

6. Kip Wiley, *Living Where the Earth Shakes: A History of the California Seismic Safety Commission*, Seismic Safety Commission, December 2000, https://ssc.ca.gov/wp-content/uploads/sites/9/2020/08/cssc_history.pdf.

7. D. J. Alesh and W. J. Petak, *The Politics and Economics of Earthquake Hazard Mitigation*, Program on Environment and Behavior Monograph #43 (Boulder: Institute of Behavioral Science, University of Colorado, 1986), 19.

8. Ibid., 9.

9. Ibid., 40.

10. Ibid., 48–49.

11. Ibid., 138.

12. Ibid., 57–79.

13. *Status of the Unreinforced Masonry Building Law*, Report SSC 2005-02, Seismic Safety Commission, June 9, 2005, https://ssc.ca.gov/wp-content/uploads/sites/9/2020/08/cssc_2005-02_urm.pdf.

14. "L'abri privé—Aide-mémoire," République et Canton de Genève, May 23, 2018, https://www.ge.ch/document/brochure-destinee-aux-proprietaires-abris-prives/telecharger.

15. "J'ai un abri de protection civile, quelles sont mes obligations?," République et Canton de Genève, October 11, 2021, https://www.ge.ch/j-ai-abri-protection-civile-quelles-sont-mes-obligations.

16. "Construction et entretien des abris," Ville de Genève, May 26, 2021, https://www.geneve.ch/fr/themes/securite-prevention/protection-civile/abris-constructions/abris-publics/construction-entretien.

17. Jo Fahy, "The Forgotten Underground World of Swiss Bunkers," Swissinfo.ch, September 1, 2016, https://www.swissinfo.ch/eng/in-case-of-emergency_the-forgotten-underground-world-of-swiss-bunkers/42395820.

18. Daniele Mariani, "Bunkers for All," Swissinfo.ch, July 3, 2009, https://www.swissinfo.ch/eng/prepared-for-anything_bunkers-for-all/995134.

19. Ibid.

20. Sweden's Civil Contingencies Agenda map of its 65,000 nuclear shelters can be accessed by clicking on the link Öppna verktyget—Skyddsrumskarta on the following webpage: https://www.msb.se/sv/verktyg--tjanster/skyddsrumskarta/.

21. Mariani, "Bunkers for All."

22. Fahy, "The Forgotten Underground World of Swiss Bunkers."

TRUTH AND LIES ABOUT RESILIENCE

1. Apparently, a term (from Latin *resilio*, "to spring back") first introduced to describe elastic recovery, on page 44 of R. Mallet, *On the Physical Conditions Involved in the Construction of Artillery: An Investigation of the Relative and Absolute Values of the Materials Principally Employed and of Some Hitherto Unexplained Causes of the Destruction of the Cannon in Service* (London: Longman, Brown, Green, Longmans, and Roberts, 1856). Without any reference to recovery, earlier mention of "resilience" (as the property of "resisting bodies in motion") is found in Thomas Tredgold, "XXX-VII. On the Transverse Strength and Resilience of Timber," *The Philosophical Magazine*, 51:239 (1818), 214–16, doi: 10.1080/14786441808637536.

2. S. Levin, "Ecological Resilience," *Encyclopedia Britannica*, December 29, 2015, https://www.britannica.com/science/ecological-resilience.

3. J. Fleming and R. J. Ledogar, "Resilience, an Evolving Concept: A Review of Literature Relevant to Aboriginal Research," *Pimatisiwin* 6, no. 2 (2008): 7–23, https://www.ncbi.nlm.nih.gov/pmc/articles/PMC2956753/.

4. "Bills Focus: Resilient Effort in Detroit," Buffalo Bills, www.buffalobills.com, October 5, 2014, https://www.buffalobills.com/video/bills-focus-resilient-effort-in-detroit-13930057.

5. Allen Foster, "The Best Chocolate Truffle," *Chicago Tribune*, January 28, 2020, https://www.chicagotribune.com/consumer-reviews/sns-bestreviews-kitchen-the-best-chocolate-truffle-20200129-cbchsfqv6bca5oazzjyc6noxf4-story.html.

6. "What Is Resilient Flooring?," Congoleum Flooring, https://www.congoleum .com/what-is-resilient/.

7. First seen in an article formerly available at: https://www.thethings.com/stunning -photos-of-brand-new-non-american-sports-cars/ describing the Porsche 911, a description reused by some Porsche dealers at the time; similar examples can be found for other brands, such as: Victor Troia, "The Resilient Roadster: How the Mazda Miata Continues to Dominate Its Class for Over 30 Years," Hotcars.com, February 12, 2022, https://www.hotcars.com/the-resilient-roadster-how-the-mazda-miata-continues -to-dominate-its-class-for-over-30-years/.

8. Natalie Clarkson, "Introducing *Resilient Lady*: Virgin Voyages' Third Ship Is an Ode to Womankind," Virgin, March 8, 2021, https://www.virginvoyages.com/ship #resilient-lady.

9. US Department of State, "What Is Resilience?," https://2009-2017.state.gov/m/ med/dsmp/c44950.htm (retrieved 9/9/2021).

10. G. P. Cimellaro, D. Solari, and M. Bruneau, "Physical Infrastructure Interdependency and Regional Resilience Index after 2011 Tohoku Earthquake in Japan," *Earthquake Engineering and Structural Dynamics* 43, no. 12 (2014): 1763–84, doi: 10.1002/ eqe.2422.

11. "List of disasters in Canada," Wikipedia, https://en.wikipedia.org/wiki/List_of _disasters_in_Canada.

12. "Ice hockey at the Olympic Games," Wikipedia, https://en.wikipedia.org/wiki/Ice _hockey_at_the_Olympic_Games.

13. "CCM (ice hockey)," Wikipedia, https://en.wikipedia.org/wiki/CCM_(ice_hock ey),=.

14. "Bauer Hockey," Wikipedia, https://en.wikipedia.org/wiki/Bauer_Hockey.

15. Shanna McCarriston, "How the Coronavirus Is Creating an NHL Stick Shortage," CBS Sports, February 11, 2020, https://www.cbssports.com/nhl/news/how-the -coronavirus-is-creating-an-nhl-stick-shortage/.

16. Tommy McArdle, "Coronavirus' Impact on Life in China Could Potentially Cause NHL Stick Shortage," per Report, *Sporting News*, February 9, 2020, https://www.sport ingnews.com/us/nhl/news/coronavirus-impact-on-life-in-china-could-potentially -cause-nhl-stick-shortage-per-report/17zx9izdxer9q1dd69zndoy2ed.

17. "Just-in-Time Manufacturing: The Path to Efficiency," Kanbanize, https:// kanbanize.com/lean-management/pull/just-in-time-production.

18. George Horwich, "Economic Lessons of the Kobe Earthquake," *Economic Development and Cultural Change* 48, no. 3 (2000): 521–42, https://doi.org/10.1086/452609.

19. "Factbox: Japan's Recovery from the 1995 Kobe Earthquake," Reuters, March 25, 2011, https://www.reuters.com/article/us-japan-quake-kobe-recovery/factbox-japans -recovery-from-the-1995-kobe-earthquake-idUSTRE72O13G20110325.

20. "Henry Ford," Wikipedia, https://en.wikipedia.org/wiki/Henry_Ford#The_five -dollar_wage.

21. Sarah Cwiek, "The Middle Class Took Off 100 Years Ago . . . Thanks to Henry Ford?," NPR, January 27, 2014, https://www.npr.org/2014/01/27/267145552/the -middle-class-took-off-100-years-ago-thanks-to-henry-ford.

22. Jeff Rindskopf, "30 Iconic U.S. Brands That Aren't Made in America," *Cheapism*, April 5, 2021, https://blog.cheapism.com/american-products-not-made-in-america-16404/#slide=5.

23. "Mattel Recalls 800,000 Toys Worldwide Because of Lead Paint," Wave 3 News September 5, 2007, https://www.wave3.com/story/7026748/mattel-recalls-800000-toys-worldwide-because-of-lead-paint/.

24. Tom Jurkowsky, "Coronoavirus Pandemic Proof That It's Time to Cut Our Dependency on China for Drugs," *Capital Gazette*, March 22, 2020, https://www.capitalgazette.com/opinion/columns/ac-ce-column-tom-jurkowsky-20200322-ougrysd64rggzogoahcz76luxu-story.html.

25. Guy Taylor, "'Wake-Up call': Chinese Control of U.S. Pharmaceutical Supplies Sparks Growing Concern," *Washington Times*, March 17, 2020, https://www.washingtontimes.com/news/2020/mar/17/china-threatens-restrict-critical-drug-exports-us/.

26. Phil Stewart and Mike Stone, "U.S. Military Comes to Grips with Over-Reliance on Chinese Imports," Reuters, October 2, 2018, https://www.reuters.com/article/us-usa-military-china/u-s-military-comes-to-grips-with-over-reliance-on-chinese-imports-idUSKCN1MC275.

27. M. Bruneau and G. MacRae, *Reconstructing Christchurch: A Seismic Shift in Building Structural Systems* (Christchurch, New Zealand: Quake Center, University of Canterbury, 2017).

28. Olivia Carville, "Church Leaders Back Bishop," *Stuff*, April 3, 2012, http://www.stuff.co.nz/the-press/news/6682160/Church-leaders-back-bishop.

29. "Christchurch Cathedral," Wikipedia, https://en.wikipedia.org/wiki/Christ Church_Cathedral#Proposed_demolition.

30. Carlie Gates, "Final Cathedral Design Options Unveiled," *Stuff*, April 10, 2013, http://www.stuff.co.nz/the-press/news/city-centre/8504585/Final-Cathedral-design-options-unveiled.

31. Scott Lewis, "The 10 Largest Base-Isolated Buildings in the World," ENR, July 17, 2017, https://www.enr.com/articles/42366-the-10-largest-base-isolated-buildings-in-the-world.

32. Andreas Illmer, "New Zealand: Fixing the Ruined Christchurch Cathedral That's Frozen in Time" (video), December 11, 2020, BBC News, World, https://www.bbc.com/news/av/world-55256198.

33. "Christchurch Cathedral Reinstatement Project, Concept Design and Project Fact Sheet," October 2020, https://christchurchcathedraltrust-my.sharepoint.com/:b:/g/personal/amandas_reinstate_org_nz/
EQkqEFZcL-1FkXUPTPznF5YBPiK0GIHxCNafikkY_Mnv-g?e=fpaf9E.

34. Nick Perry, "10 Years After Quake, Christ Church Cathedral Finally Rising," AP News, February 20, 2021, https://apnews.com/article/world-news-earthquakes-christchurch-new-zealand-10711b47a4b90b03a6bb21d80f48af17.

35. Christopher Helman and Hank Tucker, "The War in Afghanistan Cost America $300 Million per Day for 20 Years, with Big Bills Yet to Come," *Forbes*, August 16, 2021, https://www.forbes.com/sites/hanktucker/2021/08/16/the-war-in-afghanistan-cost-america-300-million-per-day-for-20-years-with-big-bills-yet-to-come/?sh=6dfe5d187f8d.

36. Multi-Hazard Mitigation Council, *Natural Hazard Mitigation Saves: 2019 Report* (Washington, DC: National Institute of Building Sciences, 2019), www.nibs.org; "Natural Hazard Mitigation Saves Interim Report," FEMA Fact Sheet, Federal Insurance and Mitigation Administration, June 2018, https://www.fema.gov/sites/default/files/2020-07/fema_mitsaves-factsheet_2018.pdf; "Mitigation Saves: Mitigation Saves up to $13 per $1 Invested," National Institute of Building Sciences, http://2021.nibs.org/files/pdfs/ms_v4_overview.pdf.

37. "Green Building Standards," EPA, https://www.epa.gov/smartgrowth/green-building-standards.

38. "The Original E.P.C.O.T Project, A Comprehensive History of Walt Disney's Last Dream," The Original E.P.C.O.T., https://sites.google.com/site/theoriginalepcot.

39. "Walt Disney's Quotes," The Original E.P.C.O.T., https://sites.google.com/site/theoriginalepcot/overview/walt-disney.

40. Wade Shepard, "Should We Build Cities from Scratch?," *Guardian*, July 10, 2019, https://www.theguardian.com/cities/2019/jul/10/should-we-build-cities-from-scratch.

41. Jacqui Palumbo, "New 'Future City' to Rise in Southwest China," CNN, February 9, 2021, https://www.cnn.com/style/article/chengdu-future-city-design-competition-masterplan/index.html.

42. "Chengdu Future Science and Technology City Launch Area Masterplan and Architecture Design," OMA, https://www.oma.com/projects/chengdu-future-science-and-technology-city-launch-area-masterplan-and-architecture-design.

43. Christchurch Central Recovery Plan (Multiple volumes), Department of the Prime Minister and Cabinet, New Zealand government, https://ceraarchive.dpmc.govt.nz/documents/christchurch-central-recovery-plan.

44. Liz McDonald, "Christchurch Nine Years On: The Reshaping of a City," *Stuff*, February 22, 2020, https://www.stuff.co.nz/national/119421238/christchurch-nine-years-on-the-reshaping-of-a-city.

45. M. Bruneau, S. E. Chang, R. T. Eguchi G. C. Lee, T. D. O'Rourke, A. M. Reinhorn, M. Shinozuka, K. Tierney, W. A. Wallace, and D. Von Winterfeldt, "A Framework to Quantitatively Assess and Enhance the Seismic Resilience of Communities," *Earthquake Spectra* 19, no. 4 (2003): 733–52.

46. See http://www.clowndoctors.org.nz/.

DOLLARS ARE FREQUENT FLYER MILES

1. James J. Nagle, "Trading Stamps: A Long History," *New York Times*, December 26, 1971, https://www.nytimes.com/1971/12/26/archives/trading-stamps-a-long-history-premiums-said-to-date-back-in-us-to.html. "S&H Green Stamps," Wikipedia, https://en.wikipedia.org/wiki/S%26H_Green_Stamps.

2. Dorri Partain, "Remember This? Trading Stamps," *Northeast News*, January 13, 2021, http://northeastnews.net/pages/remember-this-trading-stamps/.

3. "Trading Stamp," Wikipedia, https://en.wikipedia.org/wiki/; Sam Moore, "Trading Stamps Were More Than Currency for Some," *Farm and Dairy*, August 2, 2012, https://www.farmanddairy.com/columns/trading-stamps-were-more-than-currency-for-some/39813.html.

4. "Loyalty Program," Wikipedia, https://en.wikipedia.org/wiki/Loyalty_program; Max Friend, "The History and Future of Loyalty Programs," Business and Tech, https://www.futureofbusinessandtech.com/loyalty-and-rewards/the-history-and-future-of-loyalty-programs/#.

5. "List of Airline Bankruptcies in the United States," Wikipedia, https://en.wikipedia.org/wiki/List_of_airline_bankruptcies_in_the_United_States#Chapter_11; Airlines for America, U.S. Bankruptcies and Services Cessations, https://web.archive.org/web/20160528074046/http://airlines.org/data/u-s-bankruptcies-and-services-cessations/.

6. "Chapter 11—Bankruptcy Basics," Administrative Office of the US Courts, Federal Judiciary, https://www.uscourts.gov/services-forms/bankruptcy/bankruptcy-basics/chapter-11-bankruptcy-basics.

7. Leslie Josephs, "How the Sept. 11 Terrorist Attacks Forever Changed Air Travel," CNBC, September 11, 2021, https://www.cnbc.com/2021/09/11/how-9/11-forever-changed-air-travel.html; Dave Carpenter, "Judge Approves Termination of United Airlines Pension Plans," *Seattle Times*, May 10, 2005, https://www.seattletimes.com/business/judge-approves-termination-of-united-airlines-pension-plans/; "Why PBGC Cannot Restore the United Airlines' Pension Plans," The Pension Benefit Guaranty Corporation, June 2017, https://www.pbgc.gov/wr/large/united/united-airlines-plan-restoration (the PBGC is a US government agency); Chris Isidore, "Delta to Dump Pension Plans," CNN Money, June 16, 2006, https://money.cnn.com/2006/06/16/news/companies/delta_pensions/; Steve Karnowski, Associated Press, "Retirees Lose Out as Bankrupt Companies Cut Pensions," *The Ledger*, September 27, 2005, https://www.theledger.com/article/LK/20050927/News/608119486/LL.

8. *Bankruptcy and Pension Problems Are Symptoms of Underlying Structural Issues*, GAO Report GAO-05-945 to Congressional Committees (Washington, DC: United States Government Accountability Office, September 2005), https://www.govinfo.gov/content/pkg/GAOREPORTS-GAO-05-945/html/GAOREPORTS-GAO-05-945.htm.

9. Airlines for America, Data & Statistics, U.S. Airline Mergers and Acquisitions, Jun 5, 2020, http://airlines.org/dataset/u-s-airline-mergers-and-acquisitions/.

10. Susan Carey, "United's Merger Turbulence Hits Elite Frequent Fliers," *Wall Street Journal*, May 24, 2012, https://www.wsj.com/articles/SB10001424052702304451104577390140073664500; "2011 Mileage Plus and OnePass Elite Program Developments," FlyerTalk.com, https://www.flyertalk.com/forum/united-mileage-plus-pre-merger/1148667-2011-mileage-plus-onepass-elite-program-developments-14.html.

11. Susanna Kim, "Chicago 'Million Miler' Sues United Airlines in Class Action for 'Immorally' Taking Away Perks," ABC News, May 29, 2012, https://abcnews.go.com/Business/chicago-million-miler-united-airlines-immorally-benefits-files/story?id=16443814.

12. Ibid.

13. "MileagePlus," Wikipedia, https://en.wikipedia.org/wiki/MileagePlus#History; United States Court of Appeals, Seventh Circuit. George Lagen, on behalf of himself and all others similarly situated, Plaintiff–Appellant, v. United Continental Holdings, Inc., and United Airlines, Inc., Defendants–Appellees. No. 14–1375. Decided: December 22, 2014, https://caselaw.findlaw.com/us-7th-circuit/1687822.html.

14. Seth Miller, "United's Somewhat Pyrrhic Court Victory," *Wandering Aramean*, https://blog.wandr.me/2014/12/united-million-miler-court-victory/.

15. David Sims, "American Factory Grapples with the Notion of Freedom," *Atlantic*, August 30, 2019, https://www.theatlantic.com/entertainment/archive/2019/08/american-factory-review-julia-reichert-steven-bognar/597067/.

16. "Items of General Interest, Frequent Flyer Miles Attributable to Business or Official Travel," Internal Revenue Service, 2002-18, https://www.irs.gov/pub/irs-drop/a-02-18.pdf.

17. Eric Rosen, "40 years of Miles: The History of Frequent Flyer Programs," The Points Guy, May 20, 2021, https://thepointsguy.com/guide/evolution-frequent-flyer-programs/.

18. "Money Miles, Beginner's Guide: Why Are Points and Miles Referred to as Currency?" Money Miles, July 25, 2016 https://monkeymiles.boardingarea.com/flexible-flexible-currencies/#_ga=2.171867049.867067196.1569179037-1939142508.1569179037.

19. "The State and National Banking Eras: A Chapter in the History of Central Banking" (Philadelphia: Federal Reserve Bank of Philadelphia, December 2016), 5, https://fraser.stlouisfed.org/files/docs/publications/education/frbphi-chapters-state-national-banking-eras.pdf.

20. "Major Foreign Holders of Treasury Securities (in billions of dollars)," US Treasury, https://ticdata.treasury.gov/Publish/mfh.txt.

21. Chris Thomas, "The World's 12 Greatest Currency Failures," Gold IRA Guide, September 26, 2014, https://goldiraguide.org/the-worlds-greatest-currency-failures/.

22. Ibid.

23. S. Hanke and A. Kwok, "On the Measurement of Zimbabwe's Hyperinflation," *Cato Journal* 29, no. 2 (2009): 355, table 1, https://www.cato.org/sites/cato.org/files/serials/files/cato-journal/2009/5/cj29n2-8.pdf.

24. Michael J Boyle, "The Great Recession," The Investopedia Team, October 23, 2020, https://www.investopedia.com/terms/g/great-recession.asp.

25. "What's a Money Line? A Beginner's Guide to Betting Football in Vegas," *STN Blog*, https://www.stationcasinosblog.com/2019/10/whats-a-money-line-a-beginners-guide-to-betting-football-in-vegas/.

26. Alexandra Licata, "42 States Have or Are Moving Towards Legalizing Sports Betting—Here Are the States Where Sports Betting Is Legal," *Insider*, August 2, 2019, https://www.businessinsider.com/states-where-sports-betting-legal-usa-2019-7.

27. Jason Fernando, "Futures," Investopedia, August 9, 2021, https://www.investopedia.com/terms/f/futures.asp.

28. "What's a Money Line?"

29. Fernando, "Futures."

30. Julia Kagan, "Reinsurer," Investopedia, February 27, 2021, https://www.investopedia.com/terms/r/reinsurer.asp.

31. "Gross Reinsurance Premiums of $42,951,000,000 USD (life and non-life reinsurance combined)," in *Transforming Tomorrow Together: Annual Report 2020* (Zurich: Swiss Re, 2020), https://reports.swissre.com/2020/assets/pdfs/2020_financial_report_swissre_ar20.pdf.

32. "Gross Reinsurance Premiums of 37,321,000,000 Euros," in *Annual Report 2020* (Munich: Munich Re, 2020), https://www.munichre.com/en/company/investors/reports-and-presentations/annual-report.html.

33. Chris B. Murphy, "Catastrophe Bond," Investopedia, May 29, 2020, https://www.investopedia.com/terms/c/catastrophebond.asp.

34. Which is attributed to *Dirty Harry*, but is actually not the real quote, See, "'Do You Feel Lucky, Punk?': From *Dirty Harry* to *Star Wars*, the Famous Movie Quotes That Fans Always Get Wrong," *Daily Mail*, August 28, 2013, https://www.dailymail.co.uk/news/article-2403742/Do-feel-lucky-punk--Famous-movie-quotes-wrong-revealed.html and Wikipedia, https://en.wikipedia.org/wiki/Dirty_Harry.

WE'RE ALL COOKED

1. "Stats and Facts," The City of Windsor, Ontario, Canada, https://www.citywindsor.ca/residents/planning/plans-and-community-information/about-windsor/pages/stats-and-facts.aspx.

2. "Alert, Nunavut," Wikipedia, https://en.wikipedia.org/wiki/Alert,_Nunavut.

3. Ted Thornhill, "Four Months of Darkness a Year, Temperatures That Freeze Eyeballs and the Nearest Town Is 340 Miles Away: Inside the Most Northerly Settlement in the World," *Daily Mail*, February 9, 2017, https://www.dailymail.co.uk/travel/travel_news/article-4207688/Inside-Alert-northerly-settlement-world.html.

4. "QuickFacts, Key West City, Florida," US Census Bureau, https://www.census.gov/quickfacts/keywestcityflorida.

5. "Wildlife," Aroostook County Tourism, https://visitaroostook.com/story/wildlife.

6. "How Climate Change Is Affecting Maple Syrup Production," Bascom Family Farms Blog, 2016, https://web.archive.org/web/20200927162512/https://www.maplesource.com/blog/how-climate-change-is-impacting-maple-syrup-production/#.X3C8633LeUk and https://mapleresearch.org/pub/climatechangevideo2018/.

7. Livia Albeck-Ripka, "Climate Change Brought a Lobster Boom. Now It Could Cause a Bust," *New York Times*, 21 June 2018, https://www.nytimes.com/2018/06/21/climate/maine-lobsters.html.

8. "Canada Has the Third-Largest Proven Oil Reserve in the World, Most of Which Is in the Oil Sands," Government of Canada, Natural Resources Canada, Offshore Technology, https://www.nrcan.gc.ca/energy/energy-sources-distribution/crude-oil/oil-resources/18085.

9. "Hibernia Oil and Gas Field Project, Newfoundland, Canada," Offshore Technology, https://www.offshore-technology.com/projects/hibernia-oil-gas-field-project/.

10. Tzeporah Berman, "Canada's Most Shameful Environmental Secret Must Not Remain Hidden," *Guardian*, November 14, 2017, https://www.theguardian.com/commentisfree/2017/nov/14/canadas-shameful-environmental-secret-tar-sands-tailings-ponds.

11. "Oil Sands, A Strategic Resource for Canada, North America and the Global Market—GHG Emissions," Government of Canada, 2013, https://www.nrcan.gc.ca/sites/www.nrcan.gc.ca/files/energy/pdf/eneene/pubpub/pdf/12-0614-OS-GHG%20Emissions_eu-eng.pdf.

12. Ibid.

13. Kathleen Harris, "Liberals Promise to Halve Tax Rate for Clean Tech Companies as Part of Long-Range Climate Action Plan," CBC News, September 24, 2019, https://www.cbc.ca/news/politics/liberals-climate-change-action-plan-2050-1.5295027.

14. Richard N. Ostling, "Power, Glory—and Politics," *Time*, February 17, 1986, http://content.time.com/time/subscriber/article/0,33009,960674,00.html.

15. Emily Johnson, "A Theme Park, a Scandal, and the Faded Ruins of a Televangelism Empire," *Religion and Politics*, October 28, 2014, https://religionandpolitics.org/2014/10/28/a-theme-park-a-scandal-and-the-faded-ruins-of-a-televangelism-empire/.

16. "#TBT: The Downfall of Heritage USA," CHS Today, August 22, 2019, https://chstoday.6amcity.com/heritage-usa-fort-mill-sc/.

17. Sherryl Connelly, "The Story of Televangelists Jim and Tammy Faye Bakker's Fall from Grace," *New York Daily News*, August 5, 2017, https://www.nydailynews.com/news/national/televangelists-jim-tammy-faye-bakker-fall-grace-article-1.3387060.

18. Catherine Bowler, "Blessed: A History of the American Prosperity Gospel," PhD diss., Duke University; Terry Mattingly, "Jim Bakker Does U-turn on 'Prosperity Gospel' from Prison," *Tampa Bay Times*, October 11, 2005, https://www.tampabay.com/archive/1992/08/15/jim-bakker-does-u-turn-on-prosperity-gospel-from-prison/.

19. "Prosperity theology," Wikipedia, https://en.wikipedia.org/wiki/; Strange Fires, "*I Was Wrong*: Excerpt from Jim Bakker's Autobiographical Book," https://www.spiritwatch.org/firejbwrong.htm.

20. "Evangelicalism," Wikipedia, https://en.wikipedia.org/wiki/Evangelicalism.

21. Tim Funk, "Jessica Hahn, Woman at Center of Televangelist's Fall 30 Years Ago, Confronts Her Past," *Charlotte Observer*, December 16, 2017, https://www.charlotteobserver.com/living/religion/article189940794.html; Charles E. Shepard, "Jim Bakker Resigns from PTL, Jerry Falwell Assumes Leadership," *Charlotte Observer*, March 20, 1987, https://www.charlotteobserver.com/news/state/article200303464.html; "Judge Orders Ex-PTL Secretary to Repay Hush Money to Ministry," *New York Times*, July 21, 1988, https://www.nytimes.com/1988/07/21/us/judge-orders-ex-ptl-secretary-to-repay-hush-money-to-ministry.html.

22. Connelly, "The Story of Televangelists Jim and Tammy Faye Bakker's Fall."

23. Ibid; "Jim Bakker," Wikipedia, https://en.wikipedia.org/wiki/Jim_Bakker; Kelsey McKinney, "The Second Coming of Televangelist Jim Bakker," Buzz Feed News, May 19, 2017, https://www.buzzfeednews.com/article/kelseymckinney/second-coming-of-televangelist-jim-bakker; https://store.jimbakkershow.com/.

24. "*Jim Bakker Show*," Morningside Church, https://jimbakkershow.morningsidechurchinc.com/news/introducing-the-new-fuel-less-generator/; Kylie Mohr, "Apocalypse Chow: We Tried Televangelist Jim Bakker's 'Survival Food,'" NPR, December 3, 2015, https://www.npr.org/sections/thesalt/2015/12/03/456677535/apocalypse-chow-we-tried-televangelist-jim-bakkers-survival-food.

25. Genevieve Carlton, "Legendary Pastors Who Fell from Grace," Ranker.com, October 17, 2018, https://www.ranker.com/list/pastors-that-fell-from-grace/genevieve-carlton.

26. Wikipedia, "List of Federal Political Sex Scandals in the United States," https://en.wikipedia.org/wiki/List_of_federal_political_sex_scandals_in_the_United_States.

27. "Scientific Consensus: Earth's Climate Is Warming," NASA's Jet Propulsion Laboratory, California Institute of Technology, September 28, 2021, https://climate.nasa.gov/scientific-consensus/.

28. List of Worldwide Scientific Organizations, Governor's Office of Planning and Research, State of California, 2021, http://www.opr.ca.gov/facts/list-of-scientific-organizations.html.

29. Independent Comprehensive Review Panel (ICRP), James Webb Space Telescope (JWST), *Final Report* (October 29, 2010), 10, https://www.nasa.gov/pdf/499224main_JWST-ICRP_Report-FINAL.pdf.

30. Ibid., 32.

31. Daniel Clery, "NASA Is Planning Four of the largest Space Telescopes Ever. But Which One Will Fly?," *Science*, December 13, 2018, https://www.sciencemag.org/news/2018/12/nasa-planning-four-largest-space-telescopes-ever-which-one-will-fly.

32. Sari Zeidler, "The End of the Galaxy as We Know It?," CNN, May 31, 2012, https://lightyears.blogs.cnn.com/2012/05/31/the-end-of-the-galaxy-as-we-know-it/.

33. David Appell, "The Sun Will Eventually Engulf Earth—Maybe," *Scientific American*, September 1, 2008, https://www.scientificamerican.com/article/the-sun-will-eventually-engulf-earth-maybe/.

34. David Verbeek and Helene Fouquet, "Can We Get to Space without Damaging the Earth through Huge Carbon Emissions?," *Los Angeles Times*, January 30, 2020, https://www.latimes.com/business/story/2020-01-30/space-launch-carbon-emissions.

35. Mark Ward, "No Such Thing as a Free Launch," *New Scientist*, June 7,1996, https://www.newscientist.com/article/mg15020332-000-no-such-thing-as-a-free-launch/.

36. Leonard David, "How Much Air Pollution Is Produced by Rockets?," *Scientific American*, November 29, 2017, https://www.scientificamerican.com/article/how-much-air-pollution-is-produced-by-rockets/.

37. Lindsey Konkel, "Space Shuttle Blast-Offs Spewed Metals, Chemicals into Wildlife Refuge," *Scientific American*, May 15, 2014, https://www.scientificamerican.com/article/space-shuttle-blast-offs-spewed-metals-chemicals-into-wildlife-refuge/.

38. Florian Kordina, "How Much Do Rockets Pollute?," Every Day Astronaut, March 20, 2020, https://everydayastronaut.com/rocket-pollution/.

39. Highlights from the 2016 Annual Meeting in San Francisco, American Association of Geographers, http://www.aag.org/cs/annualmeeting/annual_meeting_archives/2016_san_francisco/2016_san_francisco_highlights, and American Association of Geographers 2019 Annual Meeting, University of Colorado, Boulder, April 2019, https://www.colorado.edu/geography/2019/04/19/american-association-geographers-2019-annual-meeting.

40. ACS Meetings & Expos, Meeting Registration Statistics (2008–2021), American Chemical Society, https://www.acs.org/content/acs/en/meetings/national-meeting/exhibitors/registration-statistics.html.

41. "Basic Facts about the COP24 Conference in Katowice," United Nations Climate Change Conference, December 2, 2018, https://cop24.gov.pl/news/news-details/news/basic-facts-about-the-cop24-conference-in-katowice/.

42. Nick Davies, "The Inconvenient Truth about The Carbon Offset Industry," *Guardian*, June 16, 2007, https://www.theguardian.com/environment/2007/jun/16/climate change.climatechange.
43. Natacha Larnaud, "Does Planting a Tree Really Offset Your Carbon Footprint?," CBS News, January 28, 2020, https://www.cbsnews.com/news/planting-a-tree-offset -your-carbon-footprint/.
44. Davies, "The Inconvenient Truth."
45. Fiona Harvey, "Businesses and Experts Reveal Plans for Carbon Offset Regulator," *Guardian*, July 8, 2021, https://www.theguardian.com/environment/2021/jul/08/ business-and-finance-group-reveals-plans-for-carbon-offset-regulator.
46. Oscar Reyes, "Beyond Carbon Markets," UN Chronicle, United Nations, https:// www.un.org/en/chronicle/article/beyond-carbon-markets.
47. Sean Fleming, "What Is Carbon Offsetting?," World Economic Forum, June 14, 2019, https://www.weforum.org/agenda/2019/06/what-is-carbon-offsetting/; Alia Al Ghussain, "The Biggest Problem with Carbon Offsetting Is That It Doesn't Really Work," Greenpeace, May 20, 2020, https://www.greenpeace.org.uk/news/the-biggest -problem-with-carbon-offsetting-is-that-it-doesnt-really-work/.
48. Isabelle Porter, "À Québec, l'auto passe avant l'autobus," *Le Devoir*, May 6, 2017, https://www.ledevoir.com/politique/ville-de-quebec/498154/ville-de-quebec-l-auto -passe-avant-l-autobus; "Régis Labeaume prend l'autobus pour la 1re fois en 30 ans," Ici Radio Canada, Radio-Canada, September 22, 2014, https://ici.radio-canada.ca/ nouvelle/685760/rtc-autobus-transport-voiture-labeaume-quebec.
49. Raphaël Gendron-Martin, "Cure de jouvence pour LOVE à Las Vegas," *Le Journal de Montréal*, June 21, 2016, https://www.journaldemontreal.com/2016/06/21/ cure-de-jouvence-pour-love-a-las-vegas; Brendan Kelly, "Cirque du Soleil Refreshes Beatles Show *Love* for 10th Anniversary," *Montreal Gazette*, February 14, 2016, https://montrealgazette.com/entertainment/cirque-du-soleil-refreshes-beatles-show -love-for-10th-anniversary.
50. Allan Kozinn, "In Las Vegas, All You Need Is '*Love*'—and 8 Million Beatles Fans," *New York Times*, August 2, 2016, https://www.nytimes.com/2016/08/02/arts/music/ beatles-love-cirque-du-soleil.html?searchResultPosition=1.
51. "359 Scientists Support Student Demonstrations for Climate Action," Pact for Transition, https://www.lepacte.ca/en/medias-en/.
52. Nathalie Petrowski, "Ils ont signé le pacte. Et vous ?," *La Presse*, http://mi.lapresse .ca/screens/9f4595c2-adeb-4ab0-a030-b4970a6491ef__7C___0.html.
53. Taylor Nicole Rogers, "Meet Secretive Cirque du Soleil billionaire Guy Laliberté, a Space Tourist and Former Street Performer Who Was Arrested for Growing Cannabis on His Private Island," *Insider*, March 2, 2020, https://www.businessinsider.com/ cirque-du-soleil-billionaire-space-tourist-guy-laliberte-career-net-worth.
54. Clara Moskowitz, "Circus Billionaire Says Space Trip Worth Every Penny," Space .com, October 6, 2009, https://www.space.com/7375-circus-billionaire-space-trip -worth-penny.html.
55. Asher Weinstein, "The Upsides of Downsizing," *Green American*, https://www .greenamerica.org/consume-less-live-more/upsides-downsizing.

56. "What Is the Tiny House Movement?.," The Tiny Life, https://thetinylife.com/what-is-the-tiny-house-movement/; "Tiny house movement," Wikipedia, https://en.wikipedia.org/wiki/Tiny_house_movement.

57. Julia Wells, "President Obama Buys Home on Edgartown Great Pond," *Vineyard Gazette*, December 4, 2019, https://vineyardgazette.com/news/2019/12/04/president-obama-buys-home-edgartown-great-pond.

58. "Climate Change and President Obama's Action Plan," White House, https://obamawhitehouse.archives.gov/president-obama-climate-action-plan.

59. "Calculate How Many Solar Panels You Need for Your Home," Solar Reviews, September 1, 2021, https://www.solarreviews.com/blog/how-many-solar-panels-do-i-need-to-run-my-house; and also extrapolating from the twenty-five to forty panels needed for an average 1,500-square-foot house see The Solar Nerd, https://www.thesolarnerd.com/blog/how-many-solar-panels-1500-sqft-house/.

60. See it at https://virtualglobetrotting.com/map/al-gores-house/view/google/.

61. Emily Atkin, "Al Gore's Carbon Footprint Doesn't Matter," *New Republic*, August 7, 2017, https://newrepublic.com/article/144199/al-gores-carbon-footprint-doesnt-matter.

62. See it at https://virtualglobetrotting.com/map/al-gores-house-2/view/google/.

63. World Population Review, https://worldpopulationreview.com/states/hawaii-population/.

64. "Increased Food Security and Food Self-Sufficiency Strategy," Office of Planning, Department of Business Economic Development & Tourism and Department of Agriculture, State of Hawaii, October 2012, http://files.hawaii.gov/dbedt/op/spb/increased_food_security_and_food_self_sufficiency_strategy.pdf.

65. Ken Meter and Megan Phillips Goldenberg, "Building Food Security in Alaska," Crossroads Resource Center, July 28, 2014, https://akfoodpolicycouncil.files.wordpress.com/2013/07/14-09-17_building-food-security-in-ak_exec-summary-recommendations.pdf.

66. World Population Review, https://worldpopulationreview.com/canadian-provinces/british-columbia-population/.

67. Wolf Depner, "British Columbia's Population Stands at 5.1 Million Heading into 2020," *Victoria News*, December 31, 2019, https://www.vicnews.com/news/british-columbias-population-stands-at-5-1-million-heading-into-2020/.

68. "B.C.'s Food Self-Reliance, Can B.C.'s Farmers Feed Our Growing Population?," B.C. Ministry of Agriculture and Lands, 2006, https://foodsecurecanada.org/sites/foodsecurecanada.org/files/BCFoodSelfReliance_Report.pdf.

69. Caroline Saunders, Andres Barber, and Lars-Christian Sorenson, "Food Miles, Carbon Footprinting and Their Potential Impact on Trade," Australian Agricultural and Resource Economics Society, 2009 Conference (53rd), February 11–13, 2009, Cairns, Australia, https://researcharchive.lincoln.ac.nz/bitstream/handle/10182/4317/food_miles.pdf; Summer Worsley, "British or New Zealand Lamb: Which Is More Sustainable?," Eco Beyond, https://www.ecoandbeyond.co/articles/british-new-zealand-lamb/.

70. "Discover the Grass Fed Difference," Lamb Talk, New Zealand Spring Lamb, 2021, https://www.nzspringlamb.com/discover-the-grass-fed-difference/.

71. "Food and Climate Change," David Suzuki Foundation, https://davidsuzuki.org/queen-of-green/food-climate-change/.

72. John Misachi, "Which States Grow Coffee?," World Facts, World Atlas, February 25, 2019, https://www.worldatlas.com/articles/which-states-grow-coffee.html.

73. "America's Coffee Obsession: Fun Facts That Prove We're Hooked," Kitchen Daily, *Huffpost*, November 2, 2011, https://www.huffpost.com/entry/americas-coffee-obsession_n_987885.

74. Daniel Workman, "Coffee Imports by Country," World's Top Exports, June 2021, http://www.worldstopexports.com/coffee-imports-by-country/.

75. Nikko Mills, "Examining the Carbon Footprint of Coffee," The Eco Guide, March 3, 2016, https://theecoguide.org/examining-carbon-footprint-coffee.

76. Niko Kommenda, "How Your Flight Emits as Much CO2 as Many People Do in a Year," *Guardian*, July 19, 2019, https://www.theguardian.com/environment/ng-interactive/2019/jul/19/carbon-calculator-how-taking-one-flight-emits-as-much-as-many-people-do-in-a-year.

77. Cathy Buyck, "Norwegian Now Non-U.S. Leader in Transatlantic NYC Market," AIN online, October 11, 2018, https://www.ainonline.com/aviation-news/air-transport/2018-10-09/norwegian-now-non-us-leader-transatlantic-nyc-market.

78. Also by comparing nutritional content per United States Department of Agriculture, Food Data Central, "Beverages, Coffee, Brewed, Prepared with Tap Water," FDC ID: 171890 NDB Number:14209, April 1, 2018 (https://fdc.nal.usda.gov/fdc-app.html#/food-details/171890/nutrients) against "Daily Value on the New Nutrition and Supplement Facts Labels," US Food and Drug Administration, May 5, 2020 (https://www.fda.gov/food/new-nutrition-facts-label/daily-value-new-nutrition-and-supplement-facts-labels); Malia Frey, "Coffee Nutrition Facts and Health Benefits," Very Well Fit, August 19, 2021, https://www.verywellfit.com/how-to-make-low-caloriecoffee-drinks-3495233.

79. "America's Coffee Obsession."

80. Radoslav Danilak, "Why Energy Is a Big and Rapidly Growing Problem for Data Centers," Forbes Technology Council, December 15, 2017, https://www.forbes.com/sites/forbestechcouncil/2017/12/15/why-energy-is-a-big-and-rapidly-growing-problem-for-data-centers/?sh=6729beb85a30; Ali Marashi, "Improving Data Center Power Consumption & Energy Efficiency," Vxchnge, February 12, 2020, https://www.vxchnge.com/blog/growing-energy-demands-of-data-centers.

81. Mary Branscombe, "Google's Solar Deal for Nevada Data Center Would Be Largest of Its Kind," DataCenter Knowledge, January 15, 2020, https://www.datacenterknowledge.com/google-alphabet/google-s-solar-deal-nevada-data-center-would-be-largest-its-kind.

82. Rebecca Lindsey and LuAnn Dahlman, "Climate Change: Global Temperature," NOAA, March 15, 2021, https://www.climate.gov/news-features/understanding-climate/climate-change-global-temperature.

83. "Retreat of glaciers since 1850," Wikipedia, https://en.wikipedia.org/wiki/Retreat_of_glaciers_since_1850#Europe.

84. Alan Buis, "Why a Growing Greenland Glacier Doesn't Mean Good News for Global Warming," NASA's Jet Propulsion Laboratory, October 21, 2019, https://climate

.nasa.gov/blog/2925/why-a-growing-greenland-glacier-doesnt-mean-good-news-for
-global-warming/.

85. "Predictions of Future Global Climate, Center for Science Education," University Corporation for Atmospheric Research, 2021, https://scied.ucar.edu/longcontent/predictions-future-global-climate.

86. "Warming Projections, Global Temperature Increase by 2100," Climate Action Tracker, May 2021 update, https://climateactiontracker.org/global/cat-thermometer/.

87. Ski Resorts Worldwide, https://www.skiresort.info/ski-resorts/.

88. Rebecca Lindsey, "Climate Change: Global Sea Level," NOAA, January 25, 2021, https://www.climate.gov/news-features/understanding-climate/climate-change -global-sea-level.

89. William B. Stronge, "Economic Impact of Beach Tourism: Florida and Palm Beach County," Florida Shore and Beach Preservation Association, https://www.fsbpa .com/13AnnualConfPresentations/StrongeW.pdf.

90. See Statistica.com, https://www.statista.com/statistics/500066/most-least-concer ned-about-climate-change-globally-by-country/.

91. Matthew Houser, "How Many Americans Believe in Climate Change? Probably More Than You Think, Research in Indiana Suggests," *The Conversation*, September 5, 2019, https://theconversation.com/how-many-americans-believe-in-climate-change -probably-more-than-you-think-research-in-indiana-suggests-118501.

92. Annenberg Public Policy Center of the University of Pennsylvania, "Do Most Americans Believe in Human-Caused Climate Change?" ScienceDaily, www.science daily.com/releases/2019/05/190509133848.htm (accessed September 17, 2021).

93. A. Gustafson, P. Bergquist, A. Leiserowitz, and E. Maibach, "A Growing Majority of Americans Think Global Warming Is Happening and Are Worried," Yale University and George Mason University Program on Climate Change Communication, https://climatecommunication.yale.edu/publications/a-growing-majority-of-america ns-think-global-warming-is-happening-and-are-worried/.

94. "USA Flash report, Sales Volume, 2019," Marklines, https://www.marklines.com/ en/statistics/flash_sales/salesfig_usa_2019.

95. "Average Fuel Efficiency of U.S. Light Duty Vehicles," Bureau of Transportation Statistics, US Department of Transportation, https://www.bts.gov/content/average -fuel-efficiency-us-light-duty-vehicles.

96. Central Pacific Hurricane Center, NOAA, https://www.nhc.noaa.gov/outreach/ history/; "List of Florida hurricanes," Wikipedia, https://en.wikipedia.org/wiki/ List_of_Florida_hurricanes#2000%E2%80%93present.

97. US Climate Data, https://www.usclimatedata.com/climate/california/united-states/ 3174.

98. Marc Reisner, *Cadillac Desert: The American West and Its Disappearing Water* (New York: Viking, 1986).

99. April Glaser, "The Hopelessness of Wildfire Season," *Slate*, October 29, 2019, https://slate.com/business/2019/10/california-wildfire-season-pge-liability-new -normal-hopeless.html.

100. PG&E Reorganization Information, https://www.pge.com/en_US/about-pge/ company-information/reorganization.page?WT.pgeac=Reorganization_Footer.

101. Kavya Balaraman, "Judge Approves PG&E Wildfire Settlements, Bringing Utility Closer to Exiting Bankruptcy," Utility Dive, December 18, 2019, https://www.utility dive.com/news/bankruptcy-judge-approves-pge-wildfire-settlements-utility-reorgan ization/569304/.

102. Kavya Balaraman, "CPUC Imposes Largest Ever Penalty of $1.9B on PG&E for Northern California Wildfires," Utility Dive, May 11, 2020, https://www.utilitydive .com/news/cpuc-imposes-largest-ever-penalty-of-19b-on-pge-for-northern-califor nia/577625/.

103. Dan Whitcomb, "PG&E Pleads Guilty to 84 Counts of Involuntary Man-slaughter in California Wildfire," Reuters, June 16, 2020, https://www.reuters.com/ article/us-california-wildfires-pg-e/pge-pleads-guilty-to-84-counts-of-involuntary -manslaughter-in-california-wildfire-idUSKBN23N35T.

104. Nathanael Johnson, "What's Driving California's Emissions? You Guessed It: Cars.," Grist, October 8, 2019, https://grist.org/article/whats-driving -californias-emissions-you-guessed-it-cars/.

105. Carolina Cavazos Guerra, "The City of Boulder: An Example of a Sustainable Community," Institute for Advanced Sustainability Studies, March 10, 2015, https:// www.iass-potsdam.de/en/blog/2015/10/city-boulder-example-sustainable-community.

106. Alex Burness, "Xcel: No More Coal-Burning at Valmont Plant in Boulder," Daily Camera, April 8, 2017, https://www.dailycamera.com/2017/04/08/xcel-no-more-coal -burning-at-valmont-plant-in-boulder/.

107. "History of Paris," Wikipedia, https://en.wikipedia.org/wiki/History_of_Paris.

108. Elaine Sciolino, The Seine: The River That Made Paris (New York: W.W. Norton, 2020), ch. 10.

109. Lucy Williamson, "The Thames Is Actually One of the Cleanest Rivers in the World—Here's Why It Looks So Dirty," MyLondon, May 14, 2021, https://www .mylondon.news/news/zone-1-news/thames-actually-one-cleanest-rivers-20601885, and also, intermittently: http://www.bbc.com/earth/story/20151111-how-the-river -thames-was-brought-back-from-the-dead.

110. "Danube Is Most Polluted River in Europe, Shows New Global Study," Kafkadesk, May 30, 2019, https://kafkadesk.org/2019/05/30/danube-is-most-polluted-river-in -europe-shows-new-global-study/.

111. Timothy Aeppel, "A Dirty Problem Meanders from East to West. Elbe River: How Possible for Foes to Cooperate In Costly Cleanup?," Christian Science Monitor, September 13, 1988, https://www.csmonitor.com/1988/0913/omud.html.

112. Michael Rotman, "Cuyahoga River Fire," Cleveland Historical, September 22, 2010, https://clevelandhistorical.org/items/show/63.

113. "The Modern Environmental Movement," PBS, https://www.pbs.org/wgbh/ americanexperience/features/earth-days-modern-environmental-movement/.

114. "Cuyahoga Fire," Ohio History Central, Ohio History Connection, https://ohio historycentral.org/w/Cuyahoga_River_Fire.

115. Laura Johnston, "Cuyahoga River Fish Safe to Eat, Ohio EPA Says," Cleveland .com, June 10, 2019, https://www.cleveland.com/news/2019/03/cuyahoga-river-fish -safe-to-eat-ohio-epa-says.html.

116. E. K. Silbergeld, and J. P. Graham, The Cuyahoga Is Still Burning," *Environmental Health Perspectives*, 116, no. 4 (2008): A150, https://doi.org/10.1289/ehp.11419, or https://www.ncbi.nlm.nih.gov/pmc/articles/PMC2290999/.

117. Rebecca N. Bushon and G. F. Koltun, *Microbiological Water Quality in Relation to Water-Contact Recreation, Cuyahoga River, Cuyahoga Valley National Park, Ohio, 2000 and 2002*, USGA Report WRIR 03-4333, https://pubs.usgs.gov/wri/wri034333/.

118. "The Largest Cleanup in History," The Ocean Cleanup, https://www.theocean cleanup.com/.

119. John Simmons, "Breaking Down the Way Plastics Break Down," *Klipsun Magazine*, May 31, 2018, https://klipsunmagazine.com/breaking-down-the-way-plastics -break-down-d6b7f688111a.

120. Melissa Kurtz, "Plastics Sustainability Part 2—'Are Biodegradable Plastics Really Sustainable,'" The Madison Group, https://www.madisongroup.com/publications/Plas tics_Sustainability_Part%202_Are_Biodegradable_Plastics_Really_Sustainable.pdf.

121. Daniel Morales, "The Shift toward Bioplastics: Insights from Lego and Current Opportunities," PreScouter, April 2018, https://www.prescouter.com/2018/04/shift -toward-bioplastics-insights-from-lego-and-current-opportunities/.

122. "Northeast of U.S. Quivers in Rare Quake," CNN, April 20, 2002, http://www .cnn.com/2002/US/04/20/new.england.tremors/index.html.

ELBOW ROOM

1. "How Fast Do Termites Eat Wood?," Terminix, March 5, 2015, https://www.termi-nix.com/termite-control/termite-facts/.

2. Bec Crew, "Long Live The Morbidly Obese Termite Queen, and Her Terrifying Army of Sweat-Licking Babies," *Scientific American*, November 5, 2014, https://blogs .scientificamerican.com/running-ponies/long-live-the-morbidly-obese-termite-queen -and-her-terrifying-army-of-sweat-licking-babies/.

3. "What Is the Role of a Termite Queen in a Termite Colony?," Orkin, https://www .orkin.com/termites/colony/queen-termite/.

4. "Are Cows the Cause of Global Warming?," Time for Change, https://timefor change.org/are-cows-cause-of-global-warming-meat-methane-CO2.

5. "The Wood-Burning Cook Stove In My Kitchen," Choosing Voluntary Simplicity, http://www.choosingvoluntarysimplicity.com/the-wood-burning-cook-stove-in-my -kitchen/.

6. Pippa Neill, "The Great Wood Burning Stove Debate," *Air Quality News Magazine*, February 5 2021, https://airqualitynews.com/2021/02/05/the-great-wood-burning -stove-debate/.

7. "EPA Certified Wood Stoves," US Environmental Protection Agency, https:// www.epa.gov/burnwise/epa-certified-wood-stoves.

8. "Residential Wood Burning, Environmental Impact and Sustainable Solutions," Environmental Action Germany, Deutsche Umwelthife, https://www.clean-heat.eu/ en/actions/info-material/download/background-paper-residential-wood-burning-3 .html.

9. Zoë Chafe, Michael Brauer, Marie-Eve Héroux, Zbigniew Klimont, Timo Lanki, Raimo O. Salonen, and Kirk R. Smith, "Residential Heating With Wood and Coal:

Health Impacts and Policy Options in Europe and North America," World Health Organization, 2015, https://www.euro.who.int/__data/assets/pdf_file/0009/271836/ResidentialHeatingWoodCoalHealthImpacts.pdf; "Air pollution: How It Affects Our Health," European Environmental Agency, December 14, 2020, https://www.eea.europa.eu/themes/air/health-impacts-of-air-pollution/health-impacts-of-air-pollution; "What Is Particle Pollution," US Environmental Protection Agency, https://www.epa.gov/pmcourse/what-particle-pollution.

10. "The Impact of Solar Panel Manufacturing," Collaboratory for a Regenerative Environment (CoRE), https://www.corebuffalo.org/impact-of-solar-panel-manufacturing; Vasilis M. Fthenakis, "Overview of Potential Hazards," in *Practical Handbook of Photovoltaics*, ed. Augustin McEvoy, Tom Markvart, and Luis Castañer, second edition (New York: Academic Press, 2012), 1083–96, https://doi.org/10.1016/B978-0-12-385934-1.00036-2.

11. Ariana Eunjung Cha, "Solar Energy Firms Leave Waste Behind in China," *Washington Post Foreign Service*, March 9, 2008, https://www.washingtonpost.com/wp-dyn/content/article/2008/03/08/AR2008030802595_pf.html.

12. "Planet of the Humans," Wikipedia: https://en.wikipedia.org/wiki/Planet_of_the_Humans.

13. MIT, "Carbon Footprint of Best Conserving Americans Is Still Double Global Average." *ScienceDaily*, April 29, 2008, https://www.sciencedaily.com/releases/2008/04/080428120658.htm.

14. "If Carbon Markets Boom, Who Will Benefit? Meet the Trillion Dollar Club," Redd, November 4, 2011, https://redd-monitor.org/2011/11/04/if-carbon-markets-boom-who-will-benefit-meet-the-trillion-dollar-club/.

15. Michaela Mujica-Steiner, "Déjà Vu at U.N. Climate Talks," *Boulder Weekly*, December 13, 2018, https://www.boulderweekly.com/opinion/deja-vu-at-u-n-climate-talks/.

16. US Youth Delegation to UN Climate Talks, "Resilience and Reparations," Sustain US, https://sustainus.org/climate/cop24/.

17. Paul Danish, "An Open Letter to Michaela Mujica-Steiner," *Boulder Weekly*, December 20, 2018, https://www.boulderweekly.com/opinion/an-open-letter-to-michaela-mujica-steiner/.

18. G. S. Okin, "Environmental Impacts of Food Consumption by Dogs and Cats," *PLoS ONE* 12, no. 8 (2017): e0181301. https://doi.org/10.1371/journal.pone.0181301.

19. "Rabbits in Australia," Wikipedia, https://en.wikipedia.org/wiki/Rabbits_in_Australia.

20. Centre for Invasive Species Solutions, "Rabbit Biology, Ecology and Distribution FactSheet," PestSmart, https://pestsmart.org.au/toolkit-resource/rabbit-biology-ecology-and-distribution.

21. Ping Zhou, "Australia's Massive Feral Rabbit Problem," Thoughtco, November 22, 2019, https://www.thoughtco.com/feral-rabbits-in-australia-1434350.

22. "Biological Control of Rabbits," Australia's National Science Agency, https://www.csiro.au/en/research/animals/pests/biological-control-of-rabbits.

23. The Voluntary Human Extinction Movement, VHEMT, http://www.vhemt.org/.

24. "Voluntary Human Extinction Movement," Wikipedia, https://en.wikipedia.org/wiki/Voluntary_Human_Extinction_Movement.

25. "People's Temple," Wikipedia, https://en.wikipedia.org/wiki/Peoples_Temple.
26. Simon Davis, "'Save the Planet, Kill Yourself': The Contentious History of the Church of Euthanasia," *Vice*, October 23, 2015, https://www.vice.com/en_us/article/bnppam/save-the-planet-kill-yourself-the-contentious-history-of-the-church-of-euthanasia-1022.
27. "FAQ, Church of Euthanasia," https://www.churchofeuthanasia.org/coefaq.html #5.
28. "*Idiocracy*," Wikipedia, https://en.wikipedia.org/wiki/Idiocracy.
29. "Vatican City Population 2021," World Population Review, 2021, http://worldpopulationreview.com/countries/vatican-city-population/.
30. "Vatican City," Wikipedia, https://en.wikipedia.org/wiki/Vatican_City#Wine_consumption.
31. "Homeless Woman Gives Birth Near Vatican, Gets Generous Offer," CBS News, January 20, 2016, https://www.cbsnews.com/news/vatican-homeless-woman-baby-st-peters-square/; "First Vatican City Baby Is Named Pias," Evening News San Jose California, June 19, 1929, https://news.google.com/newspapers?dat=19290614&hl=en&id=_FIiAAAAIBAJ&nid=1977&pg=3076,7997824&sjid=Q6QFAAAAIBAJ.
32. "Pope Francis: 'A society that does not welcome life stops living,'" *Vatican News*, May 14, 2021, https://www.vaticannews.va/en/pope/news/2021-05/pope-francis-general-state-of-births-italy.html.
33. Gretchen C. Daily, Anne H. Ehrlich, and Paul R. Ehrlich, "Optimum Human Population Size," *Population and Environment: A Journal of Interdisciplinary Studies* 15, no. 6, July 1994.
34. "World Population Projected to Reach 9.8 Billion in 2050, and 11.2 Billion in 2100," United Nations Department of Economic and Social Affairs, 21 June 2017, https://www.un.org/development/desa/en/news/population/world-population-prospects-2017.html.
35. "World Population Prospects 2019," United Nations, https://population.un.org/wpp/Graphs/Probabilistic/POP/TOT/.
36. "Residential water use in the U.S. and Canada," Wikipedia, https://en.wikipedia.org/wiki/Residential_water_use_in_the_U.S._and_Canada.
37. "Per Capita Consumption of Poultry and Livestock, 1965 to Forecast 2022, in Pounds," National Chicken Council, 2021, https://www.nationalchickencouncil.org/about-the-industry/statistics/per-capita-consumption-of-poultry-and-livestock-1965-to-estimated-2012-in-pounds/.
38. Benjamin Ravin, "Average Transaction Price for New Car More Than $35K in 2018, Analysts Say," Michigan Live, January 29, 2019, https://www.mlive.com/auto/index.ssf/2018/10/new_car_prices_up_2_percent_th.html.
39. "Oil and Petroleum Products Explained: Use of Oil," US Energy Information Administration, May 10, 2021, https://www.eia.gov/energyexplained/index.php?page=oil_use.
40. Dan Kopf, "The World Is Running out of Japanese People," Quartz, June 4, 2018, https://qz.com/1295721/the-japanese-population-is-shrinking-faster-than-every-other-big-country/.

41. Chrystia Freeland, "The Problems of a Graying Population," *New York Times*, July 28, 2011, https://www.nytimes.com/2011/07/29/world/americas/29iht-letter29.html.

42. G. Seetharaman, "Child Policies Across the World: From Restrictions to Incentives," *Economic Times*, April 16, 2017, https://economictimes.indiatimes.com/news/politics-and-nation/child-policies-across-the-world-from-those-imposing-restrictions-to-others-offering-incentives-to-have-more-kids/articleshow/58199912.cms.

43. "Zero Population Growth," Wikipedia, https://en.wikipedia.org/wiki/Zero_population_growth.

44. Max Roser, "Fertility Rate," OurWorldData.org, 2014, https://ourworldindata.org/fertility-rate.

45. Jesse Bering, "God's Little Rabbits: Religious People Out-Reproduce Secular Ones by a Landslide," *Scientific American*, December 22, 2010, https://blogs.scientific american.com/bering-in-mind/gods-little-rabbits-religious-people-out-reproduce-secular-ones-by-a-landslide/.

46. Guillaume Vandenbroucke, "The Link between Fertility and Income," Federal Reserve Bank of St. Louis, December 13, 2016, https://www.stlouisfed.org/on-the-economy/2016/december/link-fertility-income. A thorough analysis of birth rates by countries, GNP, religion, contraception availability, schooling, participation in work force, and many other categories is available at https://ourworldindata.org/fertility-rate.

47. Elizabeth Howton, "Nearly Half the World Lives on Less Than $5.50 a Day," The World Bank, October 17, 2018, https://www.worldbank.org/en/news/press-release/2018/10/17/nearly-half-the-world-lives-on-less-than-550-a-day.

48. M. Myrskylä, H. P. Kohler, and F. Billari, "Advances in Development Reverse Fertility Declines," *Nature* 460 (2009): 741–43, https://doi.org/10.1038/nature08230 and https://www.nature.com/articles/nature08230.

49. Corinne Purtill and Dan Kopf, "The Reason the Richest Women in the US Are the Ones Having the Most Kids," Quartz, November 11, 2017, https://qz.com/1125805/the-reason-the-richest-women-in-the-us-are-the-ones-having-the-most-kids/.

50. "Nutrition," Bill & Melinda Gates Foundation, https://www.gatesfoundation.org/What-We-Do/Global-Development/Nutrition.

51. "Africa: Make Girls' Access to Education a Reality," Human Rights Watch, June 16, 2017, https://www.hrw.org/news/2017/06/16/africa-make-girls-access-education-reality.

52. "One Child Policy," Wikipedia, https://en.wikipedia.org/wiki/One-child_policy.

53. Baron Laudermilk, "China's One-Child Policy: Urban and Rural Pressures, Anxieties, and Problems," *World Report News*, September 13, 2011, http://www.worldreportnews.com/far-and-south-east-asiaaustralia-archived/chinas-one-child-policy-urban-and-rural-pressures-anxieties-and-problems.

54. Abigail Haworth, "Breaking China's One-Child Law," *Marie Claire*, November 15, 2010, https://www.marieclaire.com/culture/news/a5563/chinas-one-child-law/.

55. Mei Fong, "Chinese Women Have Already Voted against Beijing's Natalist Hopes," *Foreign Policy*, June 4, 2021, https://foreignpolicy.com/2021/06/04/china-women-two-child-policy-birth-rates/.

56. Sui-Lee Wee, "China Says It Will Allow Couples to Have 3 Children, Up from 2," *New York Times*, May 31, 2021, https://www.nytimes.com/2021/05/31/world/asia/china-three-child-policy.html.

57. "Two-Child Policy," Wikipedia, https://en.wikipedia.org/wiki/Two-child_policy.

58. University of Adelaide, "Reducing Population Is No Environmental 'Quick Fix,'" *Science Daily*, October 27, 2014, https://www.sciencedaily.com/releases/2014/10/141027181959.htm.

59. George Fink, *Stress of War, Conflict and Disaster* (New York: Academic Press, 2010).

60. "Estimates of Historical World Population," Wikipedia, https://en.wikipedia.org/wiki/Estimates_of_historical_world_population.

61. "List of Wars and Anthropogenic Disasters by Death Toll," Wikipedia, https://en.wikipedia.org/wiki/List_of_wars_and_anthropogenic_disasters_by_death_toll.

62. "Baby Boom," Wikipedia, https://en.wikipedia.org/wiki/Baby_boom.

63. University of Adelaide. "Reducing population."

64. "Black Death," Wikipedia, https://en.wikipedia.org/wiki/Black_Death.

65. Ole Benedictow, "The Black Death: The Greatest Catastrophe Ever," *History Today* 55, no. 3 (March 2005), https://www.historytoday.com/archive/black-death-greatest-catastrophe-ever.

66. J. M. Bennett and C. W. Hollister, *Medieval Europe: A Short History* (New York: McGraw-Hill, 2006), 326.

67. "Black Death Jewish Persecutions," Wikipedia, https://en.wikipedia.org/wiki/Black_Death_Jewish_persecutions.

68. J. P. Glutting, "Barcelona in the Plague Years," *Barcelona Metropolitan*, https://www.barcelona-metropolitan.com/features/the-plague-years/.

69. "Waldemar Haffkine," Wikipedia, https://en.wikipedia.org/wiki/Waldemar_Haffkine#Anti-plague_vaccine.

70. "Emerging Pandemic Threats, and Emerging Pandemic Threats 2 Program," USAID, July 12, 2021, https://www.usaid.gov/news-information/fact-sheets/emerging-pandemic-threats-program and https://www.usaid.gov/news-information/fact-sheets/emerging-pandemic-threats-2-program.

71. Gavin Yamey, "Op-Ed: The Odds of a Devastating Pandemic Just Went Up," Global Health Institute, Duke University, February 9, 2018, https://globalhealth.duke.edu/media/news/op-ed-odds-devastating-pandemic-just-went.

72. Jolene Creighton, "We're Officially on the Path to a Global Pandemic," Noescope, February 12, 2018, https://futurism.com/officially-path-global-pandemic.

73. Lena H. Sun, "This Mock Pandemic Killed 150 Million People. Next Time It Might Not Be a Drill," *Washington Post*, May 30, 2018, https://www.washingtonpost.com/news/to-your-health/wp/2018/05/30/this-mock-pandemic-killed-150-million-people-next-time-it-might-not-be-a-drill/?noredirect=on&utm_term=.d1b79a2a749b.

74. "Here's How Long the Stock Market Has Historically Taken to Recover from Drops," Four Pillar Freedom, https://fourpillarfreedom.com/heres-how-long-the-stock-market-has-historically-taken-to-recover-from-drops/.

NUCLEAR HOLOCAUST

1. "Be My Yoko Ono," https://en.wikipedia.org/wiki/Be_My_Yoko_Ono.

2. "The Mindset Lists of American History," The Mindset List, http://themindsetlist .com/lists/.

3. "The Animated History of the USSR," https://www.youtube.com/watch?v=CJVD qlWJ7vY.

4. "Pre-emptive Nuclear Strike," Wikipedia, https://en.wikipedia.org/wiki/Pre-emptive_nuclear_strike.

5. "Reagan Remark Stirs European Furor," *Washington Post*, October 21, 1981, https://www.washingtonpost.com/archive/politics/1981/10/21/reagan-remark-stirs -european-furor/2a057366-f45b-40e3-8daf-77d8739c5cf3/?utm_term=.b92e55230da0.

6. "Ronald Reagan," Wikipedia, https://en.wikipedia.org/wiki/Ronald_Reagan.

7. "LittleBoy," Wikipedia, https://en.wikipedia.org/wiki/Little_Boy.

8. "B53. Nuclear Bomb," Wikipedia, https://en.wikipedia.org/wiki/B53_nuclear _bomb.

9. "Nuclear Weapon Yield," Wikipedia, https://en.wikipedia.org/wiki/Nuclear_weap on_yield.

10. "Tsar Bomba," Wikipedia, https://en.wikipedia.org/wiki/Tsar_Bomba#Test.

11. "World Nuclear Forces," in *SIPRI Yearbook 2017* (Stockholm, Sweden: Stockholm International Peace Research Institute, 2017), https://www.sipri.org/year book/2017/11.

12. "Firewalking," Wikipedia, https://en.wikipedia.org/wiki/Firewalking.

13. R. Karl Zipf Jr. and Kenneth L. Cashdollar, *Explosions and Refuge Chambers*, Docket Number 125 (Washington, DC: National Institute for Occupational Safety and Health, US Department of Health and Human Services), https://www.cdc.gov/niosh/ docket/archive/docket125.html.

14. "Effects of Nuclear Explosions," Wikipedia, https://en.wikipedia.org/wiki/Effects _of_nuclear_explosions.

15. "Nuclear Bomb Explosion Steps," Fox News, March 8, 2001, https://www.foxnews .com/story/nuclear-bomb-explosion-steps.

16. Ruth W. Shnider, *Compilation of Nuclear Test Flash Blindness and Retinal Burn Data and Analytic Expressions for Calculating Safe Separation Distances* (Washington, DC: URS Research Co, Defense Technical Information Center), https://apps.dtic.mil/ sti/pdfs/AD0742837.pdf.

17. "Effects of Nuclear Explosions," Wikipedia, https://en.wikipedia.org/wiki/Effects _of_nuclear_explosions.

18. "Atom Bomb Testing—the House in the Middle," https://www.youtube.com/ watch?v=lrYjVv9SyMQ.

19. "Duck and Cover (1951) Bert the Turtle," https://www.youtube.com/watch?v=IK qXu-5jw60.

20. "Kuwaiti Oil Fires," Wikipedia, https://en.wikipedia.org/wiki/Kuwaiti_oil_fires.

21. "Nuclear Winter," Wikipedia, https://en.wikipedia.org/wiki/Nuclear_winter.

22. "Nanjing Massacre," Wikipedia, https://en.wikipedia.org/wiki/Nanjing_Massacre.

23. Ye Zaiqi and Wu Xiaoling, "Feature: Japan's War Crime in Nanjing Massacre Lingers On as Horrible Memories for Asians in U.S.," Xinhuanet.com, December 11,

2017, http://www.xinhuanet.com//english/2017-12/11/c_136817747.htm; Joe Ren-
ouard, "Japan, China, and the Strains of Historical Memory: 80 Years after the Nanjing
Massacre, Historical Issues Continue to Haunt China-Japan Relations," *The Diplomat*,
December 26, 2017, https://thediplomat.com/2017/12/japan-china-and-the-strains-of
-historical-memory/.

24. Michael Kort, "Was the US Justified in Dropping Atomic Bombs on Hiroshima
and Nagasaki during the Second World War?," History Extra, *BBC History Magazine*, August 2015, https://www.historyextra.com/period/second-world-war/was-the
-us-justified-in-dropping-atomic-bombs-on-hiroshima-and-nagasaki-during-the
-second-world-war-you-debate/; Nathan J. Robinson, "How to Justify Hiroshima,"
Current Affairs Magazine, May 11, 2016, https://www.currentaffairs.org/2016/05/how
-to-justify-hiroshima.

25. Nancy Bartlit and Richard Yalman, "Japanese Mass Suicides," Atomic Heritage
Foundation, July 28, 2016, https://www.atomicheritage.org/history/japanese-mass
-suicides.

26. "Atomic Bombings of Hiroshima and Nagasaki," Wikipedia, https://en.wikipedia
.org/wiki/Atomic_bombings_of_Hiroshima_and_Nagasaki#Leaflets.

27. Ibid.

28. "Surrender of Japan," Wikipedia, https://en.wikipedia.org/wiki/Surrender_of
_Japan.

29. "Kyūjō incident," Wikipedia, https://en.wikipedia.org/wiki/Kyūjō_incident.

30. "Cold War History," History.com, October 27, 2009, https://www.history.com/
topics/cold-war/cold-war-history.

31. Christopher Chantrill, "What Is the Spending on Defense?, USgovernmentspend
ing.com, https://www.usgovernmentspending.com/defense_spending.

32. Mark Atwood Lawrence, "After WWII: A Soviet View of U.S. Intentions,"
NotEvenPast.com, November 20, 2014, https://notevenpast.org/after-wwii-a-soviet
-view-of-u-s-intentions/.

33. "Operation Crossroads," Wikipedia, https://en.wikipedia.org/wiki/Operation
_Crossroads.

34. Stephanie Pappas, "Hydrogen Bomb vs. Atomic Bomb: What's the Difference?,"
Live Science, September 22, 2017, https://www.livescience.com/53280-hydrogen
-bomb-vs-atomic-bomb.html.

35. "Nuclear Arms Race," Wikipedia, https://en.wikipedia.org/wiki/Nuclear_arms
_race.

36. "Mutual Assured Destruction," Wikipedia, https://en.wikipedia.org/wiki/Mutual
_assured_destruction.

37. See graph at https://en.wikipedia.org/wiki/Nuclear_arms_race#/media/File:US
_and_USSR_nuclear_stockpiles.svg.

38. "List of Nuclear Close Calls," Wikipedia, https://en.wikipedia.org/wiki/List_of
_nuclear_close_calls.

39. Douglas Birch, "The U.S.S.R. and U.S. Came Closer to Nuclear War Than We
Thought," *Atlantic*, May 28, 2013, https://www.theatlantic.com/international/archive/
2013/05/the-ussr-and-us-came-closer-to-nuclear-war-than-we-thought/276290/.

40. "START I," Wikipedia, https://en.wikipedia.org/wiki/START_I.

41. "Anti-Ballistic Missile Treaty," Wikipedia, https://en.wikipedia.org/wiki/Anti-Ballistic_Missile_Treaty.

42. Dave Majumdar, "Russia's Nuclear Weapons Buildup Is Aimed at Beating U.S. Missile Defenses," *National Interest*, March 1, 2018, https://nationalinterest.org/blog/the-buzz/russias-nuclear-weapons-buildup-aimed-beating-us-missile-24716.

43. Julian E. Barnes and David E. Sanger, "Russia Deploys Hypersonic Weapon, Potentially Renewing Arms Race," *New York Times*, December 28, 2019, https://www.nytimes.com/2019/12/27/us/politics/russia-hypersonic-weapon.html.

44. Ed Pilkington and Martin Pengelly, "'Let It Be an Arms Race': Donald Trump Appears to Double Down on Nuclear Expansion," *Guardian*, December 24, 2016, https://www.theguardian.com/us-news/2016/dec/23/donald-trump-nuclear-weapons-arms-race.

45. Jen Judson, "Pentagon's Major Hypersonic Glide Body Flight Test Deemed Success," *Defense News*, March 20, 2020, https://www.defensenews.com/smr/army-modernization/2020/03/20/pentagons-major-hypersonic-glide-body-flight-test-deemed-success/.

46. Julian Borger, "US Staged 'Limited' Nuclear Battle against Russia in War Game," *Guardian*, February 24, 2020, https://www.theguardian.com/world/2020/feb/24/limited-nuclear-war-game-us-russia.

47. "A Winnable Nuclear War? New Pentagon Document Shows US Military Thinks So," RT News, June 19, 2019, https://www.rt.com/usa/462251-nuclear-doctrine-document-pentagon/.

48. Michael Mazza and Henry Sokolski, "China's Nuclear Arms Are a Riddle Wrapped in a Mystery," *Foreign Policy*, March 13, 2020, https://foreignpolicy.com/2020/03/13/china-nuclear-arms-race-mystery/.

49. Franz-Stefan Gady, "Russia Reveals 'Unstoppable' Nuclear-Powered Cruise Missile," *The Diplomat*, March 2, 2018, https://thediplomat.com/2018/03/russia-reveals-unstoppable-nuclear-powered-cruise-missile/.

50. Zachary Cohen, "New Satellite Images Show Russia May Be Preparing to Test Nuclear Powered 'Skyfall' Missile," CNN, August 18, 2021, https://www.cnn.com/2021/08/18/politics/russia-skyfall-missile-test-satellite-images/index.html.

51. Yuval Noah Harari, *Sapiens: A Brief History of Humankind* (New York: HarperCollins, 2015).

52. "War Studies," Wikipedia, https://en.wikipedia.org/wiki/War_studies.

ADDENDUM

1. Chris Isidore, "Ford Will Start Shipping Explorers without all the Parts—and Add Them Later," CNN Business, March 14, 2022, https://www.cnn.com/2022/03/14/business/ford-gm-eliminate-options-chip-shortage/index.html.

2. "Supply Chain Issues Causing Retail Shortages in Spring Apparel," *Daily Report*, https://www.businessreport.com/business/supply-chain-issues-causing-retail-shortages-in-spring-apparel.

3. "Russian Forces Capture Chernobyl Nuclear Power Plant, Says Ukrainian PM," Radio Free Europe, Radio Liberty's Ukrainian Service, February 24, 2022, https://www.rferl.org/a/ukraine-invasion-russian-forces-chernobyl-/31721240.html.

4. Ahmet Üzümcü, Goran Svilanović, Maximilian Hoell, "Nuclear Dangers of Russia's War against Ukraine," European Leadership Network, https://www.european leadershipnetwork.org/commentary/nuclear-dangers-of-russias-war-against-ukraine -implications-for-multilateral-nuclear-diplomacy-and-recommendations-for-risk -reduction/.

Index

border security, 237–38
brain, 207–22; segmentation of, 207–9
breast cancer, incidence of, 177–78
bridges: collapse of, 100–105; construction of, 243–45; earthquakes and, 36, 157; load factors and, 141–42; personal experience of, 106–7
Brooklyn Bridge, 106–7
Broughton suspension bridge, 103
Brown, Michael, 238
Browning, Iben, 26–27
Buddhism, 45
builders, master, 100–105
building codes: calculation of, 175–76, 179–80, 241; for earthquakes, 241–50; education on, 143–44; for hurricanes, 223–24; politicians and, 228; and resilience, 286–87; resistance to, 249–52.
See also construction
Building Officials and Codes Administrators (BOCA), 246, 249

California: and dam safety, 226–33; and earthquake construction, 248–49, 255, 273–75; and earthquake preparedness, 272–73; and earthquake zoning, 133–36; and floods, 68–70; and infrastructure hardening, 156–57; moving away from, reasons for, 136
Camp Fire, 72
Canada: and border crossings, 237–38; and climate change, 302–3; and food supply, 309; La Grande 4, 259–61
carbon credits, 307–8, 320–21
cascades, 114–16
Cassandra complex, 170
Catastrophe Bonds, 299–300
cathedral collapses, 242–43
causation, beliefs about, 11, 46–47, 73, 85, 119
cellars, 73
censorship, 211
certainty, nature of, 182, 189–90

Champagne, Dominic, 308
charitable donations, guilt and, 216–17, 307
charlatans, and earthquake predictions, 23–27
Chaudière River, 63–64
Chemical Facility Anti-Terrorism Standards (CFATS), 112
chemical industry, regulation of, 111–12
Chernobyl nuclear disaster, 91–92
China: and climate change, 303; and manufacturing, 280–83; and nuclear power, 95; and nuclear weapons, 341; and population issues, 327–29; and resilient cities, 288
Christchurch earthquake, 27–28, 180, 252–56, 283–84, 286, 290; personal experience of, 291–92
Churchill, Winston, 226
Church of Euthanasia, 325
cinders, 78, 82
cities, resilient, 287–88
Clade X exercise, 199
Clary, Charles W. III, 223–24
climate change, 302–17, 334, 342; severity of, 310–14
cloud computing, and carbon emissions, 310
Clown Doctors of New Zealand, 291–92
coercive population control, 328–29, 342
coffee, and carbon emissions, 309–10
Cold War, 340–42
collective resource problems, 112–14
commons, tragedy of, 112–14, 117–18
comparisons, incomplete, 162–63
cone of uncertainty, 169–70
confidence level in predictions, 176
confirmation bias, 210–11
conspiracy thinking, 210; challenges to, reactions to, 211–12; and earthquake predictions, 25
Constitution, Tenth Amendment, 241, 245
construction: of bridges, 105; for earthquakes, 32–34, 42–43, 139, 145–

disaster recovery, 288–91; just in time approach and, 280–83; and research, 305–6; SimCity and, 204

E-Defense, 35

Eiffel, Gustave, 243

elastic rebound theory, 18

Eldfell volcano, 80–81

end of the world syndrome, 266–70

engineers: and codes, 246–47; and construction, 243–45; and continuing education, 234–35; and denial, 141–44; and fatalism, 257–58; and lives versus property, 253; master, 100–105; as silent heroes, 273; standards for, 97–99; and statistics, 178–79; as Third Little Pig, 9

Enjanced Fujita Scale, 74–75

environmental issues: climate change, 302–17; pollution, 314–16

Environmental Protection Agency, 315

EPCOT, 287–88

epistemic uncertainty, 180

Etna, 80

evacuation: for floods, 65; for hurricanes, 54–55; for volcanoes, 79, 82

experts: blaming, 218–20; and continuing education, 234–35

eye, of hurricane, 53

Eyjafjallajokull, 87

farming: and carbon emissions, 309; and cascades, 116

fatalism, 257–61; fake, 258–59

Federal Aviation Administration, 148

Federal Crop Insurance, 62

Federal Emergency Management Agency (FEMA), 6, 64, 67–68, 174, 238, 250–51

Federal Energy Regulatory Commission (FERC), 231

fires, 72, 242; climate change and, 313; earthquakes and, 43; Japan and, 7–8; nuclear explosion and, 337

flash floods, 65, 104

floodplains, 60–62; definition of, 60; relocating from, 64–65

floods, 59–71; and bridges, 104; Japan and, 7; personal experience of, 70–71

Ford, Henry, 281–82

foreshocks, 22

frequent flyer miles, 293–97

Fukushima nuclear disaster, 8, 40, 91–95

future: modeling, 297–98. *See also* predictions

Gable, Clark, 129

gambler's fallacy, 160–62

La Garita Caldera, 89

Gasconade Bridge, 103–4

Gates, Bill, 198

Gates (Bill and Melinda) Foundation, 328

General Services Administration, 235, 247

Geo-Disaster (movie), 129

geysers, 86

globalization, and supply chains, 280–83

global warming, 302–17, 334, 342; severity of, 310–14

Godzilla, 8

Goebbels, Joseph, 155, 209–10

Golden Gate Bridge, 126, 243

Gore, Al, 308

government: and currency, 295–96; size of, 233–34; trust in, 236–37; types of, and mitigation, 224–28

La Grande 4, 259–61

graphs, misleading, 164

Great East Japan earthquake, 36–41

Great Hanshin earthquake, 4, 10, 32–34, 281; personal experience of, 44–45

The Great Los Angeles Earthquake (movie), 124

Great Pacific Garbage Patch, 315

green energy, 308

greenhouse gas emissions: Canada and, 303; offsets to, 307; reduction proposals, 307–10, 312, 319–20; scientific research and, 306–7

Nest, 97–98
Netherlands, 66
New Madrid earthquakes, 26, 250
New Orleans, 66–68
Newton, Isaac, 248
New Zealand: airport security in, 158–59. *See also* Christchurch earthquake
Nile River, 60
Noah, 60
North American Craton, 179–81
Northridge earthquake, 4, 28–29, 173
not invented here, 141–44
nuclear bombs: effects of, 336–38; Japan and, 8; national stockpiles of, 335; and tornadoes, 76
nuclear disasters, 8, 40, 91–92
nuclear holocaust, 333–45, 350
nuclear industry: drills in, 197; risks in, 188
nuclear power: Fukushima disaster and, 93–94; predictions on, 269
nuclear shelters, in Switzerland, 275–77
nuclear winter, 337

Obama, Barack, 308
odds, 174–78; misconceptions on, 160–62
offshoring, 282
oil extraction, Canada and, 302–3
Ono, Yoko, 220
optical illusions, 209
Order of the Solar Temple, 212
organized crime, 6, 35
Oroville Dam, 226–28, 232
Ortelius, Abraham, 16
Oruanui eruption, 88–89
Otis, E. G., 95

Pacific Gas & Electric (PG&E), 313
pandemics, 119–21; and population, 330–31
Pandora (movie), 126–27
pareidolia, 24
Parkfield Experiment, 20–21
Paterson, James, 165

Pele, 85
perception, 209–13
persuasion, issues in, 321
pets: and carbon emissions, 321–22; personal experience of, 332
Plague, 330–31
plastic crisis, 315–16
plate tectonics, 17, 86
Plato, 249
polemology, 343
politicians, 223–40; and dams, 228–36; and mitigation, 223–24; nature of, 224–28; and predictions, 268; and reconstruction, 290
pollution, 314–16
Pompeii, 81–83
Ponzi schemes, 297
population issues, 318–32, 342; growth, 318–24; pets, 321–22; reduction proposals, 327–31; zero growth, 326–27; zero population, 323–26
Poseidon, 12, 60
power disruptions, 101–2; burying lines and, 157–58; cascades and, 114–16
predictions: and chaotic systems, 189; of COVID-19, 197–200; of earthquakes, 19–23; failed, 195–201; personal experience of, 204–6; pessimistic, 265–70; round numbers in, 267; term, 19; value of, 190–93
prejudice, 212–13
preparedness: attitudes toward, 347–49; cost-effectiveness of, 254–55; resistance to, ix–x, 137–38, 249–52; Switzerland and, 275–77; timeframe for, 190–93
prevention, 155–58; charity and, 217–18; versus reaction, 154–55; versus resilience, 284–85; timeframe for, 190–93
priorities: versus preparedness, 233; for public policy, 236–39; and resilience, 284–85
probability, 160–62, 174–78, 297–98

procrastination, 213–15; factors affecting, 214
professional denial, 141–44
professional sports, 182–83
projectiles, wind and, 50
public policy, 236–39
public relations, and disaster effects, 138–39
Putin, Vladimir, 350
pyramid schemes, 297
pyroclastic flow, 78, 82

Quake (movie), 128
The Quake (Skjelvet) (movie), 127
quality control, issues in, 97–99
Québec Bridge, 101–2

rabbits, in Australia, 323–24
rational decision making, 202–3; versus irrationality, 208–13
Reagan, Ronald, 226, 272, 334–35
real estate markets: costs by region, 72; earthquake zones and, 133–36; and floods, 65
regulation: and dam safety, 226–33; Japan and, 93; self, 110–12; tragedy of the commons and, 112–14. *See also* building codes
religion, 215; Bakkers and, 303–4; and fatalism, 257–58; and flood control, 63; and Madrid earthquake, 13; and population issues, 325–28
relocation: floods and, 64; volcanoes and, 82–83
resignation, versus fatalism, 258–59
resilience, 120–21; attitudes toward, 347–49; blind, 284–85; buy-in to, 283–84; as buzzword, 278–79, 291–92; Japan and, 41–44, 85; from scratch, 285–88; truth and lies about, 278–92; value of, 99
resiliency rating systems, 286–87
resistance, ix-x, xii, 137–38, 249–52, 274–75; fatalism and, 257–61; personal experience of, 256

resource wars, 343–44
rewards: and procrastination, 213–14; and risk, 186–89
Richter Magnitude scale, 15, 171
risk: aversion to, 184–85; brain and, 207–22; definitions of, 186; denial of, 136–39; versus hazard, 185–86; management of, 184–94; and reward, 186–89; statistics and, 160–83; tolerance for, 184–85; transferring consequences of, 137
risk assessment: and hurricanes, 54–55; issues in, 185; and where to live, 57
rivers, and floods, 60–61
RMS, 41
Roebling, John A., 243
Rosatom, 95
Rūaumoko, 12
Russia: and nuclear power, 95; and nuclear weapons, 335, 341–42, 350. *See also* Union of Soviet Socialist Republics

Sacramento Valley, 68–69
safe rooms, 72–73
Saffir, Herb, 47
Saffir-Simpson hurricane scale, 47–48
St. Francis Dam, 231
Sakurajima, 78–79, 90
samples, issues with, 165
Samsung Galaxy Note 7, 98
San Andreas (movie), 126
San Andreas fault, 268
San Andreas Fault, 17, 19–21, 135
San Andreas Megaquake (movie), 128
San Andreas Quake (movie), 126
Sand Palace, 56–57
Sandy, Hurricane, 58, 115, 216
San Fernando earthquake, 157, 272–74
San Francisco earthquake, 17–18, 247
San Francisco (movie), 129
Sano, Toshikata, 248
Sanriku earthquake, 8
Sauquoit Creek Bridge, 103

SBCC. *See* Southern Building Code Congress International

scapegoating, 219–21, 228

sea level, 311

seawalls, 39–40, 66–67

seismology, 15

selective exposure, 210–11

self-regulation, 110–12

September 11, 2001, 108–9, 216; and airline industry, 293–94; and airport security, 151–52

shake table, 34–35

Shankar, Ravi, 216

shaped charges, 110

Sharknado (movie), 88

sharks, 187; risks of, 178

Siemens, 93–94

significant-hazard potential, term, 230

silent heroes, xii–xiii, 253, 271–77

SimCity, 201–4

Simpson, Robert, 47

sleep aids, 349

Snider-Pellegrini, Antonio, 16

software bugs, 97–99

solar flares, 101–2

Someone Else's Fault syndrome, 218–21

sophists, 303–10

Southern Building Code Congress International (SBCC), 246–47, 249

South Fork Dam, 231

Space X, 306

Sperry, Roger Wolcott, 208

split-brain patients, 208

sports, 182–83; betting on, 298–99

Staller, Ilona, 226

statistics, 160–83; with incomplete data, 171–74; lying with, 160–66; Twain on, 298

steel buildings, standards for, 241, 244, 246

Steen, Elizabeth, 25

stock market, 297–300, 331

storm surge: definition of, 47–48. *See also* tropical storms and surges

Strauss, Joseph, 243

stress, post-disaster, 6, 290–91

subsidence, 69

suicide bombers, 108

supercell, 75

supervolcanoes, 87–89

supply chains, 280–83, 350

surveys: question spin and, 311–12; statistical errors in, 162

survival, personal experience of, 120–21

Suyehiro, Kyoji, 248

SwissRe, 41

Switzerland, 275–77

Tacoma Narrows Bridge, 104

Taipei 101, 194

Taleb, Nassim Nicholas, 196

Tangshan earthquake, 22–23

Taupo Volcano, 88–89

technological disasters, 91–107; factors in, 95–99

teenagers, brains of, 189, 207

temporal discounting, 213–14

temporary structures, and earthquakes, 193–94

10.0 Earthquake (movie), 125

10.5 Earthquake (movie), 125

Tennant Creek earthquake, 180

Tenth Amendment, 241, 245

terrorist attacks, 108–18

Therac-25, 98–99

Three Little Pigs, 9, 120; and fourth pig, 258; identification of, 347–49

Thunberg, Greta, 273, 306–7

tidal bore, 39

tidal waves. *See* tsunamis

Tohoku earthquake/tsunami, 8, 18, 84, 92–93

tornadoes, 72–77; DNA of, 73–75; incidence of, 74; personal experience of, 76–77

Toshiba, 94

Toxic Substances Control Act, 112

Tracy, Spencer, 129

traffic accidents, 162–63; versus building codes, 253–54; speed and, 163–64 tragedy of the commons, 112–14; personal experience of, 117–18

trains: bombings of, 154; timetables for, 30–31, 33

Transportation Security Administration, 151–53

Tri-State Tornado, 75

tropical storms and surges: DNA of, 46–49; effects of, 49–54; hurricanes, 46–58; Japan and, 8

Truman, Harry S., 339

trust, in institutions, 236–37

tsunamis, 30–45; Japan and, 8; nature of, 37–39; volcanoes and, 79

Twain, Mark, 298

typhoons, 8, 47

Ukraine, 350

uncertainty: cone of, 169–70; epistemic, 180; nature of, 182

UNICEF, 217

Uniform Building Code, ICBO, 247–48

Union of Soviet Socialist Republics (USSR), 109, 117–18, 334–35, 339; and Cold War, 340–42. *See also* Russia

United Nations Climate Change Conference, 307

United States: and Cold War, 340–42; and nuclear weapons, 335, 338–40; and population issues, 326–27

UN Office for Disaster Risk Reduction, 41

unreinforced masonry construction, 366n12; in California, 273–75; in Italy, 218–19

US Army Corps of Engineers, 61–62, 67

US Coast and Geodetic Survey, 268

US Geological Survey (USGS), 20–21, 23, 125

USSR. *See* Union of Soviet Socialist Republics

Vatican City, 326

Vatnajökull, 87

Verne, Jules, 85

Vesuvius, 79–83, 89

ViaRail, 30–31

Volcanic Eruption Index (VEI), 87–88

volcanoes, 78–90; DNA of, 85–86; Japan and, 8; personal experience of, 89–90

Voltaire, 14

Voluntary Human Extinction Movement (VHEMT), 324–25

voluntary population reduction, 327–28

Vulcan, 85

war: environmental/population issues and, 343–44; and population, 329–30; Russia and, 350

water disasters: dam failures and close calls, 228–31; floods, 59–71, 104; hurricanes, 46–58; tsunamis, 30–45

waterfront: attraction of, 46, 65; building codes and, 50–51

Waters, Roger, 217, 258

water supply, earthquakes and, 156–57

The Wave (movie), 127

weather prediction, issues in, 167–70

Webb Space Telescope, 305

Wegener, Alfred Lothar, 16

Westinghouse, 94

where to live, decision making on, 72, 133–36

White Island, 90

Willis, Baily, 267

wind speed, 47–48

women, education of, and birth rates, 327–28

World Health Organization, 120

Xin, 208

Y2K bug, 269–70

Yarmouth suspension bridge, 103

Yellowstone, 89

Yi, 208

Zeus, 59–60